International Rubber Conference (IRC 2018)

Transforming Technology, Driving Innovation

IOP Conference Series: Materials Science and Engineering
Volume 548

Kuala Lumpur, Malaysia
4 – 6 September 2018

ISBN: 978-1-5108-9383-2
ISSN: 1757-8981

Printed from e-media with permission by:

Curran Associates, Inc.
57 Morehouse Lane
Red Hook, NY 12571

Some format issues inherent in the e-media version may also appear in this print version.

Printed by Curran Associates, Inc. (2019)

For permission requests, please contact the Institute of Physics at the address below.

Institute of Physics
Dirac House, Temple Back
Bristol BS1 6BE UK

Phone: 44 1 17 929 7481
Fax: 44 1 17 920 0979

techtracking@iop.org

Additional copies of this publication are available from:

Curran Associates, Inc.
57 Morehouse Lane
Red Hook, NY 12571 USA
Phone: 845-758-0400
Fax: 845-758-2633
Email: curran@proceedings.com
Web: www.proceedings.com

Table of contents

Volume 548

International Rubber Conference

4–6 September 2018, Kuala Lumpur Convention Center (KLCC), Malaysia

Accepted papers received: 10 May 2019
Published online: 27 August 2019

Preface

Papers

IOP Conf. Series: Materials Science and Engineering **548** (2019) 011001 doi:10.1088/1757-899X/548/1/011001

PREFACE

This issue contains the proceedings from International Rubber Conference 2018 (IRC 2018) which was held in Kuala Lumpur Convention Center, Kuala Lumpur, Malaysia from 4-6th September 2018. The conference was organized by the Plastic and Rubber Institute of Malaysia (PRIM) under the auspices of The International Rubber Conference Organization (IRCO). With conferences theme of "Transforming Technology, Driving Innovation", The conference aimed to promote activities in various areas of rubber and polymer research by providing a forum for exchange of ideas, presentation of technical achievements and discussion on their most recent innovations, trends, concerns, issues and challenges encountered with the solutions adopted towards the new generation technology and innovation.

The conferences consisted of 5 plenary talks and 150 presentations (oral & poster) from various country around the globe which includes; China, Czech republic, Germany, India, Indonesia, Iran, Italy, Japan, the Netherlands, Philippines, Russian federation, South Korea, Thailand, Turkey, Singapore, Slovakia, United Kingdom, and United States of America. The conference encompasses all aspect of rubber technology research, development, and innovation, environment and market analysis. The conference were held in-conjunction with the 9th International Rubber Glove Conference and Exhibition (9th IRGCE) and National Symposium on Polymeric Materials 2018 (NSPM 2018). The best selected 24 papers for publication in IOP Conference Series: Materials Science and Engineering is contributed from both IRC and NSPM 2018. All papers included in the conference proceedings were peer reviewed according to IOP Publishing standards.

The success of IRC 2018 is a result of tremendous support and contribution from PRIM, key industrial player and sponsors, and of course, the enthusiastic response from all participants. The conference could not have happened without the commitment of the Organizing Committee, who helped in many ways to organize the conference. Finally, we are also very grateful to the contributors and reviewers for helping us to put together this proceeding.

IOP Conf. Series: Materials Science and Engineering **548** (2019) 011002 doi:10.1088/1757-899X/548/1/011002

TECHNICAL COMMITTEE

CHAIRMAN : Prof. Dato' Dr. Azanam Shah Hashim (UniKL)
CO-CHAIRMAN : Dr Supramaniam S. Shanmugam (MARGMA)
SECRETARIAT 1 : Assoc. Prof. Dr. Siew Kooi Ong (UniKL)
SECRETARIAT 2 : Assoc. Prof. Dr. Nadras Othman (USM)

JOURNAL PUBLICATION :

1. Prof. Dr. Azura A. Rashid (USM)
2. Azlina Pawan (MRB)

OTHER TECHNICAL COMMITTEE :

1. Dr. Faridah Hanim Ab Hanan (MRB)
2. Prof. Dr. Chee Cheong Ho (IKM)
3. Sum Teck Lim (MRPMA)
4. Dr Nor Nadiah Mohamad Yusof (UniKL)
5. Dr Amir Hashim Md Yatim (MRB)
6. Dr Yen Wan Ngeow, Roland (MRB)
7. Dr Ah Seng Tan (MARGMA)
8. Dr Devaraj Veerasamy (Suraaj Mahir Consultancy, SMC)
9. Dr Abu Amu (MREPC)
10. Shafinaz Jamaludin (MREPC)

IOP Conf. Series: Materials Science and Engineering **548** (2019) 011003 doi:10.1088/1757-899X/548/1/011003

Peer review statement

All papers published in this volume of *IOP Conference Series: Materials Science and Engineering* have been peer reviewed through processes administered by the proceedings Editors. Reviews were conducted by expert referees to the professional and scientific standards expected of a proceedings journal published by IOP Publishing.

IOP Conf. Series: Materials Science and Engineering **548** (2019) 012001 doi:10.1088/1757-899X/548/1/012001

Silanization Efficiency of Silica/Silane in Dependence of Amines in Natural Rubber-based Tire Compounds

C Hayichelaeh[1,2], L A E M Reuvekamp[2,3], W K Dierkes[2], A Blume[2], J W M Noordermeer[2*] and K Sahakaro[1*]

1. Department of Rubber Technology and Polymer Science, Faculty of Science and Technology, Prince of Songkla University, Pattani Campus, 94000 Thailand
2. Elastomer Technology and Engineering, Department of Mechanics of Solids, Surfaces and Systems (MS3), Faculty of Engineering Technology, University of Twente, P.O.Box 217, 7500AE Enschede, The Netherlands
3. Apollo Tyres Global R&D B.V., Colosseum 2, 7521PT Enschede, The Netherlands

E-mail: Sahakaro (kannika.sah@psu.ac.th) and Noordermeer (j.w.m.noordermeer@utwente.nl)

Abstract. Silica-silane technology for low rolling resistance tire compounds requires efficient bridging between the silica surface and rubber molecules through silanization and coupling reactions. The presence of diphenylguanidine (DPG) as secondary vulcanization accelerator is also needed to catalyze the silanization reaction between the alkoxy groups of silane coupling agents and the silanol groups on the silica surface. However, DPG can liberate toxic aniline under high mixing temperatures and therefore safer alternatives are required. This study investigates the influence of amines with different structures, i.e. hexylamine (HEX), octadecylamine (OCT), cyclohexylamine (CYC) and dicyclohexylamine (DIC) on the primary silanization reaction rate constant in a model system, and on interfacial compatibility of practical silica-reinforced NR compounds. Compared to the system without, the amines clearly increase the reaction rate constant for which linear aliphatic amines work better than cyclic ones. This is due to better accessibility of the amines towards the silica surface, in agreement with the values of Payne effect as observed in the rubber compounds, except for the OCT case. The lowest Payne effect of the OCT-containing rubber compound is attributed to the additional shielding effect obtained from the long alkyl-chain that leads to more hydrophobicity, resulting in good physical interaction between silica and rubber. The presence of all amines improves the cure properties in which the linear aliphatic amines give shorter cure times than the cyclic aliphatic ones. As a result of good interfacial compatibility, the OCT-containing compound which shows lowest filler-filler interaction gives good mechanical properties that are closest to the reference compound with DPG.

1. Introduction

Natural rubber (NR) is a renewable resource that is used for several applications in which tires take up a large portion of NR consumption. NR is used in many parts of the tires, especially for truck tire tread compounds while it provides low heat build-up and excellent mechanical properties derived from its ability to form crystals upon being stretched, i.e. strain-induced crystallization. Reinforcing fillers are a crucial ingredient in tire compounds, added to improve tire performances. After the emergence of the "Green Tire" introduced by Michelin for low rolling resistance tire treads, i.e. energy saving tires based on a silica/silane reinforcement system that can reduce about 25% rolling resistance compared to conventional carbon black-filled tire tread compounds [1], the silica technology was widely implemented. However, the mixing of silica into rubber compounds encounters the incompatibility problem, that can be overcome by using a silane coupling agent. Interfacial interaction of the silica/silane-reinforced rubber compounds is obtained via chemical bonding between the silica surface and rubber matrix by silane bridge formation. The reaction between silanol groups on the silica surface and alkoxy groups in the silane molecules, the so-called silanization reaction [2-4], occurs via a condensation reaction during mixing of the rubber compounds. Thereafter, the sulfur moiety in the silane molecule reacts with the rubber chains during the vulcanization reaction [4].

Diphenyl guanidine (DPG) is widely used as secondary accelerator in rubber compounds to provide a synergistic effect with the primary accelerators such as sulfenamides and thiazoles. For the silica/silane system DPG is found to also act as a silanization catalyst [5]. Further, the DPG provides a positive effect on deactivation of free silanol groups that are left over after the silanization reaction. However, the DPG can liberate toxic aniline during high temperature mixing, i.e. compounding and vulcanization [6]. Aniline has been classified as a probable carcinogenic substance [7]. Several amines have been investigated to find an alternative for DPG. The effect of different amine types with varying acid dissociation constant (pKa) values on the rate constant of the silanization reaction was studied by Mihara [8] using model compounds. It was found that the rate constant of the silanization reaction increases while the activation energy decreases with increasing pKa values from 6.5 to 11.5.

The present study covers the effect of different amine types having similar pKa values but different structures. Linear aliphatic amines, i.e. hexylamine (HEX) and octadecylamine (OCT), and cyclic aliphatic amines, i.e. cyclohexylamine (CYC) and dicyclohexylamine (DIC), are compared in both a model system and practical silica-reinforced natural rubber compounds. The silanization efficiency of the silica/silane systems having different amines is evaluated by their rate constant of the primary silanization reaction. Then, the properties of rubber compounds with these different amine types are tested.

2. Materials and methods

2.1. Materials

The chemicals used in the model compound study were ULTRASIL7005 silica and bis-(3-triethoxysilyl-propyl)tetrasulfide (TESPT) (both from Evonik, Germany), anhydrous-grade decane and diethyleneglycolmonobutylether (both from Sigma-Aldrich Chemie, Germany). Diphenyl guanidine (DPG) was obtained from Flexys, Belgium. Amines used in this study as alternatives for DPG were hexylamine (HEX), octadecylamine (OCT), cyclohexylamine (CYC) and dicyclohexylamine (DIC) (all from Sigma-Aldrich Chemie, Germany). The chemical structures of all amines used in this study are shown in Fig. 1. The ingredients used in rubber compound study were Ribbed Smoked Sheet (RSS#3)-grade Natural Rubber (NR) (locally produced in Thailand), Treated Distillate Aromatic Extract (TDAE) oil (Vivatec 500, Hansen & Rosenthal, Germany), 2,2,4-Trimethyl-1,2-dihydroquinoline (TMQ) (Flexys, Belgium), zinc oxide, stearic acid and sulfur (all from Sigma-Aldrich Chemie, Germany).

2.2. Model compound study

Mixtures containing 0.5 g of silica, 0.94 mmol of TESPT, 0.30 mmol of the respective amines and 4 ml of decane in ampoules were immersed into an oil bath at 135±2°C for different reaction times. Then, the reactions were quenched by cooling in an ice bath. All physical interactions were removed by adding 2 ml of diethyleneglycolmonobutylether. The resulting mixtures were filtered by using filter paper. The concentration of ethanol (EtOH) was determined by gas chromatography (GC) using a GC-2010 plus (Shimadzu, Kyoto, Japan). The rate constant of the primary silanization reaction was calculated by plotting $ln\ [TESPT]_t - ln\ [TESPT]_0$ against time t; where $[TESPT]_t$ is the TESPT concentration at time t and $[TESPT]_0$ is the initial TESPT concentration. The rate constant of the primary silanization reaction is obtained from the slope of the plot [2].

IOP Conf. Series: Materials Science and Engineering **548** (2019) 012001 doi:10.1088/1757-899X/548/1/012001

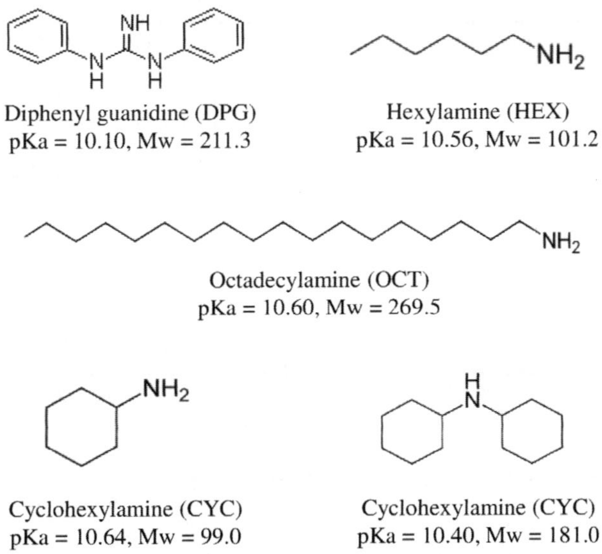

Fig. 1. Chemical structures of amines used in this study.

2.3. Rubber compound study

Rubber compounds were prepared in an internal mixer (BrabenderPlasticorder 350S) with mixing conditions of fill factor 0.7, initial mixer temperature setting of 100°C, and rotor speed of 60 rpm. The rubber formulation and mixing procedure are shown in Tables 1 and 2, respectively. The discharge temperature after the first mixing step of all mixes containing the respective amine types were in the range of 131-135°C. The rubber compounds obtained from the first mixing step were tested for the property that can be related to filler-filler interactions, i.e. the Payne effect.

Table 1. Compound formulations.

Ingredient	Quantity (phr)
NR	100
Silica	55
TESPT	5
ZnO	3
Stearic acid	1
TMQ	1
TDAE oil	8
DPG*	1.1
CBS	1.5
Sulfur	1.5

* Other types of amines were used based on molar equivalents to 1.1 phr of DPG.

Table 2. Compound mixing procedures.

Cumulative mixing time (min)	Step of mixing
1st step: Internal mixer	
0	NR
2	½ Silica + ½ TESPT + ¼ amine
7	½ Silica + ½ TESPT + ¼ amine + TDAE oil
12	ZnO + Stearic acid + TMQ
15	**Discharge**
2nd step: Two-roll mill	
0	CBS + ½ amine
2	S
5	**Mixing completed**

2.4. Testing and characterization

2.4.1 Payne effect

The storage shear moduli (G') of the rubber compounds without curatives were evaluated by using a Rubber Process Analyzer (RPA) (Alpha Technologies, Akron, USA) at a temperature of 100°C, frequency 0.5 Hz and varying strains in the range of 0.28-100%. The Payne effect was calculated from the difference in storage shear moduli at low strain (0.56%) and high strain (100%), i.e. G'(0.56%)-G'(100%).

2.4.2 Cure characteristics

The cure behaviors (i.e., scorch time and optimum cure time [$t_{c,90}$]) of the rubber compounds were tested using the RPA at 150°C, frequency 0.833 Hz, and 2.79% strain.

2.4.3 Tensile tests

Type 2 dumbbell test specimens were prepared, and the tensile properties (i.e., modulus at different strains, tensile strength, and elongation at break) were tested with a Zwick tensile tester model Z1.0/TH1S (Zwick Roell Group, Ulm, Germany) at a cross-head speed of 500 mm/min according to ASTM D412.

3. Results and discussion

3.1. Rate constant of silanization reaction

The silanization or condensation reaction takes place between silanol groups on the silica surface and alkoxy groups on the silane molecules in which the silanization efficiency depends on the adsorption of silane onto the silica surface. The amines play a role as basic catalyst for the silanization reaction promoting more nucleophilicity on the silica surface via hydrogen bonding between the amino group in the amines with silanol groups on the silica surface. Thereafter, the silica surface can easily interact with the silicon atom in the silane molecules through a pentacoordinate intermediate prior to the condensation reaction, resulting in enhancement of the silanization reaction [9-11]. Rate constants of the primary silanization reaction of the model silica/silane systems with different amine types are shown in Fig. 2, in comparison with the system without amine. Due to the enhanced silanization reaction by amines, the mixtures with amines show much higher rate constants of the primary silanization reaction than the one without amine. Compared to the one without amine, the use of DPG, HEX, OCT, CYC and DIC increases the rate constants of the primary silanization reaction by 256, 266, 194, 192 and 79%, respectively. The linear aliphatic amines enhance the reaction better than the cyclic aliphatic amines. The increase of carbon atoms in the linear aliphatic amines from 6 to 18 carbon atoms reduces the rate constant of the primary silanization reaction by 72% due to less mobility of the larger molecule and less accessibility to the silica surface. In case of cyclic aliphatic amines, a reduction by 113% of the rate constant of the primary silanization reaction is observed when the number of aliphatic rings is increased from one to two, causing steric hindrance. It is clear that the structure affects the adsorption efficiency of the amines on the silica surface. The straight alkyl-chains can adsorb easier on the silica surface than the cyclic structures. With regard to DPG that contains two bulky groups of aromatic rings in the structure but still displays a high rate constant of the primary silanization reaction, its three polar –NH groups help to enhance the adsorption efficiency onto the silica surface.

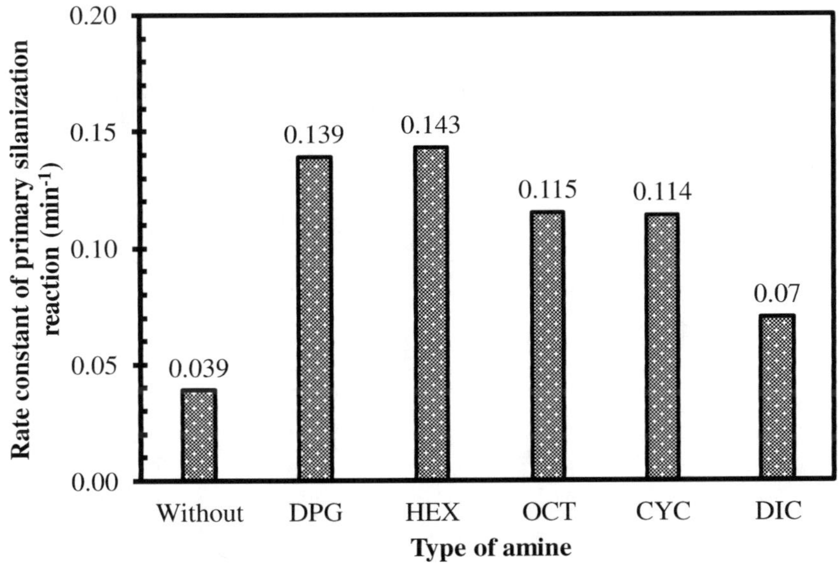

Fig. 2. Rate constant of primary silanization reaction of silica/silane system with different amine types.

3.2. Payne effect

The storage modulus as a function of strain of the silica-reinforced NR compounds with different amine types is shown in Fig. 3. The storage modulus of filled rubber compounds is influenced by four main factors, i.e. the hydrodynamic effect, the rubber network, filler-rubber interaction and filler-filler interaction. The storage modulus decreases under strain deformation due to reduction of the filler-filler interaction causing a breakdown of the filler network, which is well-known as the "Payne effect" [12]. The rubber compounds with amines show a lower filler-filler interaction than the mix without amine. As discussed earlier for the model system, all amines can promote the silanization reaction of the silica/silane system. The reaction between silane and alkoxy groups on the silica surface reduces the reactive silanol groups, resulting in weaker filler-filler interaction within the rubber matrix and a reduction in accelerator adsorption. The linear aliphatic amines that show a higher primary silanization reaction rate constant display lower filler-filler interaction in the rubber compounds compared to the cyclic amines. The CYC with one cyclic ring in the structure gives a lower filler-filler interaction than the bi-cyclic ring DIC. The Payne effect results agree well with the model study. The linear aliphatic amines that have better accessibility towards the silica surface not only enhance the silanization but also shield the free silanol groups that are left over after the silanization reaction. This shielding effect by long alkyl-chains results in more hydrophobicity of the silica surface. Therefore, the use of OCT gives the lowest filler-filler interaction in the silica-filled rubber compound.

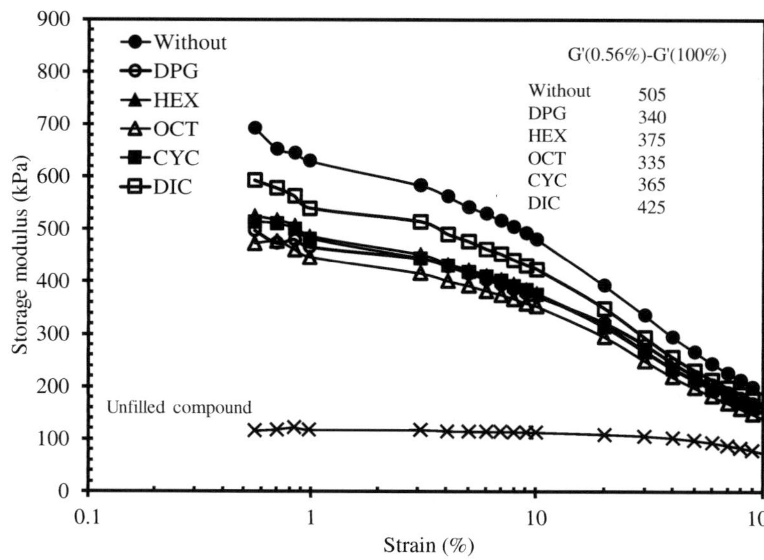

Fig. 3. Payne effect of silica-reinforced rubber compounds with different amine types.

3.3. Cure characteristics

Cure curves, scorch times and optimum cure times ($t_{c,90}$) of the rubber compounds with different amine types are shown in Fig. 4. The presence of all amine types clearly enhances the cure properties of the rubber compounds, as observed in the reduction of scorch and cure times. The amine-containing rubber compounds show reversion behavior whereas the one without amine shows a marching curve. The amines with linear aliphatic chains give shorter cure times than the cyclic structures, as the result of the higher silanization efficiency and better mobility of the secondary accelerators. Among the amines studied, the use of DPG as conventional secondary accelerator gives the fastest cure, i.e. shortest scorch and cure times. Further, the torques of the rubber compounds are affected by the different amine types. The rubber compounds with cyclic aliphatic amines show higher maximum cure torques than the mixes with linear alkyl-chains. The level of torque difference of the rubber compounds is influenced by both chemical and physical crosslinks/interactions. The OCT- and DPG-containing rubber compounds with similar Payne effect show similar cure torques, indicating similar crosslink density. The compounds with higher filler-filler interaction show higher maximum cure torque.

3.4. Tensile properties

The mechanical properties, i.e. moduli at different strains, tensile strength and elongation at break of the rubber compounds are shown in Fig. 5. The use of all amines improves the mechanical properties due to enhancement of the silanization reaction, interfacial compatibility by the shielding effect, and the vulcanization reaction. Comparing the different amine types applied as alternative for DPG, the moduli of the rubber compounds with cyclic aliphatic amines, i.e. CYC and DIC, are higher than the mixes with linear structures, i.e. HEX and OCT, corresponding to the Payne effect and crosslink density within the rubber matrix. However, the interfacial compatibility within silica-reinforced rubber compounds is a crucial factor affecting the ultimate tensile properties. The rubber compound with OCT which shows lowest filler-filler interaction, has a tensile strength and elongation at break that are closest to the reference compound with DPG.

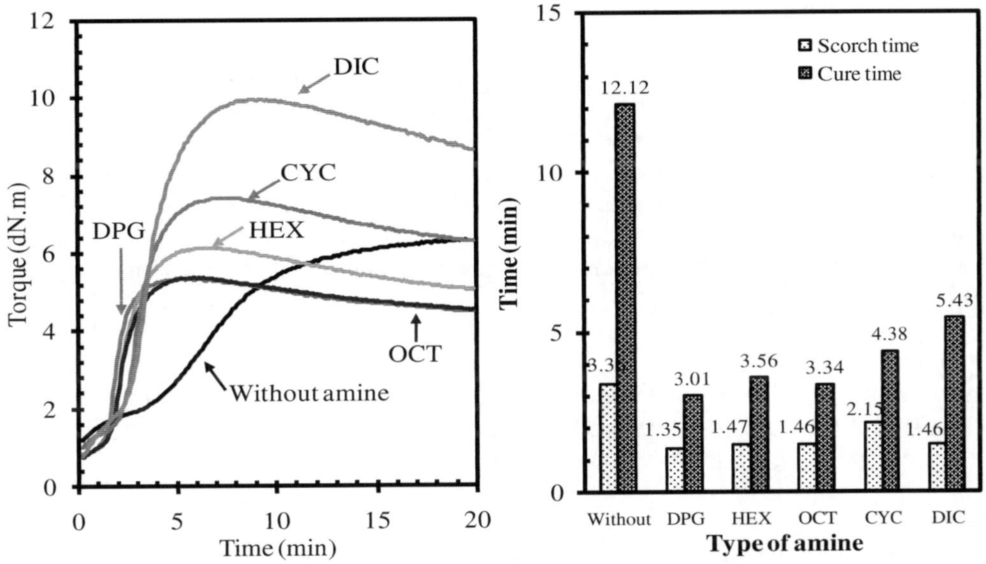

Fig. 4. Cure curves, scorch times and cure times of silica-reinforced rubber compounds with different amine types.

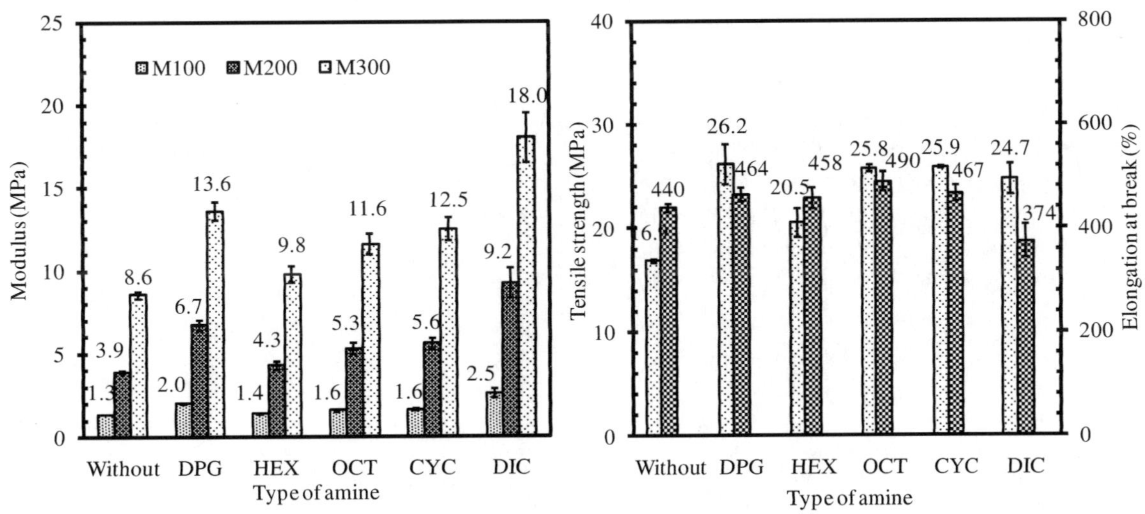

Fig. 5. Tensile properties, i.e. modulus, tensile strength and elongation at break, of silica-reinforced rubber compounds with different amine types.

4. Conclusions

The rate constant of the primary silanization reaction of a silica/silane model system with different amine types is investigated. The presence of all amine types promotes a higher rate constant of the primary silanization reaction compared to the system without amine. Linear aliphatic amines give a higher rate constant than the ones with cyclic aliphatic amines. An increase of carbon atoms in the linear amine structure from 6 to 18 decreases the rate constant. An increase of the cyclic aliphatic from 1 to 2 cyclic rings also decreases the rate constant of the reaction, caused by steric hindrance of the amine molecules that reduces the accessibility of the amines onto the silica surface. With regard to practical silica-reinforced rubber compounds, the Payne effect of the compounds with amines corresponds well with the rate constant of the primary silanization reaction, except for the linear aliphatic amines. The OCT-containing rubber compound shows the lowest Payne effect due to a shielding effect by the long alkyl-chains that promote hydrophobicity and interfacial compatibility. The use of all amines improves the cure properties, in which the linear aliphatic amines give faster cure than the cyclic ones. But, the use of CYC and DIC gives a higher maximum cure torque and torque difference, thus results in higher modulus in the rubber vulcanizates. Among the alternative amines studied, tensile strength and elongation at break of the OCT-containing compound are highest and closest to the reference compound with DPG, as influenced by optimized interfacial compatibility involving both chemical and physical interactions.

Acknowledgements

The authors gratefully acknowledge the Dutch Natural Rubber Foundation (Rubber Stichting, the Netherlands), Apollo Tyres Global R&D B.V. (the Netherlands) and the Graduate School of Prince of Songkla University (Thailand) for financial support.

References

[1] Rauline R 1992 E.P. Patent 0501227A1 (to Michelin and Cie) Feb. 12
[2] Görl U *et al* 1997 *Rubber Chem. Technol.* **70** 608
[3] Blume A 2011 *Kautsch. Gummi Kunstst.* **64** 38
[4] ten Brinke J W *et al* 2003 *Compos. Sci. Technol.* **63** 1165
[5] Mihara S *et al* 2011 U.S. Patent 7923493B2 (to The Yokohama Rubber Co., Ltd.) Apr. 12
[6] Okel T A 2011 *Rubber World* **30** 244
[7] Agency for Toxic Substances and Disease Registry 2002 http://www.atsdr.cdc.gov/toxfaqs/ tfacts171.pdf
[8] Mihara S 2009 *Reactive processing of silica-reinforced tire rubber* Ph.D. thesis, Dept. of Elastomer Technology and Engineering, Univ. of Twente, Enschede, the Netherlands.
[9] Tripp C P *et al* 1993 *J. Phys. Chem.* **97** 5693
[10] White L D *et al* 2000 *J. Colloid Interface Sci.* **232** 400
[11] Kanan S M *et al* 2002 *Langmuir* **18** 6623
[12] Payne A R 1966 *Rubber Chem. Technol.* **33** 365

IOP Conf. Series: Materials Science and Engineering **548** (2019) 012002 doi:10.1088/1757-899X/548/1/012002

Effect of Different Types of Rubbers and Carbon Black Grades on Curing Characteristics and Dynamic Properties for Anti-vibration Application

N.M. Shukri[1] **and A.R. Azura** [1]

[1] Polymer Engineering, School of Materials and Mineral Resources Engineering, Engineering Campus, Universiti Sains Malaysia, 43000 Nibong Tebal, Pulau Pinang, Malaysia

E-mail: nafeesashuk@gmail.com and srazura@usm.my

Abstract. Nowadays, the automotive industry is facing a major challenge to meet the changing demand on performance, material specifications and durability requirements. An anti-vibration part is a critical component in automotive, used to eliminate the vibration. In order to produce a good anti-vibration component, the rubber compound needs to be formulated with good damping properties. This research work focused on the effect of (unfilled) rubber types; natural rubber (SMR 60CV, SMR L, RSS) and synthetic rubber (SBR, NBR, CR, EPDM) and the effect of carbon black grades (N220, N330, N550, N774, N990). The curing characteristics and dynamic properties (dynamic stiffness, Kd) at 15, 45 and 100 Hz were analyzed. The curing characteristic results showed there is a relationship between both factors. The longer cure time for types of rubber indicates longer time is needed to break higher bond energy of the rubber chain. Meanwhile, for carbon black grades, the cure rate is gradually increased with lower carbon black surface area due to the diffusion limitation and lower chain mobility. The dynamic properties results were based on the Payne effect theory which influenced by the rubber bond structure that gave higher value of dynamic stiffness due to rubber-rubber interaction. For carbon black grades, the higher surface area of carbon black gave higher dynamic stiffness due to increased rubber-filler and filler-filler interactions.

1. Introduction

The automotive or also known as automobile industry is directly related to the rubber industry. There are various components or parts made from rubber used in automobiles including the high-technology adhesive and sealant products required by the aerospace and other demanding applications in automobiles and automotive industries. One of the most critical rubber products is mainly related to the dynamic application for example an anti-vibration component usually used for shock absorption [1].

A rubber-based anti-vibration component is widely used with the following main objectives; to eliminate harmful vibration, reduce noise and absorb shock [2]. Rubber is used as anti-vibration based on its good vibration-absorbing characteristic which reduces the effect of running engines, ensuring smooth humming for automotive anti-vibration components. The high internal isolating properties of rubber particularly contribute as anti-vibration components [3]. Some examples of the anti-vibration components for automotive application are; rubber bushing, damper, engine mounting, exhaust hanger and others. There are several commonly used rubbers for the anti-vibration application including natural rubber (NR), chloroprene rubber (CR), styrene butadiene rubber (SBR), ethylene propylene diene monomer (EPDM) [4, 5].

The most common way of reducing unwanted vibrations at a receiving structure in between the source and receiver, is either to reflect the vibration energy by significantly dissipate or changing the impedance through induction of additional damping. [6]. The dynamic motion in anti-vibration application is correlated to the dynamic properties of the rubber compound as it

IOP Conf. Series: Materials Science and Engineering **548** (2019) 012002 doi:10.1088/1757-899X/548/1/012002

responds to the periodic forces applied during application which do not cause failure. The dynamic properties of anti-vibration components are referring to the viscoelasticity properties of the rubber compound. Different types of rubbers contribute to different viscoelastic properties. Selecting a right type of rubber can produce a good anti-vibration component as it refers to the compound stiffness and viscoelastic properties [6]. The dynamic properties of rubber compounds are also contributed by the type of carbon black. The effect of different carbon black fillers on the dynamic properties of a given rubber and cure system, filler networks and strength, are the main parameters that govern the behavior of the filled rubber. The thermodynamic and kinetic of carbon black filler network formation is related to filler-filler, rubber-filler, as well as rubber-rubber interactions that also known as the Payne effect theory [7].

In automotive engineering application, rubber is an efficient material to be used as anti-vibration component due to its properties. However, designing a right rubber formulation to meet critical product specifications such as dynamic and damping properties was deemed difficult. Although there are many previous research studies on the effect of rubber and carbon black types, the studies however were found to be limited to the general mechanical properties and lesser works were reporting on the dynamic stiffness (Kd) using single-bonded rubber. Besides, the influence of the stiffness test of rubber compound towards noise and vibration has also become a concern. Due to the limitation of study on dynamic properties in designing rubber anti-vibration, this research work is conducted to study the effect of rubber types and carbon black grades on the dynamic properties for rubber anti-vibration application by using single-bonded rubber following JIS K6394 (N_2 type).

2. Materials and Methods

In this work, purchases of the natural rubbers (SMR 60 CV, SMR L RSS) were made from Mardec Berhad and the synthetic rubbers (SBR, NBR, CR, EPDM) from Imago Chemical's Sdn. Bhd. All the carbon black grades were purchased from Cabot Cans Sdn. Bhd. This research works were divided into two series; (i) unfilled and (ii) filled rubber compounds. The first series (unfilled rubber compounds) were used to study the effect of different types of rubbers (NR, SBR, NBR, CR and EPDM) on the cure characteristics and dynamic stiffness properties. The second series (filled rubber compounds) were used to analyze the effect of different types of carbon black grades (N220, N330, N550, N774 and N990) with 50 phr of carbon black loading on the cure characteristics and dynamic stiffness properties of rubber compounds. Both of these series used semi-efficient sulphur vulcanization (semi-EV) system. Materials and formulations used in this study are tabulated in Table 1 (unfilled) and Table 2 (filled).

The rubber compounds were prepared according to ASTM D3184-89 using laboratory size two-roll mill machine model XK-160. After compounding process, the curing characteristics of rubber compounds were carried out by using an oscillating die rheometer tester (ODR 2000) according to ASTM D2084 with the temperature of 165 °C. The dynamic test (dynamic stiffness, Kd) was conducted on single rubber bonded according to JIS K6394 (N_2 type) by using Servopulser testing machine at a control temperature of 23 ± 2 °C and the results were plotted at 15, 45 and 100 Hz.

IOP Conf. Series: Materials Science and Engineering **548** (2019) 012002 doi:10.1088/1757-899X/548/1/012002

Table 1. The rubber formulations for first series (unfilled) rubber compounds

Ingredients	Compounds/phr*						
	SMR 60CV	SMR L	RSS	SBR	NBR	CR	EPDM
SMR 60CV	100	-	-	-	-	-	-
SMR L	-	100	-	-	-	-	-
RSS	-	-	100	-	-	-	-
SBR	-	-	-	100	-	-	-
NBR	-	-	-	-	100	-	-
CR	-	-	-	-	-	100	-
EPDM	-	-	-	-	-	-	100
Zinc oxide	5	5	5	5	5	5	5
Stearic acid	1	1	1	1	1	1	1
Sulphur	1.5	1.5	1.5	1.5	1.5	1.5	1.5
CBS	2	2	2	2	2	2	2
TMTD	1	1	1	1	1	1	1

*phr = part per hundred rubber

Table 2. The rubber formulations for second series (filled) rubber compounds

Ingredients	Compounds/phr*				
	N220	N330	N550	N774	N990
SMR 60CV (Rubber)	100	100	100	100	100
N220	50	-	-	-	-
N330	-	50	-	-	-
N550	-	-	50	-	-
N774	-	-	-	50	-
N990	-	-	-	-	50
Zinc oxide	5	5	5	5	5
Stearic acid	1	1	1	1	1
Sulphur	1.5	1.5	1.5	1.5	1.5
CBS	2	2	2	2	2
TMTD	1	1	1	1	1

*phr = part per hundred rubber

3. Results and Discussion

In this research work, all the rubber compounds used similar types of cure system which is semi EV system sulphur vulcanization. The scorch time (t_{s2}) and cure time (t_{c90}) results were plotted in Fig. 1. The results showed that both t_{s2} and t_{c90} values gave no significant difference for NR compounds due to similar rubber structure which is polyisoprene. For synthetic rubber, CR showed the lowest t_{s2} followed by NBR, SBR and EPDM due to their average bond energy. As the vulcanization process is the breakage of rubber chain, therefore molecules with the higher bond energy need higher energy to be broken [8]. Longer cure times indicated longer times needed to break higher bond energy of the rubber chain. For carbon black grades effect, the vulcanization rate was gradually increased with the decreased surface area of carbon black. This may be due to the diffusion limitation and lower chain mobility of rubber introduced by the carbon black-rubber interactions [9].

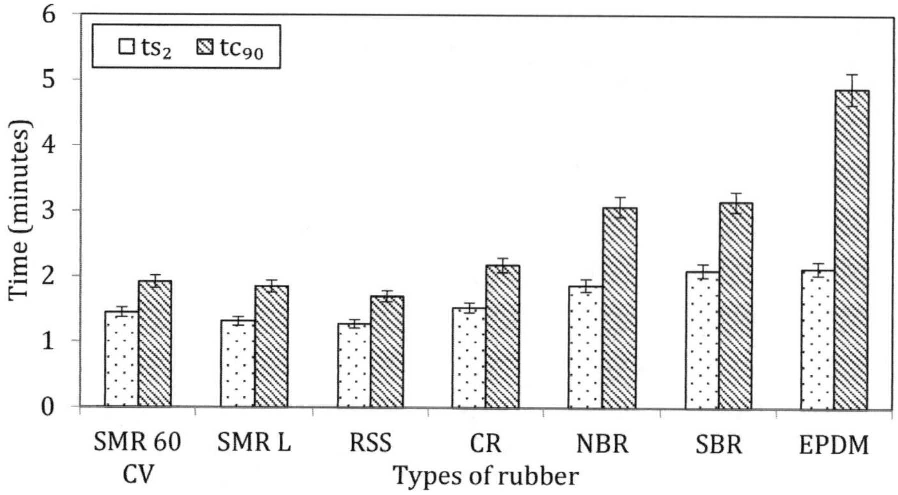

(a)

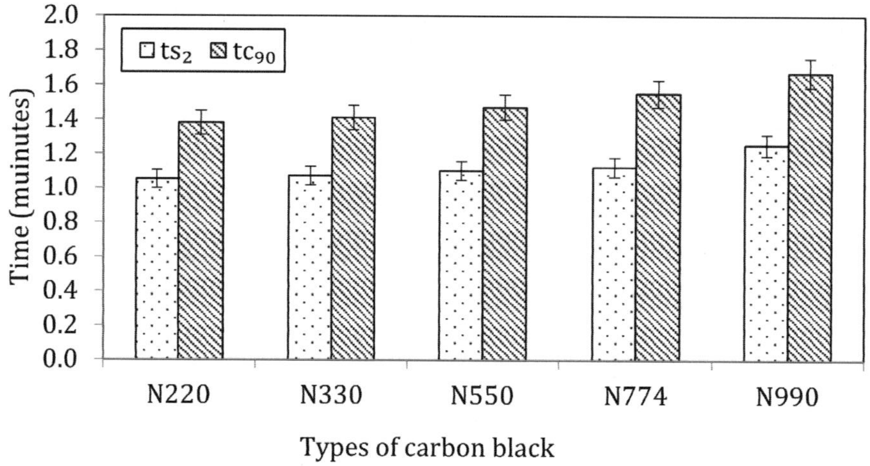

(b)

Fig. 1. The curing characteristics; (a) scorch time (t_{s2}) and (b) cure time (t_{c90}) for different types of rubbers and carbon black grades

By referring to Fig. 2, the Kd results at 15, 45 and 100 Hz were plotted for rubber type and carbon black grade effect. Higher Kd values with respect to EPDM and SBR are due to the high bond energy of diene and styrene bulky groups attached at the rubber backbone chain, requiring a higher energy for the occurrence of bonding breakage. Hence, it restricted the molecular chain orientation. While for NBR rubber, high Kd values is due to the strong C≡N bond in the rubber structure where it needed high energy to break. The C-Cl side chain bond in CR rubber has also increased the bond energy. However, in NR compounds, the polyisoprene rubbers with empirical formula $(C_5H_8)_n$ are essentially hydrocarbons consisted of only carbon and hydrogen elements which are also inherent in other rubber structures. Increasing frequency has decreased the dynamic stiffness for all rubber compounds due to the increased in rubber chain movement thus leads the molecules to slip past one another and sometimes even bond breakage occurred [10].

The highest surface area of carbon black increased the Kd values due to the increased interactions between rubber-filler and filler-filler. The rubber-filler interaction includes physical absorption of the rubber chains on the carbon black filler surfaces [11]. This interaction restricted the rubber chain mobility on the filler surfaces (Payne theory). Increment of the Kd values succeeding frequency amplification may be due to elevation in the energy absorption through a viscous flow of the rubber matrix and the interfacial interaction between the filler and rubber matrix. This absorption might be due to the overcoming inter-particles interaction by the van der Waals forces.

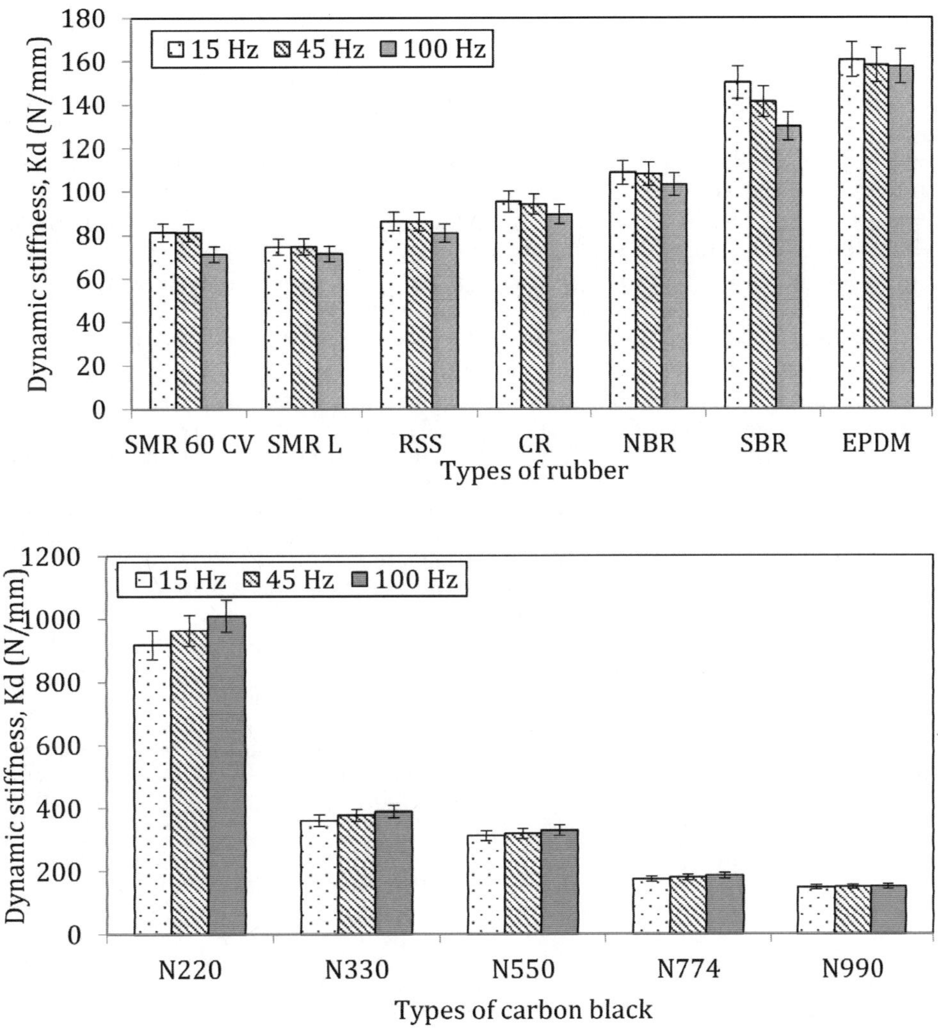

Fig. 2. The dynamic stiffness, Kd at 15, 45 and 100 Hz for different types of rubbers and carbon black grades

4. Conclusions

Based on this research work, it can be observed that different types of rubber compounds gave significant effect on the cure characteristics. Higher values of t_{s2} and t_{c90} were due to the bulky group attached at the rubber backbone chain. This bulky group tends to hinder the chain mobility, thus higher energy was needed to break the bond. The Kd test was related to the Payne effect behavior which depends on the viscoelasticity and crosslinking of the rubber itself. The differences in carbon black grades were clearly affecting the cure characteristics of the rubber compounds. The vulcanization rate was gradually increased with reduction of the carbon black surface area due to limitation in diffusion and lower chain mobility of the rubber compounds. Larger carbon black surface area would strengthen the interactions between rubber-filler and filler-filler, thus leads to increment in the Kd values.

References

[1.] Oshima H, Aono Y, Noguchi H and Shibata S 2007 Fatigue characteristics of vulcanized natural rubber for automotive engine mounting (characteristics of composition and mechanical properties) Memoirs of the faculty of Engineering, Kyushu University **67**(2)

[2.] Kurna S, Jain R, Mathur A, Parwal M and Legala A 2017 Design Optimization of Engine Mounts for Commercial Vehicle Application to Avoid Failures. SAE Technical Paper (No. 2017-26-0287)

[3.] Zhou X Q, Yu D Y, Shao X Y, Zhang S Q and Wang S 2016 Research and applications of viscoelastic vibration damping materials: a review. Compos. Structures **136**, 460-480

[4.] Phu P Q and Nguyen Khac Tien L T T H 2017 Effects of sevral factors on anti-vibration ability of EPDM rubber J. Sci Technol. **55**(1B) 202-207

[5.] Li Q, Zhao J C and Zhao B 2009 Fatigue life prediction of a rubber mount based on test of material properties and finite element analysis engineer. failure analysis **16**(7), 2304-2310

[6.] Sjoberg M 2002 On dynamic properties of rubber isolators (Doctoral dissertation, Institutionen för farkostteknik)

[7.] Wang Z, Li S, Wei D and Zhao J 2015 Mechanical properties, Payne effect, and Mullins effect of thermoplastic vulcanizates based on high-impact polystyrene and styrene-butadiene rubber compatibilized by styrene-butadiene-styrene block copolymer. J. Thermoplas. Compos Mater. **28**(8)1154-1172

[8.] Agrawal C M, Ong J L, Appleford M A and Mani G 2013 Introduction to Biomaterials: Basic Theory with Engineering Applications. Cambridge University Press.

[9.] Salehi M M, Khalkhali T, Solevman R and Barari M 2016 Effect of carbon black type on the thermal properties of fluoroelastomer European J. Adv. Eng. Technol. **3**(2) 27-30

[10.] Geethamma V G, Asaletha R, Kalarikkal N. and Thomas S 2014 Vibration and sound damping in polymers Resonance **19**(9) 821-833

[11.] Shuib R K and Pickering K L 2016 Effect of Carbon Black on the Dynamic Properties of Anisotropic Magnetorheological Elastomer J. Eng. Sci. **12** (1)

IOP Conf. Series: Materials Science and Engineering **548** (2019) 012003 doi:10.1088/1757-899X/548/1/012003

Evaluation of Vulcanization Depth of Thick Rubber Products by Terahertz Radiation

Y Yasumoto[1], Y Hirakawa[1], and T Gondo[1]

[1.] National Institute of Technology, Kurume College, 1-1-1 Komorino, Kurume 830-8555, Japan

E-mail: hirakawa@kurume-nct.ac.jp

Abstract. An evaluation of the vulcanization depth of thick rubber products was performed using terahertz (THz) time-domain spectroscopy. The sample herein was a styrene -butadiene rubber (SBR)-based cylindrical products with a 35 mm diameter and 20 mm thickness, vulcanized under 10 MPa of pressure and a temperature of 150 °C. To clarify the vulcanization effect, no carbon black (CB) was included in the sample, because its apparent THz absorbance is extremely large compared with that of polymers and other additives. To perform the THz absorbance imaging, the thick sample was sliced parallel to the upper and lower heating plates of the cure mold into six thin specimens a few millimeters thick. Herein, the samples cured for two different cure times ($T90$ and $T100$) were investigated and compared. It was found that the coefficient of variation (CV) of the THz absorbance in the sliced specimens was suitable for evaluating the cure condition in the thick sample. The THz imaging results were used to compare the crosslinking densities of the sliced specimens using the Flory-Rehner equation.

1. Introduction

THz radiation has both light-like and radio wave-like characteristics, because it lies between the microwave and infrared regions of the electromagnetic band, as shown in Fig. 1. It can penetrate most materials except metal and can be manipulated by mirrors and lenses [1]. There has been an attempt to use THz light for various industrial inspections such as paint films [2] and tablet coatings [3]. Further, THz radiation is suitable for nondestructive evaluation/inspection of materials including industrial products such as polymers and elastomers. Our research group has developed a novel technique using THz technology to evaluatie rubbers, and various applications have been reported [4-7], and the THz imaging of a carbon black dispersion over an area several millimeters square and the possibility of visualizing the cure process have been described.

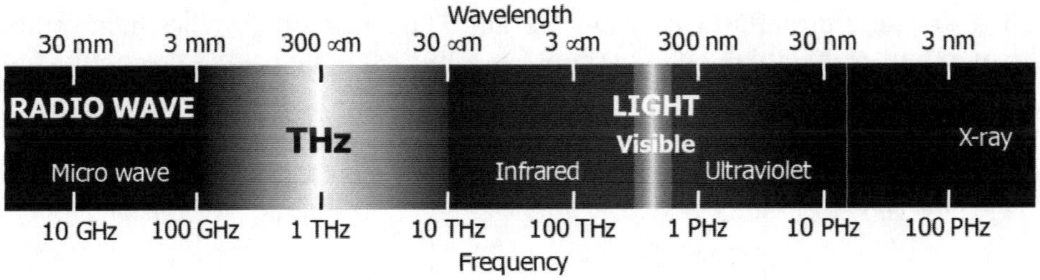

Fig. 1. THz region in the electromagnetic band

In this article, we report the estimation of the vulcanization depth of thick rubber products. The sample used was cylindrical with a thickness of 20 mm. This thick sample was sliced into several thin specimens each with a thickness of a few millimeters, and each specimen was evaluated by THz time-domain spectroscopy (THz-TDS) to obtain the THz absorbance images. It

IOP Conf. Series: Materials Science and Engineering **548** (2019) 012003 doi:10.1088/1757-899X/548/1/012003

Table 1. Compound formulation. The materials include styrene-butadiene rubber (SBR), tetramethylthiuram disulfide (TMTD), dibenzothiazole disulfide (MBTS) and n-cyclohexyl -2-benzothiazole sulfenamide (CBS).

Ingredient	Material	Quantity [phr]
Polymer	SBR	100
Accelerator activators	Zinc oxide	5
	Stearic acid	1
Vulcanizing agent	Sulfer	1
Vulcanization accelerators	TMTD, MBTS, CBS	1 each

was found that the coefficient of variation (CV) of the THz absorbance dispersion is suitable for evaluating the degree of vulcanization at each depth position. The correlation between the THz absorbance values and the crosslinking density calculated using the Flory-Rehner equation [8] was confirmed using the same samples after soaked in toluene and swollen, indicating the validity of the THz evaluation proposed herein.

2. Experiment

2.1 Samples

The used samples used herein were vulcanized rubber based on styrene-butadiene rubber (SBR, the commercial grade: JSR 1502, JSR Co.), whose compounding formulation is given in Table 1. All the reagents used were commercial grade and supplied by Sanshin Chemical Industry Co., Ltd. To clarify the cure effect, no carbon black (CB) was included in the sample, because the apparent THz absorbance of CB is extremely large compared with polymers and other additives [4]. The samples were cured under 10 MPa of pressure and a temperature of 150 °C using a special mold for thick rubber production. The cure time $T90$ ($T100$) is defined as the time required for the torque to reach 90% (100%) of difference between the minimum and the maximum vulcanization curve torque. $T90$ is so-called optimum cure time. Herein, $T90$ and $T100$ for the thick samples were estimated using the results obtained by a thin (t = 2 mm) sample having the same composition using a curemeter (JSR Trading Co., Ltd., Curelastometer W). The cured thick sample was cylindrical with a 35 mm diameter and a thickness of 20 mm, which was sliced parallel to the upper and lower heating plates of the cure mold into six thin specimens with the thickness of about 3 mm thick, as shown in Fig. 2.

2.2 Experimental setup and method

To evaluate the rubber samples in this study, the spectroscopic technique of THz-TDS [9,10] was used, which is a unique and representative technique using THz radiation that characterizes material information such as transmission, absorption, reflection, and complex physical quantities

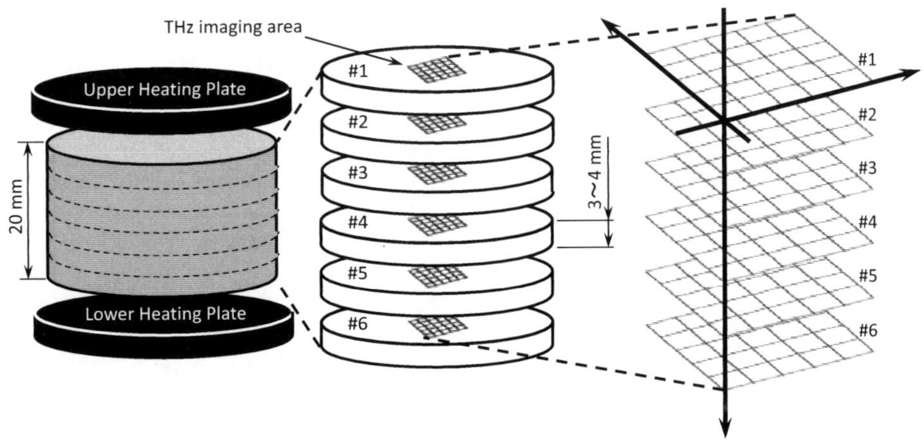

Fig. 2. Schematic of the thick sample used in this study (left), the six thin slices obtained from the thick samples (center), and the areas on each sample scanned by THz time-domain

by data processing. Figure 3 shows a schematic drawing of the THz-TDS system (Otsuka Electronics Co., Ltd., TR-100KS) used herein, in which a femtosecond fiber laser (IMRA America, Inc., Femtolite HX-150, 1620 nm, 100 fs) is used as the fundamental laser light source. In this study, the THz transmission was measured to obtain the THz absorbance. Herein, the THz radiation generated by the 4'-dimethylamino-N-methyl-4-stilbazolium tosylate (DAST) crystal was focused into the sample by the off-axis paraboloidal mirrors. The THz light that passed through the sample was detected by the photoconductive (PC) switch made of low-temperature-grown GaAs to record a real-time wave form $E(t)$ in the computer. By converting the waveform by Fourier transformation, the frequency spectrum $E(\omega)$ was derived as

$$E(\omega) = \int_{-\infty}^{\infty} E(t)e^{i\omega t}dt = |E(\omega)|e^{i\phi(\omega)},$$

where ω and ϕ are the anglular frequency and the phase of the THz radiation, respectively. The frequency spectra for the reference (i.e., blank) and the sample were then calculated to obtain the absorbance $A(\omega)$, such that

$$A(\omega) = -\log\left|\frac{E_{sample}(\omega)}{E_{ref}(\omega)}\right|,$$

where the subscripts *sample* and *ref* indicate the sample and the reference spectra, repectively.

Because the absorbance is a function of the sample thickness, the THz absorbance obtained herein was converted into the value per unit thickness (1 cm). During the two-dimensional THz absorbance imaging measurement, the sample was raster scanned in the $x-$ and $y-$ axes on the sample stage in the THz-TDS system. The scanned area was 5 mm × 5 mm with an interval of 1 mm (25 sampling points) at the central part of each specimen, as depicted in Fig.2. The THz absorbance image was then constructed by calculating the representative THz absorption $A_{\text{image}}(x,y)$ at each measured point (x,y), which was calculated by integrating the THz absorbance spectrum as

$$A_{image}(x, y) = \int_{\omega_1}^{\omega_2} A_{x,y}(\omega) d\omega / (\omega_2 - \omega_1),$$

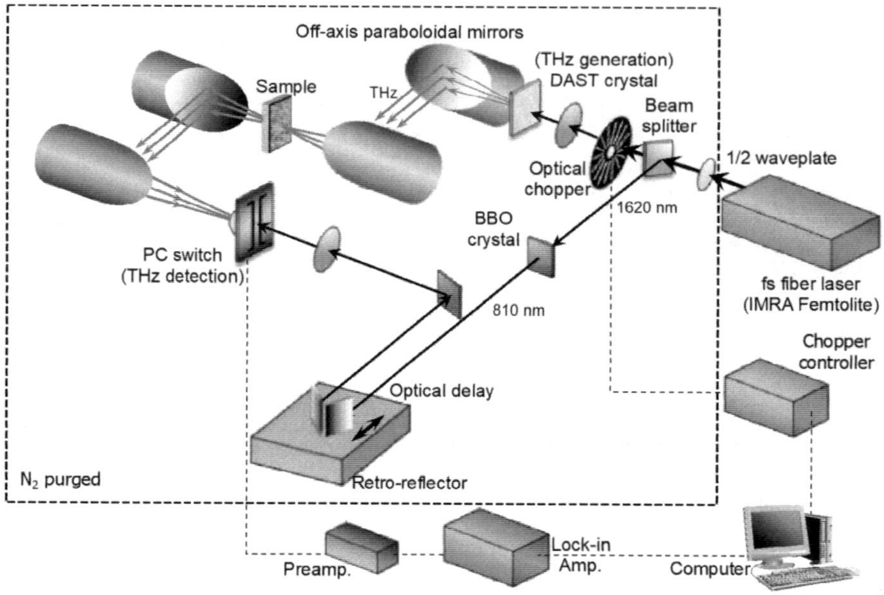

Fig. 3. Schematic of the THz time-domain spectroscopy system used in this study.

where ω_1 and ω_2 are the low- and high-end frequencies of the THz absorbance spectrum, respectively. The values for ω_1 and ω_2 were determined by the spectral region exhibiting a superior signal-to-noise ratio. In this study, the values of ω_1 and ω_2 were 0.21 and 2.7 THz, respectively.

To obtain the crosslinking densities, the used silenced samples were swelled in toluene at room temperature for 72 hours. After that, the samples were dried for 24 hours at 80 °C to vaporaize toluene inside the samples. The crosslinking density of the sliced specimens were calculated by the Flory-Rehner equation,

$$\nu = \frac{V_R + \ln(1 - V_R) + \mu {V_R}^2}{-V\left({V_R}^{\frac{1}{3}} - \frac{V_R}{2}\right)}$$

where V_R is the volume fraction of the polymer (i.e., SBR) in the swollen gel, V is the molar volume of the solvent (i.e., toluene), and μ is the Flory solvent-polymer interaction term. The obtained densities were compared with the THz results to confirm the validity of the THz evaluations.

3. Results and discussion

According to our previous experimental results [7] after the optimum cure time $T90$, the THz transmission becomes smaller as the vulcanization reaction progressed; namely, the THz absorbance of the rubbers increased with the cure process. However, the dispersion of the THz absorbance tends to gradually decrease with the cure time. Prior to imaging the slice specimens, this trend of the THz absorbance was confirmed using a thin (about 2 mm) SBR sample with the same compounding formulation shown in Table 1. In these experiments, a total of $3 \times 3 = 9$ points

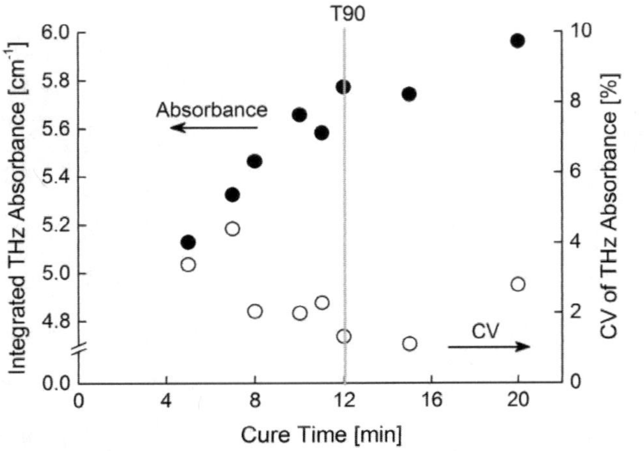

Fig. 4. Integrated THz absorbance (solid circles) and CV (open circles) of the THz absorbance dispersion of nine measured points in the thin sample. The line labeled T90 indicates the optimum cure time for the thin sample.

were evaluated by the THz-TDS system with a resolution of 1 mm. The result is summarized in Fig. 4, where the averaged THz absorbance and the coefficient of variation (CV) of the THz absorbance at the nine measured points are plotted against the cure time. The CV is a parameter indicating the rubber homogeneity. The optimum cure time of this thin sample was found to be 12 min, as indicated by the vertical line labeled T90 in Fig. 4. We found a prevailing tendency of the THz absorbance to gradually increases and of the CV to decrease with the cure time. These same parameter tendencies have been previously reported in Ref 7. This result suggests that, as the reaction proceeds, the THz radiation encounters increased disturbance in the sample such as absorption, scattering, and interference owing to unidentified chemical products or chemical structures (i.e., crosslinking) induced by the vulcanization, and the disturbances ultimately become homogeneous. Interestingly, we can observe the CV parameter to suddenly increase at 20 min (i.e., $T100$), which likely caused by polymers being cut or crosslinked by overcuring, indicating chaotic condition in the thin sample. It is well known that the crosslink density in rubber is formed by the vulcanization reaction [11], but the chemical condition of the crosslinking has not been fully revealed yet [12]. In a previous study we confirmed that the crosslink density in samples increased with the cure time and we have also found that the density curve of these samples exhibited a profile identical to the torque-cure curve obtained by a cure meter [7]. Based on these facts, it is reasonable to conclude that the observed increase of THz absorbance with the cure time can be correlated to the crosslink density in the sample, though an accurate accounting of the factors inducing the THz absorbance increase is still unclear. The possible candidates mentioned above can be considered as factors, but further detailed investigations are necessary to reveal this point.

The results of Fig. 4 confirm that the THz absorbance has a correlation with the crosslink density induced by the vulcanization reaction. As a next step, we conducted the THz imaging of the thick samples (i.e., sliced specimens shown in Fig. 2). The THz absorbance images of the thick samples at $T90$ and $T100$ are given in Fig. 5, where the dark and light contrast in the images show high and low THz absorbance, respectively. Comparing the images of the $T90$ and $T100$ samples, we can see that the latter sample exhibits darker image contrast than that of the former. This result shows that the vulcanization reaction in the sample at $T100$ may proceed farther than at $T90$, according to the result of Fig. 4. To recognize the difference between two samples clearly, the

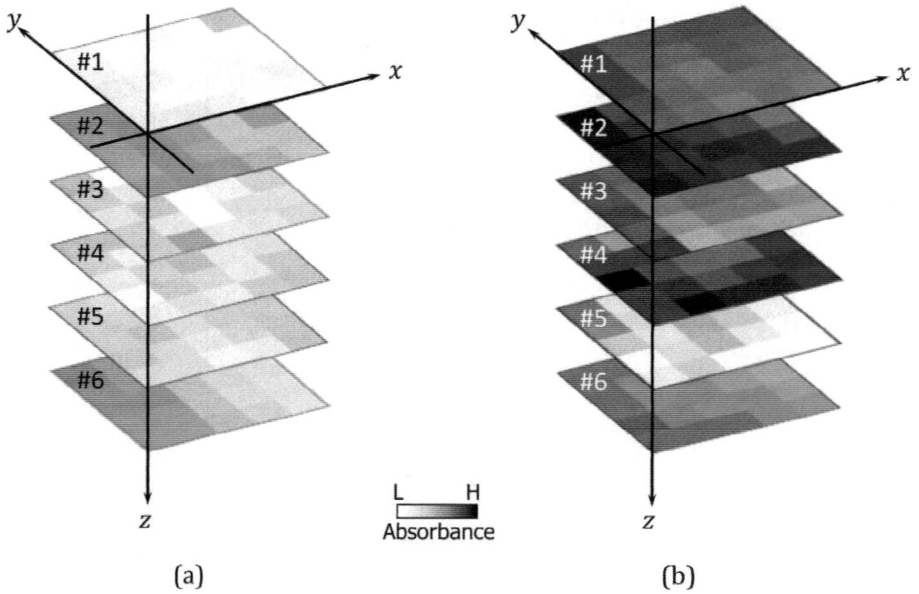

Fig. 5. Images mapping the THz absorbance for the (a) *T90* sample (b) *T100* sample.

average of the THz absorbance and the CV of the THz absorbance dispersion in one slice specimens were 2.27 and 2.29 cm^{-1} for *T*90 and *T*100 sample, respectively, and the mean CVs of *T*90 and *T*100 samples were 2.59 and 2.54 %, respectively. These results suggest that the vulcanization process in the *T*100 thick sample progressed farther and the reaction became more homogeneous than that of the T90 sample. We also note the behavior of the CV values as a function of slice position in each sample, shown in Fig. 5, where the slice specimens are placed in sequence of height in the sample from the left to the right (the slice number is omitted). It can be seen that the CV values maximum at the center of the sample (slice #3 or #4), which indicates that the cure reaction is still active at the center slice specimen of each sample. It is reasonable to believe that an increased CV values reflects a random vulcanization condition (i.e., active condition), while a lower CV value is owing to a uniform reaction (i.e., calm condition). For example, at one point in the central specimen of the thick sample the reaction just started, and at other point in the same slice the reaction has almost finished. In the sliced specimens nearer the heating plates of the cure mold (especially #1 and #6 slices), the reactions have almost finished and the condition became rather homogeneous. Comparing the maximum CV value in each thick sample, the *T*90 sample presents a larger CV peak than the *T*100 sample, which is likely owing to the same theory posited above wherein the vulcanization reaction in the *T*90 sample was more active than that of the *T*100 sample. From these experimental results and our theory, we conclude that the CV of the THz absorbance dispersion is a suitable parameter for evaluating the degree of vulcanization at each depth position.

To confirm the validation of the cure condition as estimated by the THz absorbance and its dispersion, the conventional crosslink density of the samples was calculated based on the Flory-Rehner equation using the weights and volumes of the exact samples used in the THz evaluation (Fig. 5) and the solvent used in the equilibrium swelling experiment. In this study, toluene was selected as solvent. The obtained the crosslink densities for the *T*90 and *T*100 samples are plotted in Fig. 7, where the crosslink density in each sample slice (#1—#6) of the *T*90 and *T*100 samples

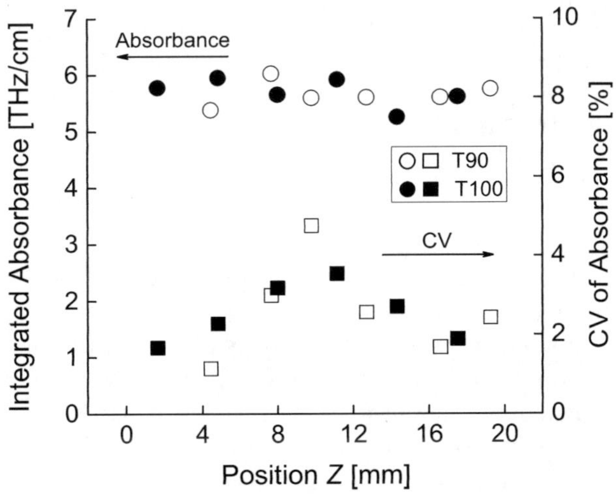

Fig. 6. Averages of the THz absorbance (circles) and CV (squares) of the absorbance dispersion in each slice.

are compared. The average densities were 0.192 and 0.221 mol/L for the *T*90 and *T*100 samples, respectively. As can be seen, the *T*100 sample had more crosslink density, indicating that more revolution of the vulcanization reaction occurred. We also note that the clear trend of increased crosslink in the samples is confirmed by both the equilibrium swelling experiment and the THz absorbance imaging, as shown in Fig. 5 and Fig. 7.

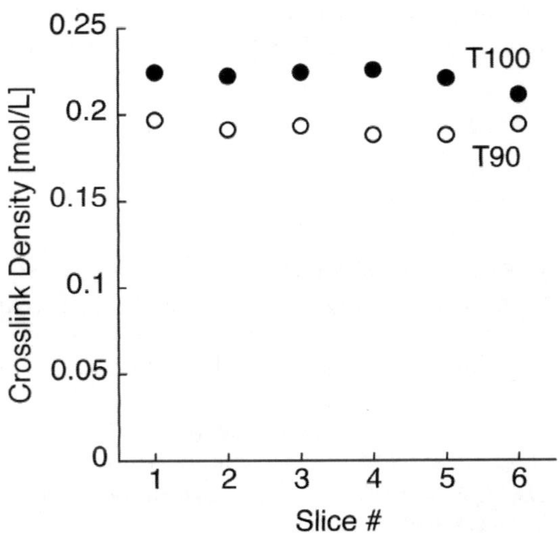

Fig. 7. Crosslink densities for the *T90* (open circles) and *T100* (solid circles) samples, as calculated by the Flory-Rehner equation.

Although the results obtained by the conventional swelling method is more obvious than that obtained by the THz experiment, the CV values of the THz absorbance also provide us information regarding the cure condition. Moreover, the THz absorbance measurements were obtained after only several hours via a nondestructive evaluation. The conventional equilibrium swelling method, however, required at least one week to obtain the final results and the samples were completely destroyed. This comparison between these two-different evaluation methods indicates that our novel method using the THz absorbance for the vulcanization reaction is sufficiently effective. Further, the experimental results in this study suggest that the THz spectroscopic method has the potential to rapidly evaluate cure reactions in thick samples with no damage. We expect that the THz technique will successfully reveal unknown aspects of various elastomer products in the near future.

4. Conclusion

An estimation of vulcanization depth of thick rubber products was performed using the THz-TDS technique. The thick sample (20 mm) was sliced parallel to the upper and lower heating plates of the cure mold into six thin specimens with a thickness of a few millimeters. Each specimen was evaluated at 25 different points to obtain the THz absorbance for the THz absorbance imaging. It was found that the THz absorbance values for each slice were almost constant, while the CV of the THz absorbance, which corresponds to the THz dispersion, had maximum value at the central slice of the thick sample. This behavior of the CV value suggests an active condition of the cure process at the center of the thick sample. We concluded that the CV parameter is suitable for evaluating the degree of vulcanization at each depth position in the sample. This study proved the effectiveness of non-destructive rapid THz estimation of the vulcanization as a function of depth in thick rubber products.

Acknowledgements

The authors thank Prof. M. Tonouchi of Osaka University, Japan, for his support on the THz systems, and Dr. H. Ohtake of IMRA America for his technical support. We thank Sara Maccagnano-Zacher, PhD, from Edanz Group (www.edanzediting.com/ac) for editing a draft of this manuscript. This study was partially supported by the 23rd and 28th Eno Foundation for the Advancement of Science and the Japan Society for the Promotion of Science (JSPS) KAKENHI (Grant Nos 24560056 and 16K05925).

References

[1] Tonouchi M 2007 *Nature Photonics* **1** 97

[2] Yasui T, Yasuda T, Sawanaka K, Araki T 2005 *App. Opt.* **44** 6849

[3] Fitzgerald AJ, Cole BE, Taday PF 2005 *J. Pharm. Sci.* **94** 177

[4] Hirakawa Y, Ohno Y, Gondoh T, Mori T, Takeya K, Tonouchi M, Ohtake H, and Hirosumi T 2011 *J. Infrared, Millimeter and Terahertz Waves* 32 1457

[5] Hirakawa Y, Kamino T, Gondoh T, Hirano S, and Noguchi T 2016, *Proceeding of EMN Meeting on Terahertz 2016* (San Sebastián) C29

[6] Hirakawa Y, Yamauchi T, Kamino T, Gondoh T, Hirano S, and Noguchi T 2017 *Technical Digests of the Conference on Lasers and Electro-Optics (CLEO) Optical Society of America* (San Jose), ATh4B

[7] Hirakawa Y, Nobutsuka A, Gondoh T, and Mori T 2015 *Proceedings of International Symposium on Frontiers in THz Technology (FTT 2015)* (Hamamatsu) Pos2.33

[8] Flory P J, Rehner J 1943 *J. Chem. Phys.* **11** 521

[9] Nishizawa S 2005 *Terahertz Time-Domain Spectroscopy* (Terahertz Optoelectronics) ed K Sakai (Heidelberg: Springer-Verlag) pp. 203-269

[10] Baxter J B, Schmuttenmaer 2007 *Time-Resolved Terahertz Spectroscopy and Terahertz Emission Spectroscopy* (Terahertz Spectroscopy) ed SL Dexheimer (Boca Raton: CRC Press) chapter 3 pp. 73-118

[11] Cpran A Y 2013 *Vulcanization* (The Science and Technology of Rubber, 4th edn) ed JE Mark, and B Erman and M Roland (Oxford: Academic Press) chapter 7 pp. 337-381

[12] Sakai Y, Usami R, Tohsan A, Junkong P, Ikeda Y 2018 *RSC Adv.* **8** 10727

IOP Conf. Series: Materials Science and Engineering **548** (2019) 012004 doi:10.1088/1757-899X/548/1/012004

Evaluation of Specialty Natural Rubbers for Engine Mount Application

S. Kamaruddin[1], A. F. Ramly[2], N. Mohd Shukri[2], A. G. Bakri[2], Q. Fourrier[2]

[1] Malaysian Rubber Board, Rubber Research Station, 47000 Sungai Buloh, Selangor Darul Ehsan, Malaysia

[2] Kumpulan Jebco (M) Sdn.Bhd., Lot 1569, Jalan Kusta, Kawasan Perindustrian Kampung Jaya, 47000 Sungai Buloh, Selangor Darul Ehsan, Malaysia.

E-mail: kshamsul@lgm.gov.my

Abstract. An engine mount secures the engine of a vehicle to its chassis. Successful application of elastomeric mounts as vehicle engine mounts has been observed over the years. Epoxidised Natural Rubber (ENR) and Deproteinised Natural Rubber (DPNR) are specialty rubbers with a potential for application in the development of engine mounts. In this study, a blend comprising natural rubber (NR) and polybutadiene rubber (BR) is used to produce the Original Equipment Manufacturer (OEM) part. Epoxidised Natural Rubber (ENR25 and ENR50) and Deproteinised Natural Rubber (DPNR) developed by Malaysian Rubber Board were used and compared with NR by evaluating both physical and mechanical properties of the rubbers with different filler loadings. The physical properties of NR exhibited excellent results opposed to other rubbers. However, a blend of ENR25 and BR reflected good compression set at 100°C for 96 hours. The performance of an engine mount comprising a blend of ENR50 and BR is promising in reference to the excellent mechanical properties over that of other rubbers. Dynamic stiffness and damping characteristics of the engine mount play an important role to achieve vibration isolation with acceptable engine motion control in a vehicle's engine mount system. It is perceived that engine mounts produced with a blend of ENR50 and BR can improve vibration effects in dynamic conditions.

1. Introduction

Generally the vehicle engine mounting system consists of an engine and several numbers of mounts connected to the vehicle structure. In the current technology of automobile industries, engine mounting systems have been primarily used to support the weight of the engine. This will ensure that the engine can be freely maintained in its specific design position. Besides supporting the engine weight, it is able to isolate the driver and from both noise and vibration generated by the engine [1]. Static and dynamic stiffnesses are very important properties to cater the performance of engine. It is known that rubbers are extensively used in many applications because of their large reversible elastic deformation, excellent damping and energy absorption characteristics [2] and it is a common material used in automotive industries. The present study was motivated by the improvement of engine mount performance, the success of which depends on the capabilities of rubber. It is well known that rubber has an ability to deform and prevent of shocks and vibrations. This latter function depends on two properties of rubber to deform: firstly, its damping effect or departure from perfect elasticity (hysteresis) which enables it to absorb the energy of shock and vibration; secondly, the fact that by choosing the correct dimensions a rubber mounting can be made as it were "impervious" to vibrations of a given frequency [3]. Specialty rubbers such as epoxidized natural rubber (ENR) 25, ENR50 and deproteinised natural rubber (DPNR) have potential to be used in the development of automotive products. These materials cover a wide range of properties desirable to meet certain requirement in engineering application reduced creep, enhancement in resilience [4] and good dynamic properties [5]. The advantages of ENR include excellent properties of damping, wet-grip or rolling resistance, oil resistance and gas permeability compared to NR [6]. In addition, DPNR has low creep and stress relaxation, low water absorption, low compression set and a more consistent modulus when subjected to conditions of variable humidity compared to NR [7, 8]. In the present investigation study, a comparative study on several types of natural and specialty rubbers was made by carrying out physical and mechanical tests. Polybutadiene Rubber (BR) was used and blended with natural and specialty rubbers to investigate the physical and mechanical properties of each rubber compound in the development of engine mount.

2. Materials and Methods

The modified rubber was compounded according to a compounding formulation for an engine mount as shown in Table 1. The compounds used in this investigation were Standard Malaysia Rubber CV 60, ENR 25, ENR 50, DPNR and Polybutadiene Rubber (BR). Eight rubber formulations containing two different loading of carbon black were mixed and cured as nominal 2mm thick sheets. NR is known for its inherent strength, its ability to withstand large deformations and resistance to crack propagation. BR, on the other hand, exhibits better abrasion resistance, resilience, and resistance to initiation of cracks. Therefore, a combination of NR and BR finds vast application in various rubber products, including the tread and sidewall of the tyre and elastomer part the engine isolate mount [9,10]. All rubber compounds were blended with BR because this synthetic rubber could improve the damping property. All the rubber compounds were prepared using a lab size two roll mill machine at temperature of 165°C.

Table 1. Rubber compound formulation

Ingredients/Name	CV60/ BR (1)	ENR25 /BR(1)	ENR50 /BR(1)	DPNR /BR(1)	CV60/ BR(2)	ENR25 /BR(2)	ENR50 /BR(2)	DPNR /BR(2)
	phr							
NR (SMR CV60)	70	-	-	-	70	-	-	-
ENR 25	-	70	-	-	-	70	-	-
ENR 50	-	-	70	-	-	-	70	-
DPNR	-	-	-	70	-	-		70
BR	30	30	30	30	30	30	30	30
Carbon black	26	26	26	26	36	36	36	36
ZnO	5	5	5	5	5	5	5	5
Stearic acid	2	2	2	2	2	2	2	2
Sulfur	1.7	1.7	1.7	1.7	1.7	1.7	1.7	1.7
CBS	1.3	1.3	1.3	1.3	1.3	1.3	1.3	1.3
TMTD	0.5	0.5	0.5	0.5	0.5	0.5	0.5	0.5
Oil	4	4	4	4	4	4	4	4
Cure Condition (Time, Temperature)	20 min, 165°C	20 min, 165°C	20 min, 165°C	20 min, 165°C	20 min, 165°C	20 min, 165°C	20 min, 165°C	20 min, 165°C

IOP Conf. Series: Materials Science and Engineering **548** (2019) 012004 doi:10.1088/1757-899X/548/1/012004

2.1. Measurement of rubber viscosity
Viscosity is a property of the fluid which opposes the relative motion between the two surfaces of the fluid that are moving at different velocities. This property is referred to evaluate of the flow of rubber compound. Viscosity of each rubber compound was measured using MV2000 by referring to the ISO 289-1 document, Part 1: Determination of Mooney viscosity.

2.2 Measurement of rubber compound hardness
Rubber hardness is essentially a measure of the reversible, elastic deformation produced by a specially-shaped indentor under a specified load. In the present investigation, three hardness buttons of each compound were cured and used for hardness tests. The test was carried out referring to BS ISO 48:2007.

2.3 Characterisation of tensile and elongation at break in aged and unaged conditions
Eighty tensile dumbbells represent of eight rubber compound formulations (Type 2, BS ISO 37:2005) were prepared for aged and unaged tensile and elongation at break tests. The dumbbells test pieces were aged at the condition of 70°C for 70 hours and the tensile strength tests were carried out according to the BS ISO37.

2.4 Characterisation of compression set property
Forty eight of test pieces were prepared according to ISO 815 (1993)/JIS K6258 (1993). Two type of compression set were carried out at temperatures of 22±3°C for 72 hours and 100°C for 90 hours respectively.

2.5 Product performance testing
An engine mount was produced using a mould with two cavities. In the moulding process, external and internal sleeve metals were used to produce an engine mount component. All the metal parts were going through metal treatment process. Four units of engine mount were cured using an injection machine and each cycle was produced using eight different rubber compounds. Static and dynamic tests were carried out on a servopulser Shimadzu testing machine. A special jig was designed and fabricated as shown in Fig. 1 and 2. The testing conditions are shown in Table 2.

Fig. 1. The overview of static compression test.

Fig. 2. Dynamic test setup on a Shimadzu testing machine

Table 2. Static and dynamic testing conditions

Static Test			
Preload (mm)	0 - 11mm (2 times)		
No. of Cycle	5		
Speed rate (mm/min)	13mm/min		
Static stiffness measuring	850N - 1150N		
Dynamic Test			
Preload (N)	1000N		
Frequency (Hz),	15	40	100
Displacement Amplitude (mm)	±1.0mm	±0.25mm	±0.05mm

The static stiffness, K_s was calculated using Equation 1.

$$K_s = \frac{(L_1 - L_2)}{(D_1 - D_2)} \quad (1)$$

L_1 and L_2 indicate the load applied region in the actual application. The calculation load region is 850N – 1150N and D_1 and D_2 are the value of displacement at the load position as shown in Fig. 3.

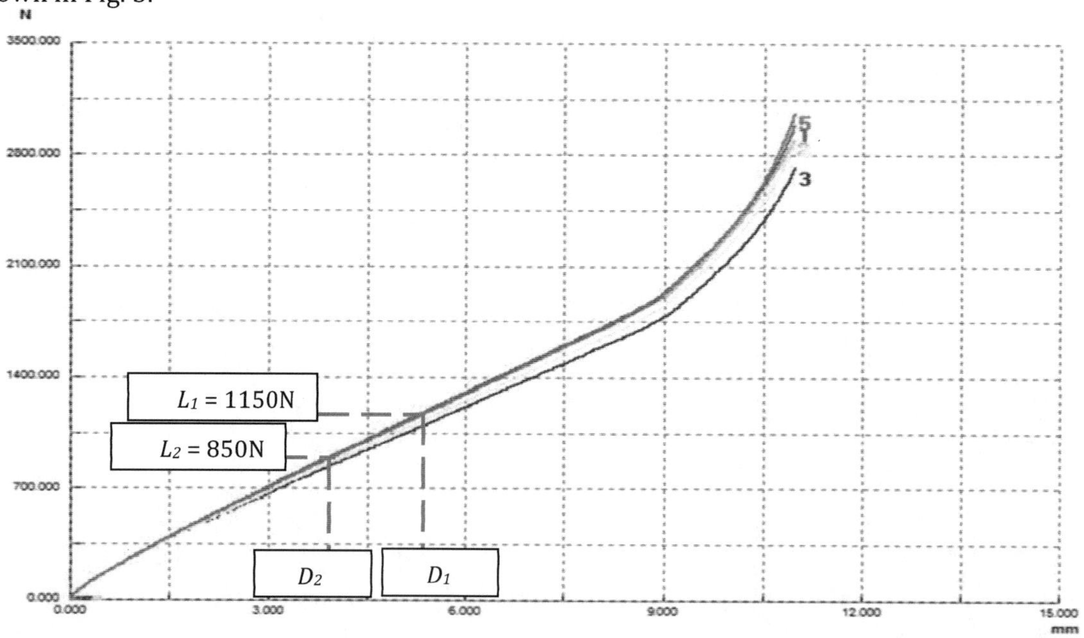

Fig. 3. Force and deflection curves for one of the static tests for 5 cycles.

The absolute spring constant or dynamic stiffness, K_d was calculated and each dynamic stiffness according to the frequency was obtained using the equation (1). Fig. 4 shows one example of testing data which was analysed to determine dynamic stiffness at frequency of 15Hz.

IOP Conf. Series: Materials Science and Engineering **548** (2019) 012004 doi:10.1088/1757-899X/548/1/012004

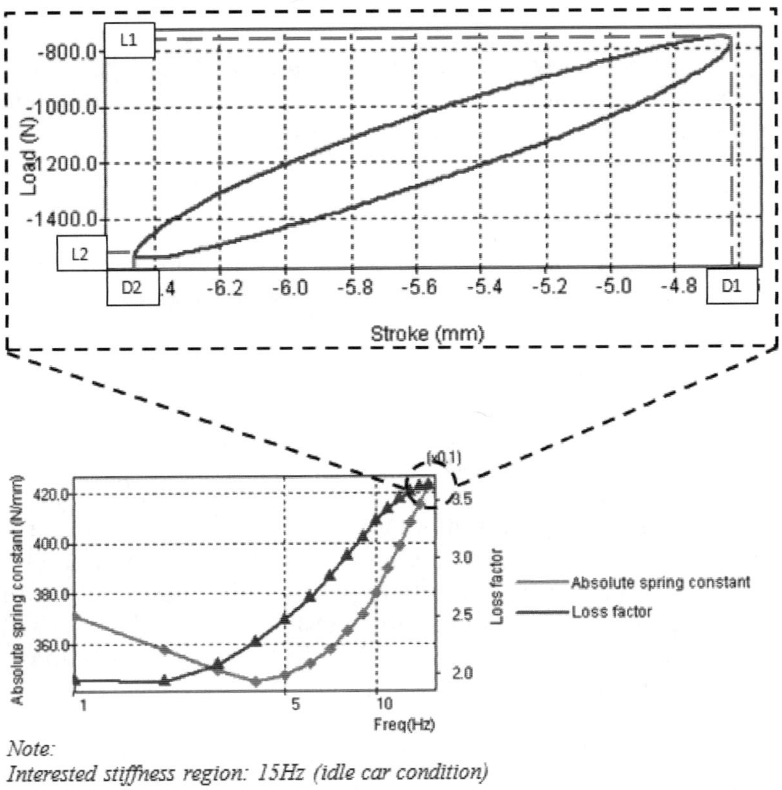

Fig. 4. A hysteresis loop of load-stroke displacement at frequency of 15 Hz

Because rubbers are not perfectly elastic, the strain during cyclic deformation always lags slightly behind the stress. If the relaxation between the force and the deformation is approximately linear the application of a sinusoidal deformation will result in a sinusoidal force of the same frequency but displaced in terms of phase by an amount termed the loss angle (Fig. 5)[11].

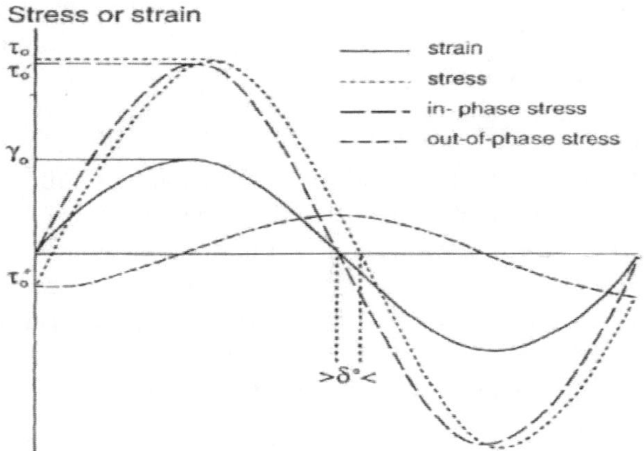

Fig. 5. Response of a linear viscoelastic material to an imposed sinusoidal shear strain of amplitude γ_o . The stress of amplitude τ_o leads the strain by the phase angle δ_o. The in-phase modulus $G' = \tau_o{}'/\gamma_o$ and the out-of-phase modulus $G'' = \tau_o{}''/\gamma_o$, where $\tau_o{}'$ and $\tau_o{}''$ are the amplitudes of the in and out-of-phase stress respectively [11].

It is convenient to consider the elastic, in-phase response and the viscous, out-of-phase in terms of two moduli. The overall response can then be expressed as a complex modulus. For instance, in shear the modulus (stress amplitude/strain amplitude),

$$G^{*}= G' + jG'' \qquad (2)$$

where G' is the in-phase, 'storage' modulus and G'', the out-of-phase, 'loss' modulus. Modulus alone is here used to refer to G*. The phase or loss angle δ is given by

$$\text{Tan } \delta= G'/G' \qquad (3)$$

3. Results and Discussion

3.1 Viscosity cure of rubber compounds

The most common parameter used to assess flow behavior is viscosity particularly in the injection mould processing of engine mount production. Rubber flow is very important for shaping process. Basically, viscosity affects both mixing and fabrication processes [12]. In the present study, the value of rubber viscosity is determined to correlate with other mechanical properties. Table 3 shows the results of Mooney viscosity for each rubber. ENR 50 shows high Mooney viscosity follow by ENR 25, DPNR and NR. The number proportional with the torque as viscosity index (Mooney viscosity) is shown in Table 3.

Table 3. Mooney viscosity of each rubber

Rubber	Mooney Viscosity (MV)
NR	57
ENR 25	88
ENR 50	91
DPNR	70

Here, the specialty rubbers have high values of viscosity compared to natural rubber. The viscosity of NR is sufficiently low and eases to incorporate with compounding ingredients. However, the viscosity of specialty rubbers is sufficiently high which can improve dispersion in the mixing process of rubber compound for engine mount [12].

3.2 Hardness of rubber compound

The results of hardness are shown in Fig. 6 according to the different rubber compounds. The results indicate that hardness increased with the increasing carbon black loading from 26phr to 36phr. ENR 50/BR showed high hardness compared to other rubber compounds followed by the hardness of DPNR. It is believed that the hardness of rubber compound can be increased using specialty rubber and this will be an alternative approach to optimise rubber hardness. The mechanism of the increase of hardness is because of the specialty rubbers have been epoxidised and deproteinised compared with NR and ease to blend completely with BR. This will helps all compounding ingredients incorporated into rubber.

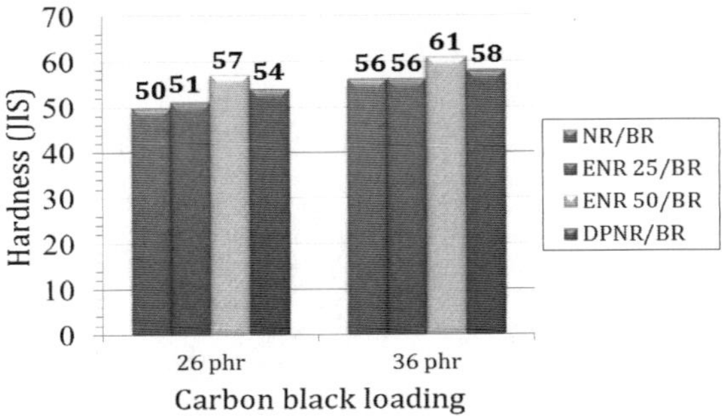

Fig. 6. Hardness value of different rubber compounds according to carbon black loading

3.3 Tensile strength

Tensile test was undertaken using all different rubber compounds. Fig. 7 shows the histogram of tensile strength versus carbon black loading for each rubber compound. The NR/BR showed high tensile strength compared to other rubber compounds. However, the tensile strength of DPNR/BR compound is higher than ENR 50/BR and ENR 25/BR compounds, respectively. Most of the tensile strength increased slightly when the loading of carbon black increased from 26phr to 36phr. The increase of carbon black gives more reinforcing filler and the degree of reinforcement into rubber.

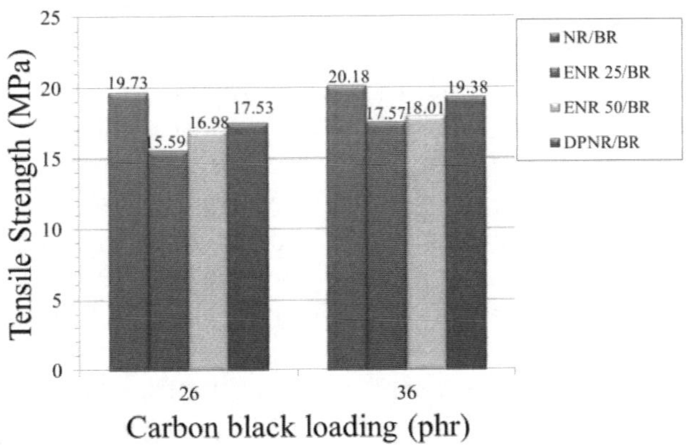

Fig. 7. Histogram of tensile strength versus carbon black loading

3.4 Elongation at break

Fig. 8 shows the histogram of elongation at break of each rubber compound versus carbon black loading. The elongation at break of DPNR/BR compound was above 464% which is higher than other rubber compounds. In contrast, ENR 50/BR compound shows the lowest elongation at break among the rubber compounds. Most of the elongation at break slightly reduced when the carbon black loading is increased from 26phr to 36phr. It is confirmed that the increase of carbon black loading give less flexibility to the rubber compounds.

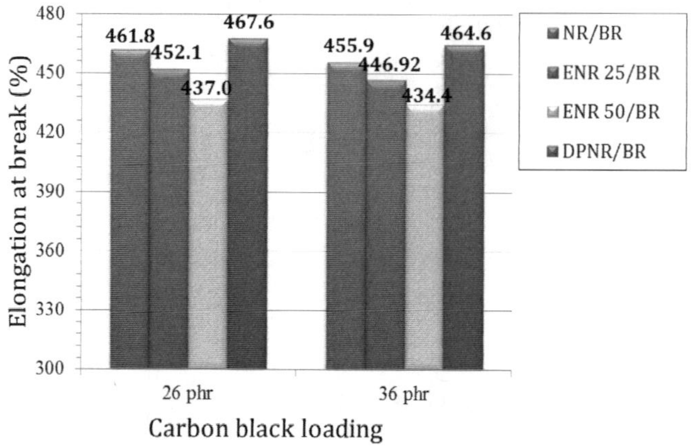

Fig. 8. Histogram of elongation at break versus carbon black loading

3.5 Tensile strength and elongation at break in normal and aged condition

Fig.s 9 and 10 show the results of tensile strength and elongation at break in both of unaged and aged conditions. The comparison of carbon black loading for these two properties was also carried out. In Fig. 8, DPNR/BR compound showed the lowest different percentage of tensile strength among other rubber compounds (T1- 26phr of carbon black) which was 3.26% between unaged and aged conditions. This is because of DPNR/BR compound has good dispersion of reinforcing filler compared with other compounds. In contrast, the different percentages of tensile strength in normal and aged conditions for specialty rubber compound (ENR 25/BR and ENR 50/BR) using 26phr of carbon black (T1) are higher than NR/BR and DPNR/BR compounds which are above 20%. Nevertheless, the different percentage of rubber compounds using 36phr of carbon black (T2) for normal and aged conditions was not very significant which less than 8%. DPNR/BR showed the lowest different percentage of tensile strength compared to other compounds. In Fig. 10, the elongation at break for rubber compound using 26phr of carbon black in unaged and aged was more significant compared with the rubber compound using 36phr of carbon black. However, the different percentage of the elongation at break for ENR50/BR rubber compound using 36phr of carbon black is higher than that of ENR 50/BR rubber compound using 26phr of carbon black. This because of the elongation at break after ageing process of ENR50/BR compounds with carbon black loading of 36phr is lower compared to ENR50/BR compounds with carbon black loading of 26phr.

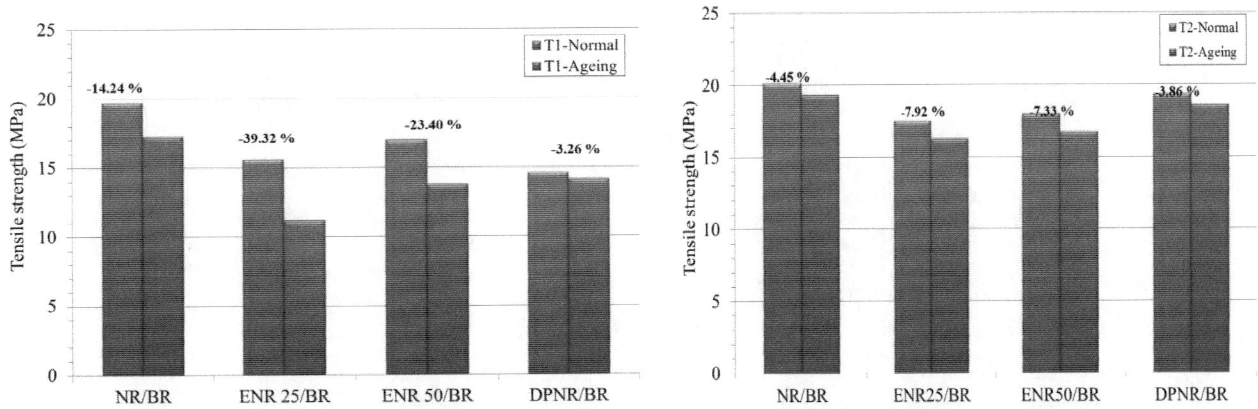

Fig. 9. Histogram of tensile strength versus rubber compound for aged and unaged (T1- 26 phr carbon black, T2- 36phr carbon black)

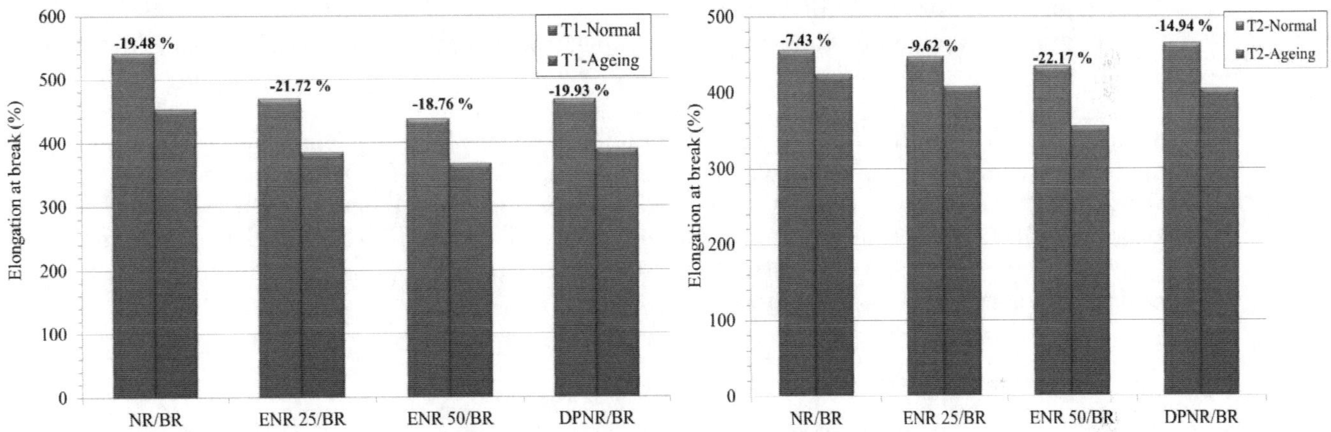

Fig. 10. Histogram of elongation at break versus rubber compound for aged and unaged (T1- 26 phr carbon black, T2- 36phr carbon black)

It is found that the ageing process of all rubber compounds using 26phr and 36phr of carbon black reduced the tensile strength and the elongation at break. The different percentages of both properties are paramount to evaluate all the rubber compounds whether their properties changed significantly.

3.6 Compression set of rubber compound

Figures 11 and 12 show the histogram of compression set versus carbon black loading for two different ageing conditions. ENR 25/BR showed the lowest percentage of compression set compared to other rubber compounds at 70°C for 22 hours as shown in Fig. 11. Fig. 12 shows DPNR/BR which is the lowest compared to other rubber compounds after aged at 100°C for 96 hours. These results showed that specialty rubbers have good compression set except ENR25/BR compound which the compression set is higher at 100°C for 96 hours. Based on the results of compression set, it is assumed that specialty rubbers have good density and an efficient cure system compared with natural rubber.

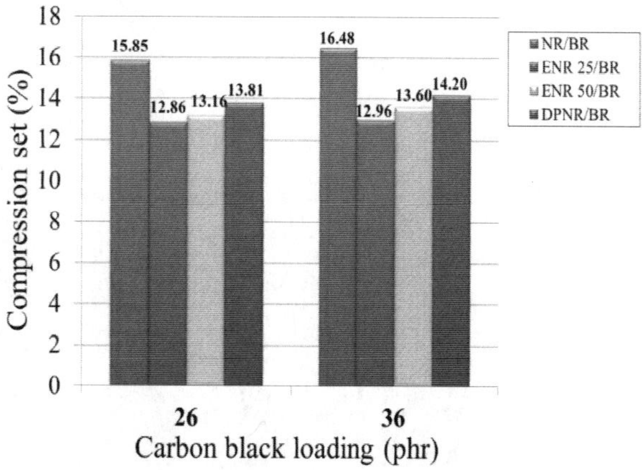

Fig. 11. Histogram of compression set results at 70°C for 22 hours.

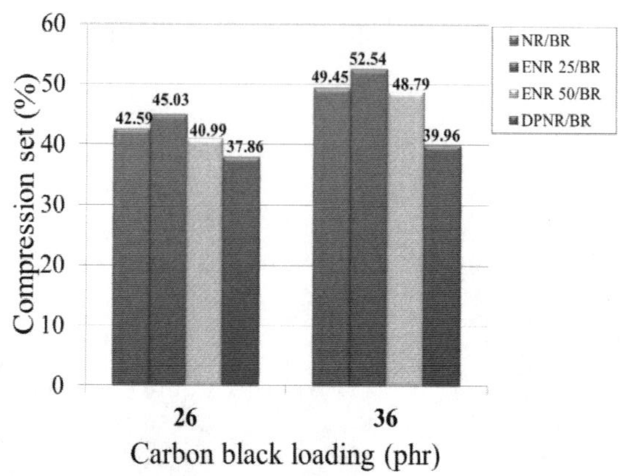

Fig. 12. Histogram of compression set results at 100°C for 96 hours.

3.7 Performance test results of engine mount

The evaluation of an engine mount using all rubber compounds is essential to ensure that the performance of the component is adequate with engine specification requirement. Fig. 13 shows static stiffness of each engine mount using four different rubber compounds. The results indicate that static stiffness of rubber compound using 36phr of carbon black (T1) is higher than rubber compound using 26phr of carbon black (T2). ENR 50/BR showed high static stiffness compared to other rubber compounds. This is because the epoxidation process of 50% epoxy moles (ENR50%) can induce more interaction with reinforcing fillers particularly carbon black and become superior against natural rubber due to random arrangement of the epoxy groups on the main chain of the polymer.

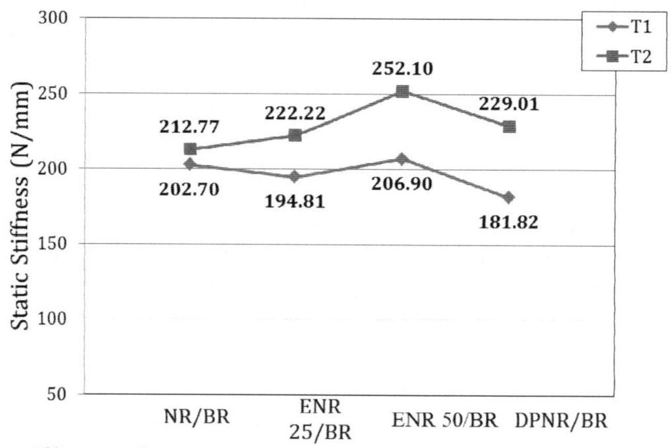

Fig. 13. Static stiffness of engine mount using four different rubber compounds.

Dynamic stiffness is also an important property in the development of engine mount. The higher the content of carbon black loading can increase static stiffness, dynamic stiffness and tan delta. The response of a mounted body to a vibration about the static deflection is controlled by the dynamic stiffness. For the filled rubber, the dynamic stiffness is determined by the slope of the force-deflection curve in retraction away from the static deflection point; the slope at this point of the force-deflection curve for loading may be substantially less, resulting in the so-called dynamic/static ratio. Fig. 14 and 15 show results of dynamic stiffness for engine mount using 26phr and 36phr of carbon black. ENR 50/BR rubber compound showed high dynamic stiffness compared to other rubber compounds at all frequencies (15Hz, 40Hz and 100Hz). This is because ENR50/BR compound showed excellent crosslink density with more interaction between rubber matrix and carbon black of 26phr and 36phr. ENR 25/BR showed high dynamic stiffness compared with DPNR/BR and CV60/BR. These results proved that the use of ENR can improve dynamic stiffness of rubber compound. ENR/BR has potential to replace CV60/BR rubber compound in the development of engine mount.

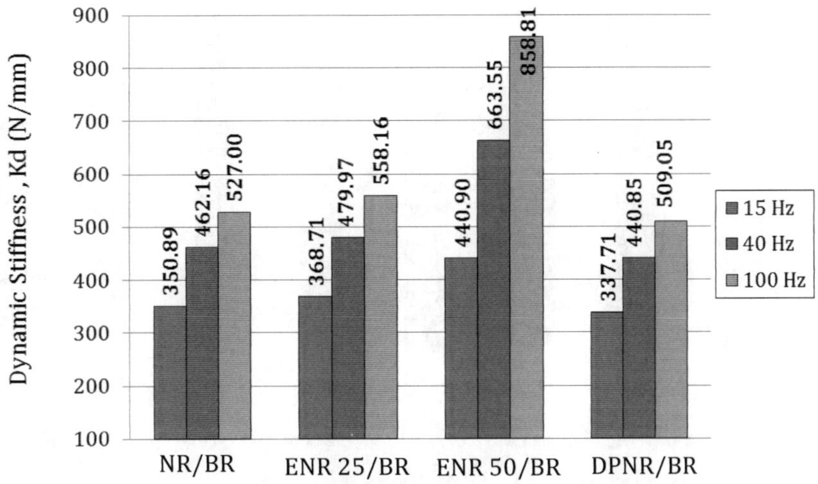

Fig. 14. Histogram of dynamic stiffness of engine mount using 26phr of carbon black for each rubber compound.

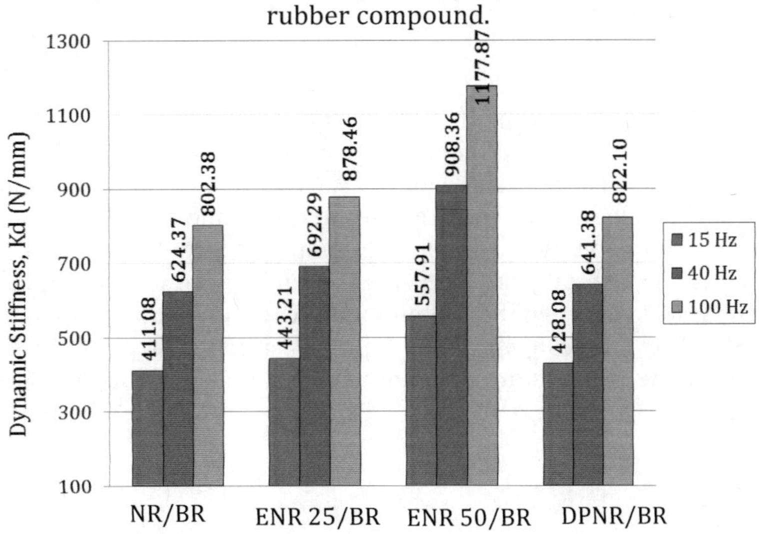

Fig. 15. Histogram of dynamic stiffness of engine mount using 36phr of carbon black for each rubber compound.

The tan delta quantifies the way in which a material absorbs and disperses energy. It expresses the out-of-phase time relationship between an impact force and the resultant force that is transmitted to the supporting body. Fig. 16 showed the histogram of tan delta of engine mount versus rubber compounds according to carbon black loading at 15Hz. ENR 50/BR shows high tan delta compared to other rubber compounds. This rubber compound demonstrates the greatest capability to dissipate energy and contribute more effective in damping properties. The higher the tan delta, the greater the damping coefficient, the more efficient the material will be in effectively accomplishing energy absorption and dispersal.

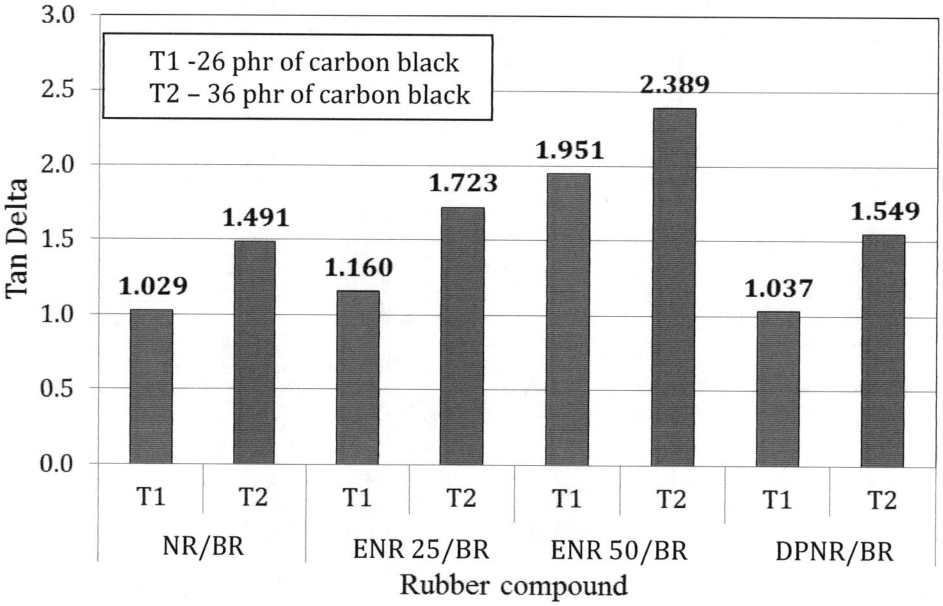

Fig. 16. Histogram of tan delta of engine mount versus rubber compound at 15Hz.

4. Conclusion

The comparison of physical and mechanical properties for BR blends with NR, ENR and DPNR has been successfully undertaken. The higher the amount of carbon black loading increases the physical and mechanical properties. The ageing process of all rubber compounds using 26phr and 36phr of carbon black reduced the tensile strength and the elongation at break. The specialty rubbers showed good compression set compared to NR. ENR 50/BR blends showed excellent physical and mechanical properties among rubber compounds. ENR 50/BR indicates high tan delta compared to other rubber compounds. Thus, it is apparent from the engineering aspect that ENR 50/BR can be considered as a potential material for automotive engineering application such as engine mount since the tan delta is significantly higher than the rest of the rubber compounds.

5. Acknowledgement

The authors gratefully acknowledge the support from management of Malaysian Rubber Board and Kumpulan Jebco (M) Sdn. Bhd. to publish this work.

IOP Conf. Series: Materials Science and Engineering **548** (2019) 012004 doi:10.1088/1757-899X/548/1/012004

References

[1] Yan Wen-Bing, Jiang Shao Zhong, Cai Yu Jun, Wen Li Zhi and Chen Yue 2009 Experimental study of engine mount system's vibration characteristic, *World Congress Software Engineering* 43-46

[2] Gent A N 2001 Engineering with rubber Hanser Fardner 18

[3] Davey A B and Payne A R 1964 Rubber in Engineering Practice, *Rubber and Plastic Research Association of Great Britain*, Applied Science Publishers Ltd. Ripple Road, Barking, Essex, England

[4] Natural Rubber Technical Information Sheet: D6 Dry Rubber Series 1976 The Malaysian Rubber Producers' Research Association

[5] Khoo T C, Abdul Rahman R. and Kamarul Baharin B 2003 Deproteinised Natural Rubber (DPNR) *Monograph No.10*, Malaysian Rubber Board

[6] Ekoprena brochure 2014 *Malaysian Rubber Board*

[7] Pureprena brochure 2014 *Malaysian Rubber Board*

[8] Nor Hidayaty K, Shamsul K, Asrul M, and Ahmad Nazir K 2011 Comparison of some properties of deproteinised natural rubber vulcanisates and observations on dynamic load analysis, *J. Rubber Res* **14**(3) 129-138

[9] Amin A F M S, Lion A, and Höfer P 2010 Temperature history effects in rubber: literature review and critical experiments, *Zammzeitschrift Fur Angewandte Mathematik und Mechanik*, **90** 347 – 369

[10] Ueda, T. Ohyama, H. Watanabe, and A. Yoshioka, 1989, Vibration isolation characteristics of chemically modified solution-polymerized rubbers for engine mount use, *J. Appl. Polym. Sci.* 44 229 – 241

[11] Lindley P B 1992 Engineering design with rubber *The Malaysian Producers' Research Association* ISSN 0956-3856

[12] Azemi Samsuri 2009 An introduction to polymer science and rubber technology *University Publication Centre (UPENA) UiTM* ISBN 978-305-288-2

IOP Conf. Series: Materials Science and Engineering **548** (2019) 012005 doi:10.1088/1757-899X/548/1/012005

Rheological Properties of Carbon Black/Silica Hybrid Filler in Acrylonitrile Butadiene Rubber

MH Mohd Hasan[1] JK Kler[1] SK Ong[1] and S Manroshan[2]

[1] Universiti Kuala Lumpur Branch Campus Malaysian Institute of Chemical and Bioengineering Technology (UniKL MICET), Lot 1988, Kawasan Perindustrian Bandar Vendor, Taboh Naning, 78000 Alor Gajah, Melaka, Malaysia

[2] Malaysian Rubber Board, RRIM Research Station, 47000 Sungai Buloh, Selangor, Malaysia

E-mail: skong@unikl.edu.my

Abstract. Filler is a common additive added into rubber vulcanizate to achieve reinforcing performance. Typically, filler is compounded via two-roll mill which results in filler agglomeration and lower reinforcing efficiency. To gain higher reinforcing efficiency, filler agglomeration must be minimized via masterbatch preparation. This study investigates colloidal stability of carbon black/silica (CB/SiO_2) hybrid filler and effectiveness between typical compounding and masterbatch process for NBR vulcanizate production. CB/ SiO_2 hybrid filler was fixed at 20 wt.% with ratios of 100/0, 75/25, 50/50, 25/75 and 0/100 were dispersed separately in sodium hydroxide (NaOH) via ball milling. The zeta potential, particle size and viscosity results showed CB and SiO_2 dispersions stabilized after 48 hours of ball milling. NBR masterbatch with 50/50 ratio has highest flow resistance. SiO_2 reinforced NBR masterbatch showed higher flow resistance compared to CB due to finer silica particle size. However, scorch and cure time for SiO_2 reinforced NBR is slower than CB reinforced NBR as silica deactivate the accelerator. CB/SiO_2 at 50/50 has highest T_{max} and ΔT while 100/0 and 0/100 has insignificant differences. As conclusion, CB and SiO_2 dispersions were stable based on zeta potential analysis and particle size reduced as ball milling time increased.

1. Introduction

Acrylonitrile butadiene rubber (NBR) is a synthetic rubber with good elongation properties, adequate tensile strength and resistant to oil.[1] Its properties affected by acrylonitrile and butadiene ratio. Increasing acrylonitrile content in NBR increase its polarity and transition glass temperature (T_g).[2]

NBR is reinforced by filler e.g., silica (SiO_2) and carbon black (CB).[3] SiO_2 is a polar filler while CB is a non-polar filler.[4] Lower filler's structure gives better load sharing within rubber vulcanizate thus having better reinforcement.[5] Typically filled compound is compounded via two-roll mill which results in filler agglomeration.[6] The agglomerated filler results in lower reinforcing efficiency. Thus, to gain higher reinforcing efficiency, filler agglomeration must be minimized.[7]

This can be achieved via masterbatch preparation; filler is dispersed in a selected medium before it is mixed with the latex. Thus, the colloid stability of the dispersed filler plays an important role.[8] The colloid stability depends on its particle size, total potential energy (sum of repulsive and attractive force), diffuse electrical double layers (surface charge) and polarity.[9] Deyaguin-Landau-Verwey-Overbeek (DLVO) theory describes stability is attained when repulsive Coulomb force is greater than attractive Van der Waals force.[10] The repulsive force determined by the magnitude of zeta potential from electrical double layers.[11]

NBR latex is a polar rubber thus poor dispersion with non-polar filler such as CB may exist and create poor reinforcement.[12] Dispersing SiO_2 in NBR latex is also challenging as both the filler and latex are polar.[13] The aim of this study is to investigate the zeta potential, particle size and viscosity of SiO_2 and CB dispersion for colloidal stability. The cure characteristics of CB/SiO_2 filled NBR masterbatch also will be studied.

2. Experimental

2.1. Material

Table 1 shows the raw materials used in this study and its respective manufacturer.

Table 1. Raw materials used in this study and its respective manufacturer

Material	Manufacturer
Silica	Evonik Industries, Germany
Carbon black	Cabot (M) Sdn. Bhd.
Nitrile butadiene latex	Rivertex Technical Fabrics Group B.V.
[a])6PPD	Flexys A.L.P.
[b])TBzTD	Bayer (M) Sdn. Bhd.
[c])TQ	Sovereign Chemical Company USA
Zinc Oxide	Bayer (M) Sdn. Bhd.
Stearic Acid	Pacific Oleochemicals
Wax Antilux 654	Rhein Chemie Rheinau Gmbh Germany
Sulphur	Bayer (M) Sdn. Bhd.

a)6PPD: N-1,3-dimethylbutyl)-N-phenyl-p-phenylenediamine
b)TBzTD: Tetrabenzylthiuram disulfide
c)TQ: Polymerized 1,2 dihydro 2,2,4- trimethyl quinolone

2.2. CB and SiO_2 dispersions in NaOH

10 wt.% of CB and SiO_2 were dispersed in NaOH separately via ball mill machine (AMACO Model No. B, Serial No. C293) for 72 hours. Particle size and zeta potential of dispersion were analysed using zeta sizer (model: Brookhaven ZetaPlus) at ball milling time interval of 0.25, 0.5, 0.75, 1, 2, 3, 4, 5, 6, 24 and 72 hours.

2.3. CB and SiO_2 masterbatch

Based on particle size and zeta potential results, the ball milling time the CB and SiO_2 dispersions were fixed to 24 hours. These dispersions then mixed with NBR latex to form NBR mixtures according to CB/ SiO_2 ratio of 100/0, 75/25, 50/50, 25/75 and 0/100. The mixtures were coagulated to produce CB/ SiO_2 masterbatch using a microwave (Model: Electrolux EMM2001W) before oven dried (Model: Memmert) until constant weight.

2.4. Rubber compounding

The rubber compounding recipes summarizes in Table 2. Compounding was carried out using two-roll mill machine (Model: SLIMC SXK-160A).

Table 2. Formulations for rubber compounding

	Formulation (phr)				
Material \ Recipe	1	2	3	4	5
Silica masterbatch	0	25	50	75	100
Carbon black masterbatch	100	75	50	25	0
6PPD	2.4	2.4	2.4	2.4	2.4
TBzTD	1.0	1.0	1.0	1.0	1.0
TQ	1.5	1.5	1.5	1.5	1.5
Zinc oxide	3.5	3.5	3.5	3.5	3.5
Stearic acid	2.4	2.4	2.4	2.4	2.4
Wax antilux 654	1.0	1.0	1.0	1.0	1.0
Sulphur	1.2	1.2	1.2	1.2	1.2

2.5. Cure characteristics
Cure characteristics were carried out on all the filled rubber compounds. The test was performed using a rheometer (Model: Monsanto 100) according to ASTM D2084-93. The rheometer was set at 150 °C for 60 minutes with 12 g of sample.

3. Results and Discussion
3.1. Zeta potential and particle size
Fig. 1 and 2 show decreasing trend of particle size for CB and SiO_2 respectively. The increasing ball milling time will produce smaller particle size of filler. More time for mechanical treatment allows filler to be further reduced from larger into smaller size.[14] The magnitude of zeta potential increased (negative charge) as the ball milling time increased for both CB and SiO_2 dispersions. Alkaline dispersion medium such as NaOH gives negative value of zeta potential due to more hydroxyl groups.[15] Dispersion with zeta potential greater than or less than +30 mV or -30 mV respectively is a stable dispersion.[16]

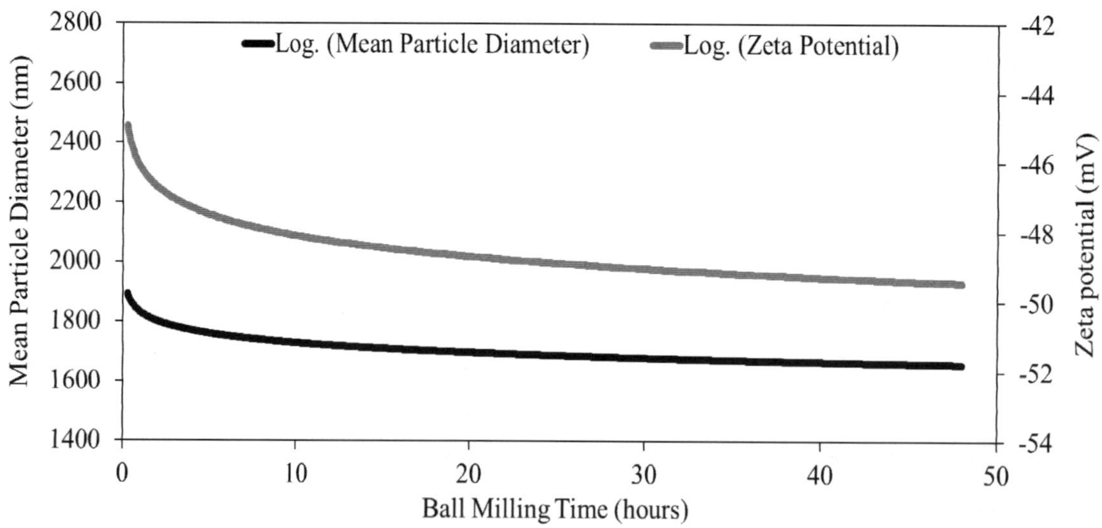

Fig. 1. Zeta potential and particle size of CB in NaOH dispersion

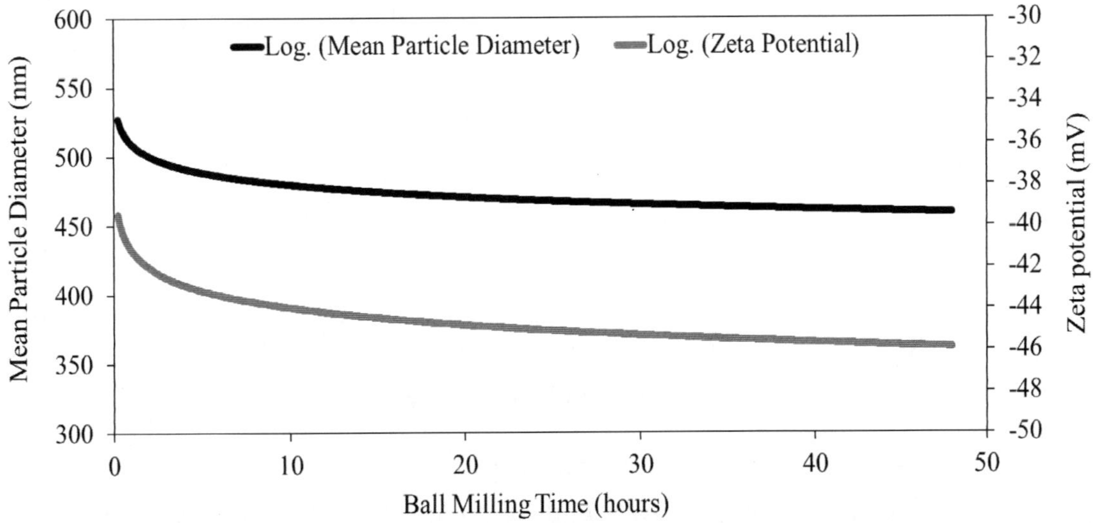

Fig. 2. Zeta potential and particle size of SiO_2 in NaOH dispersion

Despite both CB and SiO_2 has their particle size decreased as the ball milling time increased, SiO_2 has smaller size compared to CB as showed in Fig. 3. SiO_2 most likely to have less aggregation formation due to polar hydroxyl group presents on its surface.[17] CB is non-polar thus attractive Van der Waals force may cause CB to form aggregates and affecting the mean particle size.[18]

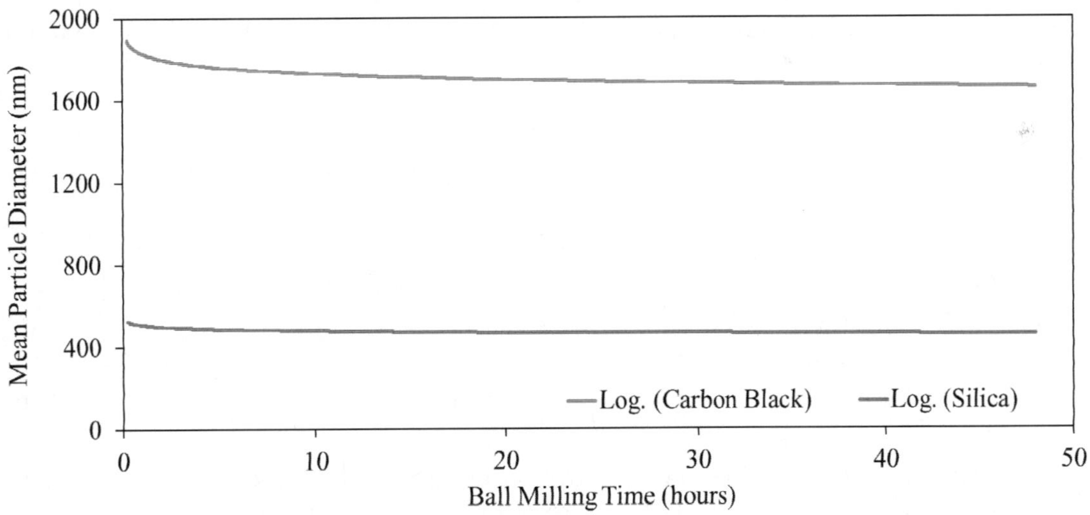

Fig. 3. Comparison of particle size of between CB and SiO_2 in NaOH dispersion

3.2. Viscosity

Fig. 4 shows viscosity of both CB and SiO_2 dispersions decreased as the ball milling time increased. CB has bigger particle size most likely due to higher filler structure as compared to SiO_2. Higher magnitude of CB dispersion zeta potential suggested higher repulsive force among the CB particles. Therefore, due to the CB's structure and its zeta potential, CB has higher viscosity.[19]

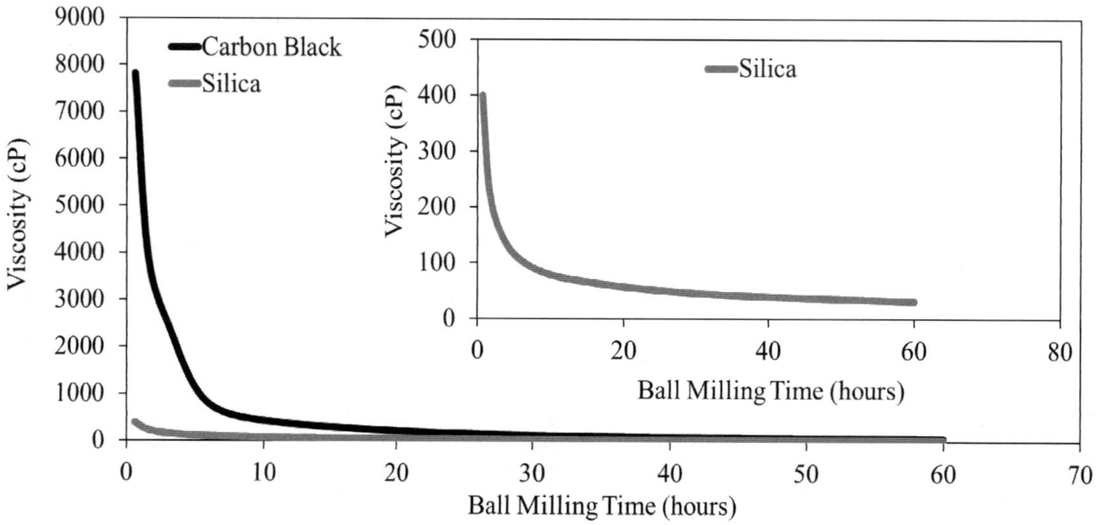

Fig. 4. Viscosity of CB and SiO_2 in NaOH dispersion as ball milling time increased

3.3. Cure characteristics

Cure characteristics of NBR masterbatches filled with CB/ SiO_2 hybrid filler at different ratio (CB/ SiO_2 of 100/0, 75/25, 50/50, 25/75 and 0/100) are shown in Table 3. Higher CB loading shorten both the scorch time and cure time of NBR masterbatch from 8.67 to 5.21 mins and 9.00 to 6.00 mins, respectively as incorporating more CB in rubber vulcanizate assist in vulcanization process.[20] Meanwhile, more SiO_2 results in longer scorch time and cure time of NBR masterbatches. It has hydroxyl group on the surface that may deactivate accelerator. Lower amount of sulfureting agent and actual accelerator making crosslinking to be slower and does affect the vulcanization process.[21]

Table 3. Cure characteristics of CB/SiO_2 filled NBR masterbatch

Characteristic	100/0 CB/SiO_2	75/25 CB/SiO_2	50/50 CB/SiO_2	25/75 CB/SiO_2	0/100 CB/SiO_2
Scorch time, t_{s2} (min)	5.21	6.11	6.72	6.25	8.67
Cure time, t_{90} (min)	6.00	7.00	7.00	7.00	9.00
Minimum torque, T_{min} (lb-in)	11.25	10.35	15.10	15.08	13.20
Maximum torque, T_{max} (lb-in)	54.91	56.73	72.10	49.10	55.70
Delta torque, ΔT (lb-in)	43.66	46.38	57.04	34.98	42.50

100% filled NBR masterbatch with CB or SiO_2 has insignificant difference in T_{max}, T_{min} and ΔT as shown in Fig. 5. The T_{max}, T_{min} and ΔT has less than 5% difference between them. Loading of hybrid filler at 50/50 has the highest T_{max} and ΔT. Increasing CB loading does not have bigger influenced on T_{min} of NBR masterbatch based on Fig. 5. Filled NBR compound with CB/SiO_2 ratio of 50/50 showed highest T_{max} and T_{min} which might due to the re-dispersion of both fillers.

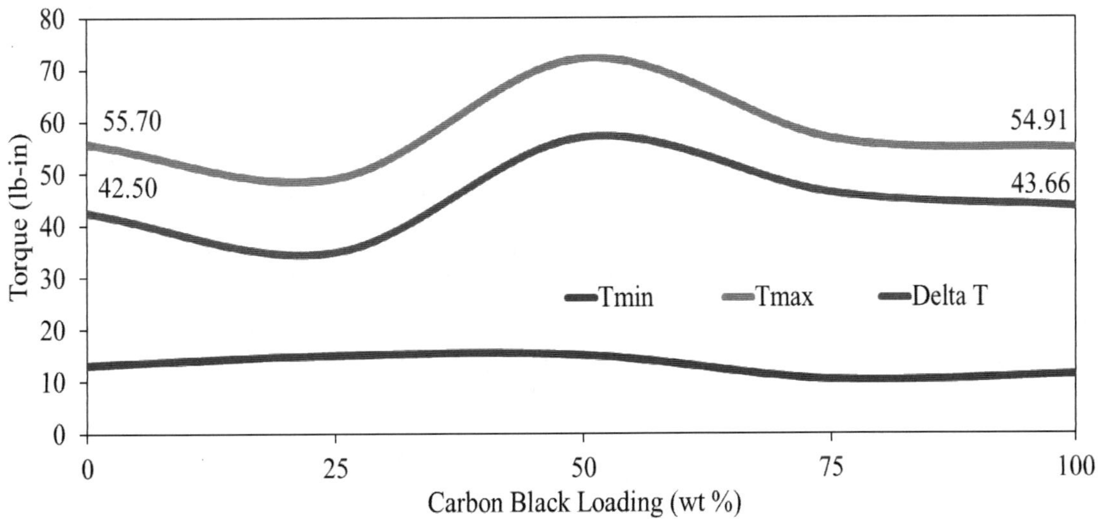

Fig. 5. T_{min}, T_{max} and ΔT of CB/SiO_2 filled NBR masterbatch at different loading ratio

4. Conclusion

Zeta potential shows both CB and SiO_2 dispersion were stable. The increasing ball milling time of CB and SiO_2 dispersion reduced the particle size. CB dispersion may have aggregation formation thus resulting higher particle size and viscosity compared to SiO_2 dispersion. Insignificant difference in T_{max} and ΔT between 100% CB and SiO_2 filled masterbatches. T_{max} and ΔT of hybrid fillers varies according to the ratio.

5. Acknowledgement

The authors acknowledge the assistance provided by Universiti Kuala Lumpur (UniKL) and Malaysian Rubber Board (MRB) in funding and facilities that resulted in this article.

References

[1] Ponnamma D *et al.*, Special Purpose Elastomers: Synthesis, Structure-Property Relationship, Compounding, Processing and Applications, in *Advances in Elastomers I: Blends and Interpenetrating Networks*, **12**, P. M. Visakh, S. Thomas, A. K. Chandra, and A. P. Mathew, Eds. 2013, pp. 47–82

[2] Yasin T, Ahmed S, Yoshii F, and Makuuchi K, 2003, Effect of acrylonitrile content on physical properties of electron beam irradiated acrylonitrile-butadiene rubber, *React. Funct. Polym.*, **57**, pp. 113–118

[3] Ismail H and Ahmad H S, 2013, The Properties of Acrylonitrile-Butadiene Rubber (NBR) Composite with Halloysite Nanotubes (HNTs) and Silica or Carbon Black, *Polym. - Plast. Technol. Eng.*, **52**, pp. 1175–1182

[4] Ulfah I M *et al.*, 2015, Influence of Carbon Black and Silica Filler on the Rheological and Mechanical Properties of Natural Rubber Compound, *Procedia Chem.*, **16**, pp. 258–264

[5] Zaaba N F, Jaafar M, and Ismail H, 2017, The Effect of Alkaline Peroxide Pre-Treatment on Properties of Peanut Shell Powder Filled Recycled Polypropylene Composites, **13**, pp. 75–87

[6] Bielinski D, Slusarski L, Dobrowolski O, Glab P, and Dryzek E, 2004, Studies of filler agglomeration by atomic force microscopy and positron anihilation spectroscopy: Part 1: Silica filled compounds, *KGK Kautschuk Gummi Kunststoffe*, **57**, pp. 579–586

[7] Reinert L, Zeiger M, Suárez S, Presser V, and Mücklich F, 2015, Dispersion analysis of carbon nanotubes, carbon onions, and nanodiamonds for their application as reinforcement phase in nickel metal matrix composites, *RSC Adv.*, **5**, pp. 95149–95159

[8] Keddie J L and Routh A F, An Introduction to Latex and the Principles of Colloidal Stability, in *Fundamentals of Latex Film Formation*, Springer, Dordrecht, 2010, pp. 1–26

[9] Ohshima H, Interaction of colloidal particles, in *Colloid and Interface Science in Pharmaceutical Research and Development*, 1st ed., H. Ohshima and K. Makino, Eds. Oxford: Elsevier B.V., 2014, pp. 1–27

[10] Horinek D, DLVO Theory, in *Encyclopedia of Applied Electrochemistry*, G. Kreysa, K. Ota, and R. Savinell, Eds. New York: Springer Reference, 2014, pp. 343–345

[11] Tadros T, General Principles of Colloid Stability and the Role of Surface Forces, in *Colloid Stability: The Role of Surface Forces, Part I*, **1**, WILEY-VCH Verlag GmbH & Co., 2007, p. 21

[12] Arayapranee W and Rempel G L, 2013, Effects of Polarity on the Filler-Rubber Interaction and Properties of Silica Filled Grafted Natural Rubber Composites, *J. Polym.*, **2013**, pp. 1–9

[13] Choi S S, 2002, Properties of silica-filled styrene-butadiene rubber compounds containing acrylonitrile-butadiene rubber: The influence of the acrylonitrile-butadiene rubber type, *J. Appl. Polym. Sci.*, **85**, pp. 385–393

[14] Mukhtar N Z F, Borhan M Z, Rusop M, and Abdullah S, 2013, Effect of Milling Time on Particle Size and Surface Morphology of Commercial Zeolite by Planetary Ball Mill, *Adv. Mater. Res.*, **795**, pp. 711–715

[15] Yukselen-Aksoy Y and Kaya A, 2011, A study of factors affecting on the zeta potential of kaolinite and quartz powder, *Environ. Earth Sci.*, **62**, pp. 697–705

[16] Nallamuthu I, Parthasarathi A, and Khanum F, 2013, Thymoquinone-loaded PLGA nanoparticles : antioxidant and anti-microbial properties, *Int. Curr. Pharm. J.*, **2**, pp. 202–207

[17] Shylesh S, Hanna D, Gomes J, Canlas C G, Head-Gordon M, and Bell A T, 2015, The role of hydroxyl group acidity on the activity of silica-supported secondary amines for the self-condensation of n-butanal, *ChemSusChem*, **8**, pp. 466–472

[18] Xu R, Wu C, and Xu H, 2007, Particle size and zeta potential of carbon black in liquid media, *Carbon N. Y.*, **45**, pp. 2806–2809

[19] Brewer S A, Carbon Black Pigments for Plastics, in *Coloring of Plastics: Fundamentals*, 2nd ed., R. A. Charvat, Ed. John Wiley & Sons, Inc, 2004, pp. 159–174

[20] Li Z H, Zhang J, and Chen S J, 2008, Effects of carbon blacks with various structures on vulcanization and reinforcement of filled ethylene-propylene-diene rubber, *Express Polym. Lett.*, **2**, pp. 695–704

[21] Chonkaew W, Mingvanish W, Kungliean U, Rochanawipart N, and Brostow W, 2018, Vulcanization Characteristics and Dynamic Mechanical Behavior of Natural Rubber Reinforced with Silane Modified Silica, *J. Nanosci. Nanotechnol.*, **10**, pp. 2018–2024

IOP Conf. Series: Materials Science and Engineering **548** (2019) 012006 doi:10.1088/1757-899X/548/1/012006

Frequency and Amplitude Dependence of Magnetorheological Elastomers Composites

N Moksin[1], R K Shuib[1], H Ismail[1] and M K Harun[1]

[1] School of Materials and Mineral Resources Engineering, USM Engineering Campus, Universiti Sains Malaysia, 14300 Nibong Tebal, Pulau Pinang, Malaysia

E-mail: raakhimi@usm.my

Abstract. In this study, anisotropic and isotropic magnetorheological elastomers (MREs) based on nickel zinc ferrite and natural rubber were prepared. The amount of nickel zinc ferrite was varied at five different levels (20, 40, 60, 80, and 100 phr) to assess its optimum content for enhancing dynamic mechanical performance of MREs. Tan δ was measured through parallel plate rheometer over frequency range of 0.1-100 Hz and strain amplitude range of 0.1-6%. The results revealed that tan δ increased with increasing waste nickel zinc ferrite content and reached maximum value at 100 phr for both isotropic and anisotropic MREs. SEM micrograph showed that anisotropic MREs had chain-like columnar structures of magnetic particles in the matrix as a consequence of an applied magnetic field during curing. For frequency sweep test, tan δ for anisotropic MRE improved 244% and for strain amplitude test 159% improvement compared to isotropic MRE.

1. Introduction

Magnetorheological elastomers (MREs) are a new group of elastomeric composites which consist of a non-magnetic elastomeric matrix with a suspension of magnetic particles. MREs offer several distinct advantages when compared with basic rubbers. The performance of MREs mainly promote by the viscoelastic properties of the rubber matrix as do basic rubber, but inclusion of magnetic particles in rubber enables additional reinforcement through magnetic particles interaction and interfacial friction between the particles and rubber matrix. Furthermore, the stiffness and dynamic mechanical properties of the MREs can be varied by application of an applied magnetic field during fabrication or in service. MREs are being used in various applications such as adaptive tuned vibration absorbers [1], auto-motive engine mounts [2] and semi active seismic dampers [3].

MREs can be fabricated into isotropic MREs and anisotropic MREs [4]. Isotropic MREs can be characterized by having uniform dispersion of magnetic particles embedded in the rubber matrix. Anisotropic MREs have a formation of chain-like structure within the rubber matrix resulting from subjecting the materials to an external magnetic field during curing. Anisotropic MREs are found to produce materials with larger stiffness and better dynamic properties [5].

Natural rubber is generally chosen as matrix material because of its associated ease of processing, and natural rubber has the highest failure strain compared with any rubber and unbeatable in term of the dynamic properties [6-8]. The secondary raw materials from the manufacturing are abandoned regardless of the large amount of ferrite (as high as 70% in nickel zinc ferrite waste) and cannot be recycled because its composition is complicated and has many kinds of impurities. Furthermore, the waste contains heavy metal elements which could cause dangerous effect to human health and environment if it is not treated properly prior to disposal [9, 10]. Nickel zinc ferrite are soft ferromagnetic materials commonly used for manufacture of electronic inductors, core material for power transformers in electronics, antennas and devices that are applied in the communication, lighting, alternative energy and automotive [10,11].

The incorporation of waste nickel zinc ferrite in MREs is an attractive approach from the view point of profit earning, recycling and sustainable environment. This work aims to assess the potential of nickel zinc ferrite as the magnetic filler particles for MREs. The MREs were prepared with different nickel zinc ferrite (0, 20, 40, 60, 80, 100 phr) in a natural rubber matrix. The dynamic properties of isotropic and anisotropic MREs were compared with unfilled rubber compound. The dynamic properties were investigated using lost tangent, commonly called tan δ under the influence of frequency and strain amplitude mode.

2. Materials and Methods

2.1 Materials

Natural rubber (grade SMR L) and other chemicals including zinc oxide, stearic acid, N-cyclohexyl-2-benzothiazole sulfonamide (CBS), tetramethylthiuram disulphide (TMTD), paraffin oil, sulphur and N-Isopropyl-N'-phenyl-P-phenylenediamine (IPPD) were purchased from Zarm & Chemical Supplier Sdn Bhd. Nickel zinc ferrite with average size of 6 μm was obtained from ACME Ferrite Products (M) Sdn. Bhd.

2.2 Preparation of nickel zinc ferrite with natural rubber MREs

Ferrite was first washed with water and gently stirred in a container. This step was repeated until no foreign particles left on the surface of the water. Then, the nickel zinc ferrite was filtered and dried at 150 °C in an oven until a constant weight was achieved. The compound formulation is given in Table 1. Formulations were compounded using a conventional laboratory two roll mill (model XK-150) according to ASTM D3184-80. The compounding began with mastication of the rubber for 2-3 minutes. Additives (other than accelerators and sulphur) were then added followed by waste nickel zinc ferrite; addition of accelerators and sulphur were delayed to the last part of the process to prevent premature vulcanization during compounding. Compounded rubber samples weighing 4 g were used to determine the crosslinking time (t_{90}) by using Mosanto Rheometer (MDR 2000) under temperature of 150 °C.

Table 1. Formulation of rubber compounds

Material	Function	Content (phr*)
Natural rubber	Matrix	100
ZnO	Activator	5
Stearic acid	Activator	1
Nickel Zinc Ferrite	Filler	0, 20, 40, 60, 80, 100
Paraffin oil	Plasticizer	4
CBS	Accelerator	2
TMTD	Accelerator	1
IPPD	Antioxidant	2
Sulphur	Crosslinking agent	1.5

*parts per hundred rubber

Compounded rubber samples weighing 45 g were placed in a mould 124 × 70 × 3 mm. The anisotropic MREs were pre-cured in a specially developed magnetic mould (as shown in Figure 1) at 80 °C for 30 minutes to align in the magnetic particles along the direction of magnetic mould

IOP Conf. Series: Materials Science and Engineering **548** (2019) 012006 doi:10.1088/1757-899X/548/1/012006

and subsequently were locked into place upon curing in a compression moulder at 150 °C under a pressure of approximately 12 MPa.

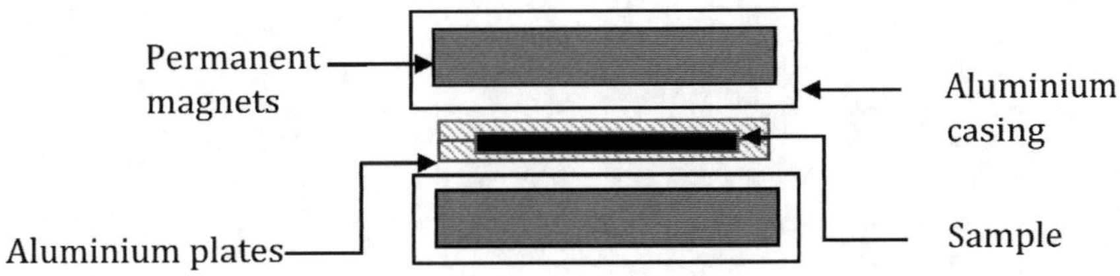

Fig.1. Sketch of specially developed magnetic mould for MRE curing

2.3 *Morphology test*
The microstructure of isotropic MRE and anisotropic MRE were observed using a field emission scanning electron microscopy (SUPRA ™ 35, V, ZEISS USA)). The samples were immersed in liquid nitrogen before being fractured using two pliers in order to bare their interior. The surfaces were coated with a thin layer of gold prior to observation at an accelerating voltage of 20 kV.

2.4 *Dynamic mechanical properties test*
The dynamic mechanical properties of rubber composites were measured using parallel and plate rheometer (MCR 301, Anton Paar Company, Germany) on a circular specimen of 20 mm diameter and 3 mm thickness in shear mode at room temperature. The dynamic mechanical properties were investigated using loss tangent, commonly called tan δ. Tan δ is the comparison of the energy loss to that stored and is obtained by dividing loss modulus (G") with the storage modulus (G'). Tan δ was measured over the frequency range of 0.1-100 Hz at a fixed strain amplitude of 2%. For the influence of strain amplitude on tan δ, the samples were measured at strain amplitude range of 0.1-6% at a fixed frequency of 10 Hz.

3. Results and discussion
3.1 *Morphology*
Figure 2 shows SEM images of isotropic and anisotropic MREs. It can be observed that isotropic MRE have a homogenous dispersion of nickel zinc ferrite particles embedded in a natural rubber matrix (Fig. 2a). Figure 2b shows anisotropic MRE cured under an applied magnetic field. It can be seen that the alignment of magnetic particles has occurred in the rubber matrix due to the influence of magnetic force to the filler particles in which the north pole of one particle attracts the south pole of its neighbor, thus resulting the formation of chain-like structures inside the rubber matrix.

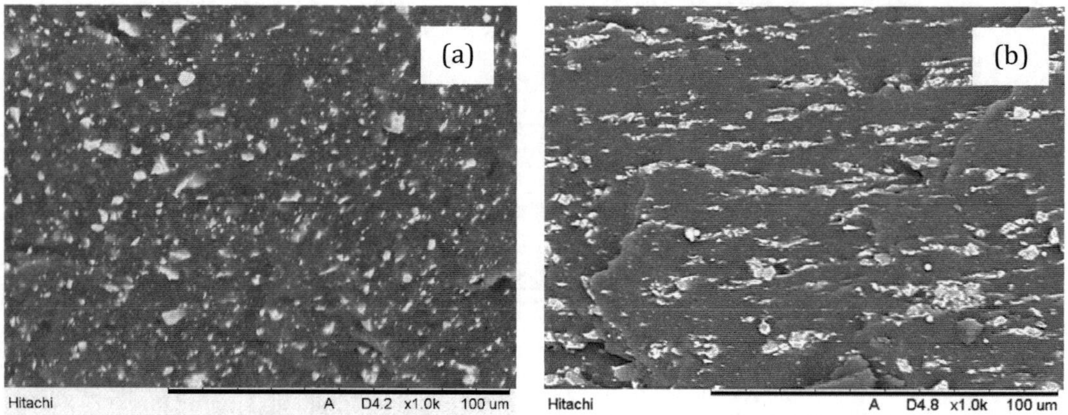

Fig. 2. SEM images of surface of; (a) isotropic MRE, (b) anisotropic MRE

3.2 Dynamic mechanical properties

3.2.1 Frequency sweep measurement

The variation of tan δ with frequency for isotropic and anisotropic MREs at different waste nickel zinc ferrite contents is shown in Figure 3. Generally, tan δ in Figures 3a and 3d gradually increased with increasing frequency for both isotropic and anisotropic MREs and slightly decreased at high frequency for anisotropic MREs. However, tan δ of anisotropic MREs was higher compared with isotropic MREs with improvement of approximately 250%, 245%, 247%, 235%, and 243% for 20, 40, 60, 80, and 100 phr of filler, respectively. It is also apparent that the tan δ for anisotropic and isotropic MREs increased with increasing nickel zinc ferrite contents. The changes in tan δ can be further explained with storage modulus (G') and loss modulus (G").

Storage modulus (G') and loss modulus (G") are plotted in Figure 3b and 3c for isotropic MREs and Figure 3e and 3f for anisotropic MREs. G' and G", similar to tan δ, were frequency dependent and the increase in tan δ as the frequency increased was mainly due to the increased in G" as opposed to G'. The increase of G" as the frequency increased for isotropic MREs could be due to increase of energy absorbed through stretching and recovery of rubber molecular chain during viscoelastic flow under dynamic loading and interfacial damping between the particles and rubber matrix [12]. For anisotropic MREs, similar mechanism would be involved with additional energy absorbed through magnetism induced damping as the formation of chain like columnar structure in anisotropic MRE improved damping [13]. This phenomenon could be attributed to separation of dipole-dipole interaction between neighbouring particles.

The increase of G" with increasing waste nickel zinc ferrite content could be explained by the increased in energy absorbed due to interfacial friction caused by the increased in the interfacial area and the volume with the increase in waste nickel zinc ferrite content. It can also be seen that G' increase with increasing waste nickel zinc ferrite content for both isotropic and anisotropic MREs. The increment of G' for isotropic MREs could be related to efficient stress transfer between homogenously distributed particles and rubber matrix which improved stiffness and strength of the material; and for anisotropic MREs, the increase of G' could be attributed to efficient reinforcing effect provided by more bulkier chains and columnar structures as the volume of the waste nickel zinc ferrite increased in the rubber matrix.

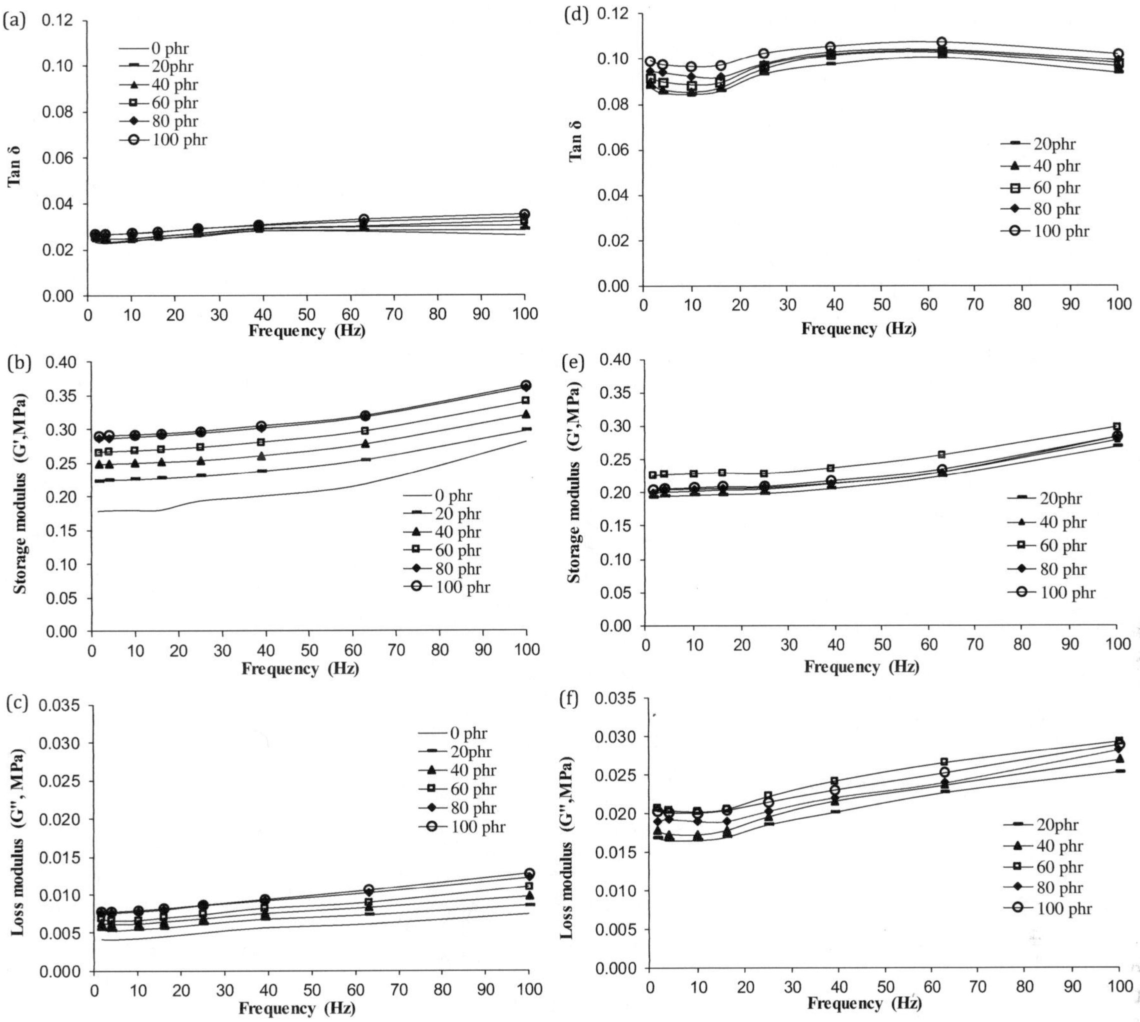

Fig. 3. (a) Tan δ, (b) G', and (c) G" of isotropic MREs and (d) Tan δ, (e) G', and (f) G" of anisotropic MREs

3.2.2 Strain amplitude measurement

The variation of tan δ, G', G" with strain amplitude for isotropic and anisotropic MREs at different waste nickel zinc ferrite contents are shown in Figure 4. The relationship between tan δ and strain amplitude for isotropic (Fig. 4a) and anisotropic (Fig. 4d) MREs is approximately linear, with tan δ increased with increasing strain amplitude up to 6%. Again, it is apparent that the changes in tan δ with increased strain amplitude is mainly due to the increase in G". Besides, it can be seen that; tan δ and G" for anisotropic MREs were higher compared to isotropic MREs at low strain amplitude up to 2.5%, and the values of tan δ for isotropic and anisotropic MREs were comparable at high strain amplitude.

The amplitude dependence for isotropic MREs is related to Payne effect [14,15]. This effect can be explained by the breakdown of filler aggregates to release trapped and occluded rubber which allows more rubber to take part in energy dissipation, filler-rubber detachment, and reformation to increase at higher strain amplitude. For anisotropic MREs similar mechanisms would be involved as well as magnetism induced damping. The higher tan δ and G" for anisotropic MRE at low strain amplitude supports that additional energy absorbed due to increase of energy required to breakdown interparticle magnetic interaction between neighboring particles.

At high strain amplitude, the magnetic interparticle interaction become weak and energy loss is largely reliant of the viscoelastic damping in the rubber matrix and the Payne effect. It can also be seen tan δ, G' and G" for isotropic MREs increased with increasing waste nickel zinc ferrite, however, for anisotropic MRE, tan δ, G' and G" increased with increasing waste nickel zinc ferrite up to 60 phr. The decrease in tan δ, G' and G" at higher filler content could be attributed to poor filler-filler interaction in a bulkier chain like columnar structures and filler-rubber interactions, it is easily disrupted as the strain amplitude increased, and the magnetism induced damping also reduced [16].

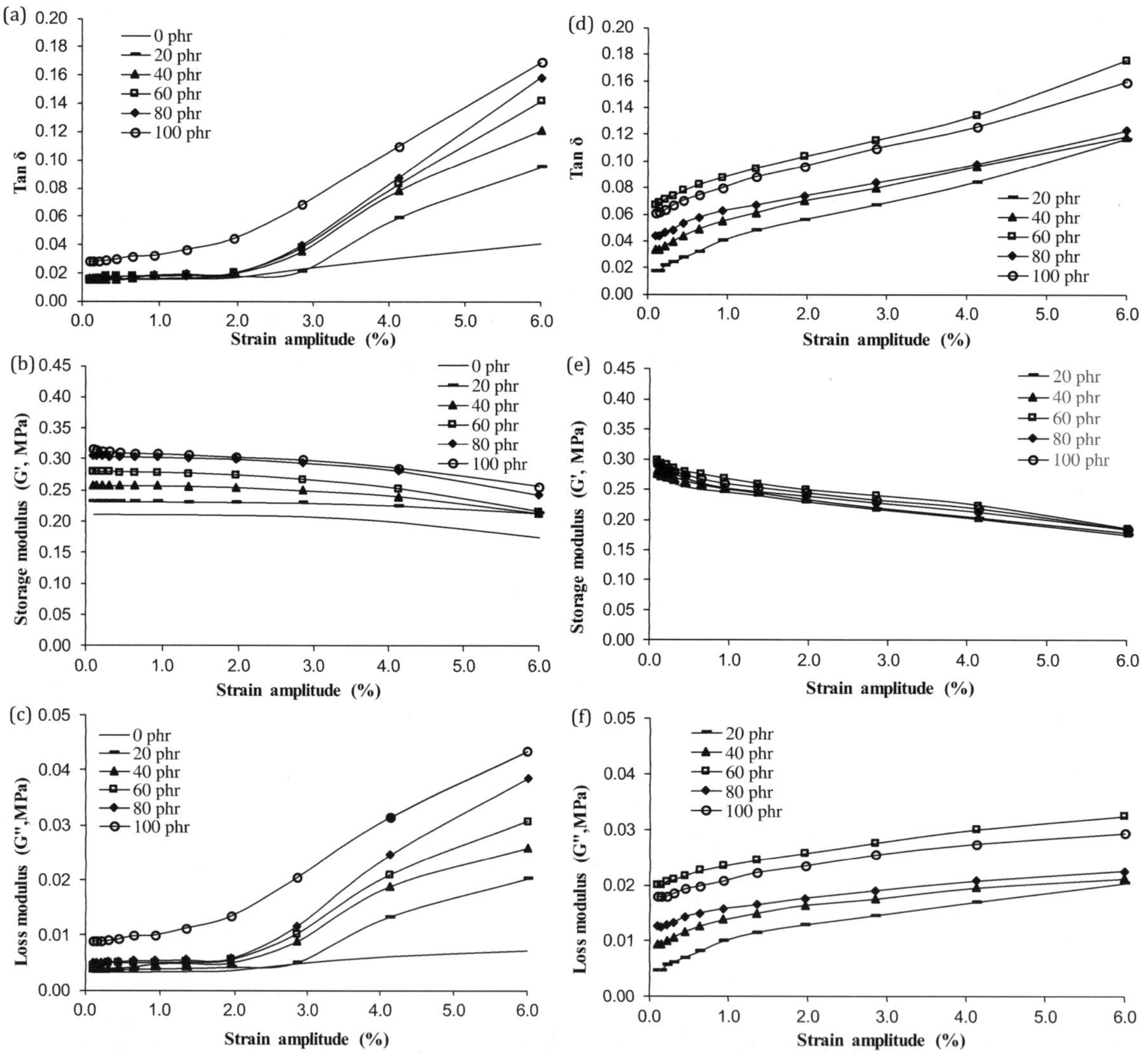

Fig. 4. (a) Tan δ, (b) G', and (c) G" of isotropic MREs and (d) Tan δ, (e) G', and (f) G" of anisotropic MREs

4. Conclusions

It was found that isotropic MREs had uniform waste nickel zinc ferrite particles distribution and alignment of magnetic particles occurred for anisotropic MREs as a consequence of curing the material under an applied magnetic field. The results revealed that tan δ was highest for anisotropic MREs with improvement of 244% compared with isotropic MREs over the whole frequency range explored and 159% improvement over the whole strain amplitude range explored, supporting that formation of chain like columnar structures contributing to higher mechanical performance which might be due to viscoelastic damping, interfacial damping as well as magnetism induced damping. The optimum filler content was found to be 100 phr for both isotropic and anisotropic MRE.

Acknowledgement

The authors gratefully acknowledge support from the Universiti Sains Malaysia short term grant 304/PBAHAN/60313045.

References

[1] Deng H X and Gong X L 2008 Application of magnetorheological elastomer to vibration absorber *Communications in nonlinear science and numerical simulation* **13(9)** 1938-1947
[2] Ginder J M, Nichols M E, Elie L D and Tardiff J L 1999 Magnetorheological elastomers: properties and applications. in Smart Structures and Materials: Smart Materials Technologies International Society for Optics and Photonics
[3] Dyke S, Spencer Jr B, Sain M and Carlson J 1996 Modeling and control of magnetorheological dampers for seismic response reduction *Smart Mater. Struct.* **5(5)** 565
[4] Boczkowska A, Awietjan S F, Pietrzko S and Kurzydłowski K J 2012 Mechanical properties of magnetorheological elastomers under shear deformation *Composites, Part B* **43(2)** 636-640
[5] Kukla M, Górecki J, Malujda I and Talaśka K 2017 The Determination of Mechanical Properties of Magnetorheological Elastomers (MREs) *Procedia Eng.* **177** 324-330
[6] Jung H S, Kwon S H, Choi H J and Jung J H 2016 Magnetic carbonyl iron/natural rubber composite elastomer and its magnetorheology *Compos. Struct.* **136** 106-112
[7] Wang Y, Hu Y, Gong X and Jiang W 2007 Preparation and properties of magnetorheological elastomers based on silicon rubber/polystyrene blend matrix. *J. Appl. Polym. Sci.* **103(5)** 3143-3149.
[8] Yunus N, Mazlan S, Aziz S and Khairi M A 2016 Investigation on magnetic field dependent modulus of epoxidized natural rubber based magnetorheological elastomer. *J. Phys.: Conf. Ser.. IOP Publishing* **776** 012024
[9] Sezgin N, Sahin M, Yalcin A and Koseoglu Y 2013 Synthesis, characterization and, the heavy metal removal efficiency of MFe2O4 (M= Ni, Cu) nanoparticles Ekoloji **22(89)** 89-96
[10] Song W H 2011 A Study on Waste-derived NiZn Soft Ferrites as EMI Suppressor. *International Journal of Electronics, Computer and Communications Technologies.* **1(2)** 1-4
[11] Hee A, Metselaar I, Johan M and Mehrali M 2011 Preparation of nickel zinc ferrite by electrophoretic deposition *J. Electrochem. Soc.* **159(1)** 18-22
[12] Khimi S R and Pickering K 2016 The effect of silane coupling agent on the dynamic mechanical properties of iron sand/natural rubber magnetorheological elastomers. *Composites, Part B* **90** 115-125
[13] Kallio M 2005 The elastic and damping properties of magnetorheological elastomers: VTT.
[14] Rendek M and Lion A 2010 Amplitude dependence of filler-reinforced rubber: Experiments, constitutive modelling and FEM–Implementation *Int. J. Solids Struct.* **47(21)** 2918-2936
[15] Sorokin V V, Ecker E, Stepanov G V and Shamonin M 2014 Experimental study of the magnetic field enhanced Payne effect in magnetorheological elastomers *Soft Matter* **10(43)** 8765-8776.
[16] Shuib R K, Pickering K L and Mace B R 2015 Dynamic properties of magnetorheological elastomers based on iron sand and natural rubber. *J. Appl. Polym. Sci.* **132(8)**

IOP Conf. Series: Materials Science and Engineering **548** (2019) 012007 doi:10.1088/1757-899X/548/1/012007

Effect of Plasticizer on Microstructure and Dynamic Mechanical Performance of Anisotropic Magnetorheological Elastomers

N H N Abdul Hadi[1] and R K Shuib[1]

[1]School of Materials and Mineral Resources Engineering, USM Engineering Campus, Universiti Sains Malaysia, 14300 Nibong Tebal, Penang, Malaysia

E-mail: raakhimi@usm.my

Abstract. Anisotropic magnetorheological elastomers (MREs) based on natural rubber and industrial waste nickel zinc ferrite were prepared. Paraffin oil was employed as plasticizer to modify the viscosity of rubber matrix. The amount of plasticizer was varied at five levels (5, 10, 20, 30 and 40 phr) to assess the ability of nickel zinc ferrite particles to align into chain like columnar structures within the rubber under an applied magnetic field. Scanning electron microscopic (SEM) and dynamic mechanical analysis over frequency range of 1-100 Hz and a strain amplitude range of 0.1-6% were examined. SEM micrographs revealed that the chain like columnar structures become longer and thicker as the viscosity of the rubber matrix decreased. It was found that 5 phr plasticizer produced the highest tan δ over the range of frequency and strain amplitude explored.
Keywords: Natural rubber; Waste nickel zinc ferrite; Magnetorheological; Plasticizer; Paraffin oil

1. Introduction

Magnetorheological elastomers (MREs) consist of magnetic particles and an elastic polymer matrix such as soft silicone elastomers, poly vinyl alcohol, gelatin, hard natural rubber and polychloroprene rubber[1] . The magnetic particles that are usually used in MREs are carbonyl iron and iron oxide [2]. MREs can be fabricated into isotropic MREs and anisotropic MREs. Isotropic MREs can be characterized by having uniform dispersion of magnetic particles embedded in the rubber matrix. Anisotropic MREs have a formation of chain like structure within the rubber matrix resulting from subjecting the materials to an external magnetic field [3]. Anisotropic MREs are found to produce materials with larger stiffness and better dynamic properties [4, 5].

Fabrication of MREs needs a lot of attention and consideration, not only in selection of the magnetic particles, but the performance of MREs also depends on matrix viscosity of rubber. MREs with different matrix viscosity are believed to have a significant effect on the ability of particles to orientate along the magnetic field direction [6]. Anna et al. found that anisotropic MREs based on silicone rubber with viscosity of 30 000 mPa.s and carbonyl iron has limited oriented structure of particle chain in the rubber matrix [7, 8]. Jung et al. reported that MREs based on natural rubber with viscosity of 81.10 mPa.s and carbonyl iron has good alignment and good dynamic properties [9, 10].

In this study, anisotropic MREs based on waste nickel zinc ferrite and natural rubbers were prepared. The viscosity of rubber matrix was modified using plasticization method [11]. The effect of matrix viscosity on magnetic particle alignment in the rubber matrix was investigated by scanning electron microscopy and dynamic mechanical testing over a range of frequency and strain amplitude. The loss tangent (tan δ) was considered as the fundamental parameter to assess dynamic performance. Tan δ gives a comparison of the energy lost to that stored; it is obtained by dividing the loss modulus (G′′) by the storage modulus (G′).

2. Materials and Methods

2.1 Materials

The anisotropic MREs sample was prepared by compounding waste nickel zinc ferrite and natural rubber. The compound formulation used in this study is listed in Table 1. Natural rubber and compound ingredients such as zinc oxide, stearic acid, paraffin oil, n-cyclohexyl-2-benzothiazole sulfonamide (CBS), tetramethylthiuram disulfide (TMTD), sulphur and n-isopropyl-n'-phenyl-p-phenylenediamine (IPPD) were purchased from Zarm & Chemical Supplier Sdn. Bhd. Waste nickel zinc ferrite was obtained from ACME Ferrite Products (M) Sdn. Bhd.

Table 1. Formulation of rubber compound

Materials	Function	Loading (phr)					
Natural rubber	Matrix	100	100	100	100	100	100
Zinc oxide	Activator/peptiser	5	5	5	5	5	5
Stearic acid	Activator/peptiser	1	1	1	1	1	1
Paraffin oil	Plasticizer	0	5	10	20	30	40
Waste Nickel Zinc Ferrite	Filler	70	70	70	70	70	70
CBS	Accelerator	2	2	2	2	2	2
TMTD	Accelerator	1	1	1	1	1	1
Sulphur	Crosslinking agent	1.5	1.5	1.5	1.5	1.5	1.5
IPPD	Antioxidant	2	2	2	2	2	2

2.2 Sample Preparation

In this study, mixing process was carried out on a conventional laboratory two roll mill size 160 x 320 mm (model XK150) according to ASTM designation D 3184-80. The compounding began with softening the rubber (mastication) about 2-3 min. The additives in Table 1 (other than accelerators and sulphur) were then added followed by nickel zinc ferrite; addition of accelerators and sulphur were delayed to the last part of the process to prevent premature vulcanization during compounding. The cure time, t_{90} at 150 °C was then determined using Monsanto Rheometer, Model MDR 2000 (moving die rheometer) according to ISO 3417. 37 g of rubber compound was placed in a mould 124x 70x 3 mm. The anisotropic MREs were pre-cured to an external magnetic field in a specially developed magnetic mould at 80 °C for 30 min. The particles were magnetized, and then form chains aligned along the field direction and subsequently were cured in a compression moulder at 150 °C under a pressure of approximately 12 MPa.

2.3 Viscosity

The viscosity test was performed using parallel plate rheometer (Physica, MCR 301). The spindle type of parallel plate rheology are plate PP25. The viscosity test was performed on a circular specimen of 20 mm diameter and 3mm thickness in shear mode at room temperature.

2.4 Dynamic Mechanical Properties Measurements

A dynamic mechanical property of rubber composites was performed using parallel plate rheometer (MCR 302, Anton Paar Company, Germany). The spindle type of parallel plate are plate PP25. The dynamic measurements were performed under room temperature through oscillatory amplitude and frequency sweep procedure. Tan δ was measured over the frequency range of 1-100 Hz at a fixed strain amplitude range of 0.1-6% at a fixed frequency of 10 Hz.

2.5 Scanning Electron Microscope (SEM)

The fracture surface of the MREs was observed by using scanning electron microscope (SEM), SUPRA ™ 35-V, ZEISS USA. The micrograph obtained from SEM can be used to observe the filler distribution in the matrix with an accelerating voltage of 20 kV. The fracture ends of the sample were mounted on aluminium stubs. Then, the sputter coated with a thin layer of gold to avoid electrostatic changing and poor resolution.

3. Results and Discussion

3.1 Viscosity of rubber matrix

Viscosity of MREs at different contents of plasticizer is shown in Fig. 1. The viscosity test measure material resistance to flow. The viscosity decreased with increasing plasticizer contents. This effect could be explained due to reduction in cohesive attraction force between polymer chains when paraffin oil molecules penetrate into the rubber matrix. Addition of plasticizers into the rubber matrix also enhance the sliding of rubber molecular chains. This effect could potentially enhance the arrangement of magnetic particles during curing under an applied magnetic field and potentially increase the magnetism induced damping of the MREs [12].

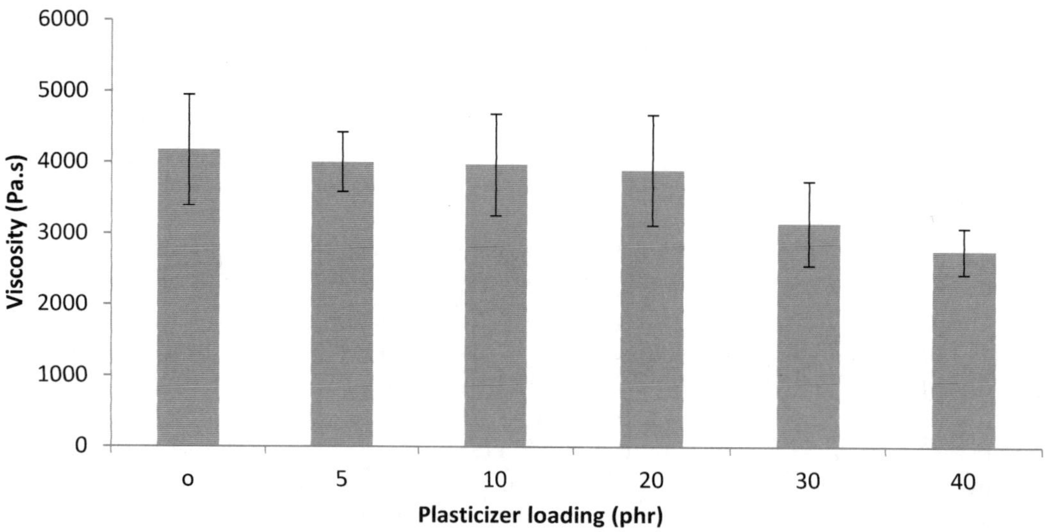

Fig. 1. Viscosity of MREs at different contents of plasticizer

3.2 Scanning electron microscope (SEM)

SEM micrographs of MREs with different plasticizer content are shown in Fig. 2. As can be seen for MRE without plasticizer, the waste nickel zinc ferrite particles was not well organized into chain like columnar structure as a consequence of curing the material under an applied magnetic field at elevated temperature. For MREs with addition of plasticizer, It can be seen that the waste nickel zinc ferrite chains get dense, longer, and the gap between the adjacent chains decreased as the plasticizer content increased from 5 to 40 phr. Thickness and length of the magnetic particle chain are tabulated in Table 2. Clearly, the thickness and length of waste nickel zinc ferrite increased as the plasticizer content increased. It is expected that increase magnetic particle interaction of waste nickel zinc ferrite by using plasticizer could further enhanced the dynamic performance of MREs. However, there are a lot of cavities remaining due to the particle pull out from rubber matrix which indicates weak interaction between waste nickel zinc ferrite particles and rubber [11].

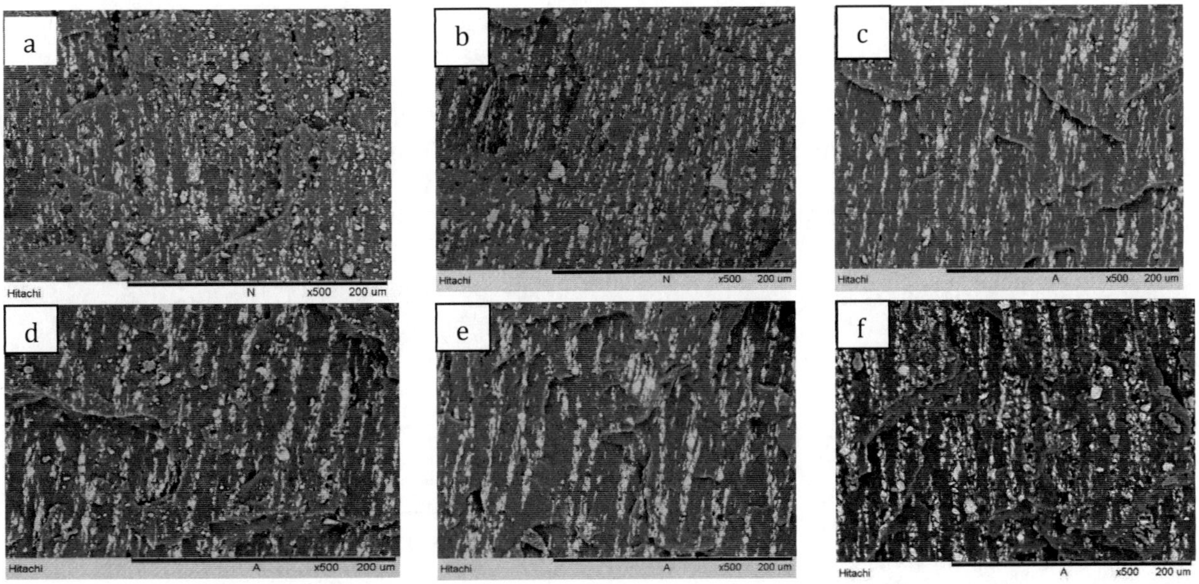

Fig. 2. SEM micrographs of fracture surface of rubber composites: (a) MRE without plasticizer and (b-f) MREs with different plasticizer contents of 5, 10, 20, 30, and 40 phr respectively

Table 2. Thickness and length of waste nickel zinc ferrite at different plasticizer content

Plasticizer content (phr)	Thickness (μm)	Length (μm)
0	2.5	39.2
5	5.4	56.2
10	6.2	149.8
20	8.1	326.2
30	10.6	614.1
40	14.2	779.6

3.3 Frequency Sweep

The variation of tan δ, G′ and G″ with frequency for MREs with different plasticizer contents is shown in Fig. 3. It can be seen that, tan δ was increased with the increase of frequency until 20 Hz and subsequently decreased until 100 Hz. It can be also seen that the tan δ appears to increase with increasing plasticizer content until it reached a maximum value at 5 phr and thereafter decreased at higher plasticizer contents (10, 20, 30, and 40 phr). The increase of tan δ for MREs with 5 phr plasticizer was mainly due to increase in G″ as opposed to G′. The increase in G″ could be explained by the increased of energy loss due to sliding of rubber molecular chains which would bring more viscous flow during deformation [13]. Furthermore, additional energy could potentially absorbed due to break down of interparticle magnetic interaction between neighbouring particles. At higher plasticizer content (from 10 to 40 phr), the decrease of tan δ could be explained by decreasing stiffness of MREs. Decreasing stiffness might be due to the increase of space in the polymer chain. The plasticizer generally has bulky structures that provide polymers with more space to move around and prevent polymers from coming close together. Therefore, the plasticizer softening the polymers and reducing the elastic modulus [14].

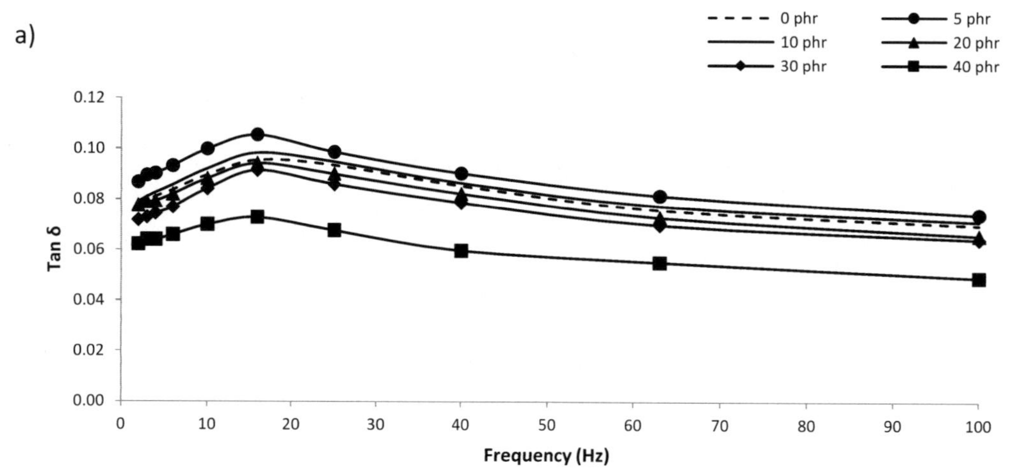

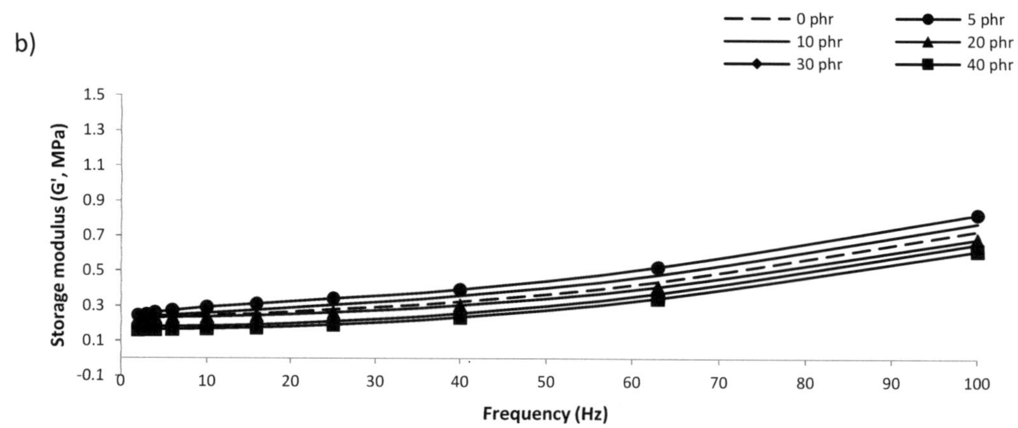

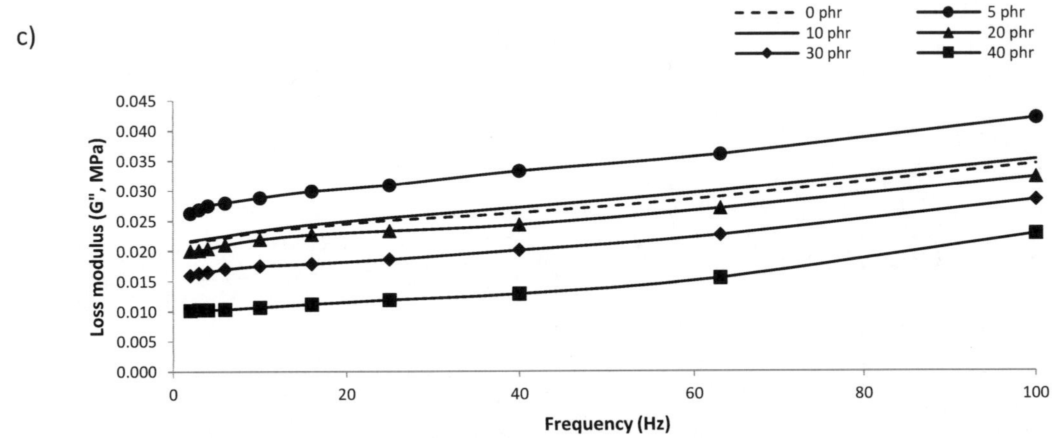

Fig. 3. Tan δ (a) storage modulus (b) and loss modulus (c) of MREs with different content of plasticizer with frequency range of 0.1 to 100 Hz

3.4 Strain Amplitude Sweep

The variation of tan δ, G′ and G″ with strain amplitude for MREs with different plasticizer contents is shown in Fig. 4. Tan δ was amplitude dependent over the whole strain amplitude explored. MREs with 5 phr plasticizer had the highest tan δ with percentage increment of 17.63% in tan δ compared to MREs without plasticizer. It is apparent that the change in tan δ with increased strain amplitude is mainly due to the decrease in G′. The amplitude dependence for MREs could be explained due to Payne effect. This effect could be related to the breakdown of filler aggregates to release trapped and occluded rubber which allows more rubber to take part in energy dissipation, filler rubber detachment and reformation that increases with increasing strain amplitude. Furthermore, tan δ, G′ and G″ for anisotropic MREs decreased with increasing plasticizer content. This is believed to occur due to poor filler-filler interaction in a bulkier chain like columnar structures and weak filler-rubber interaction. The weak filler-filler and filler-rubber interaction in anisotropic MREs can easily disrupted as the strain amplitude increased and therefore the magnetism induced damping reduced.

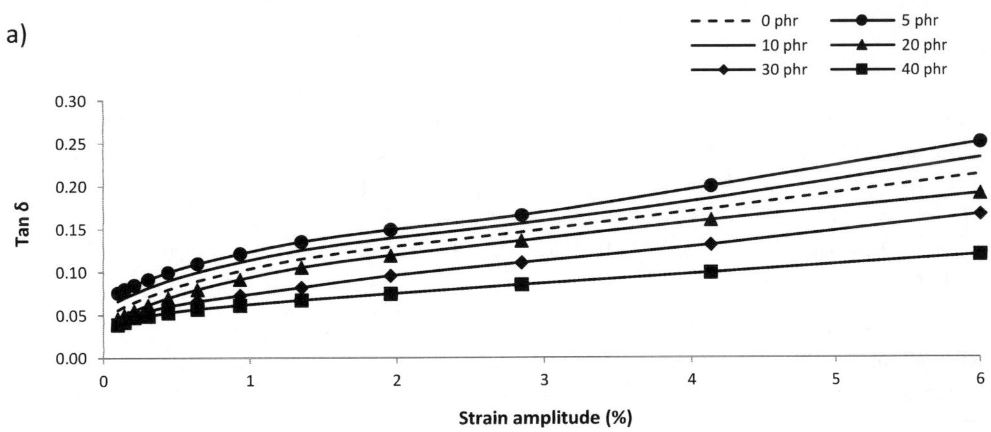

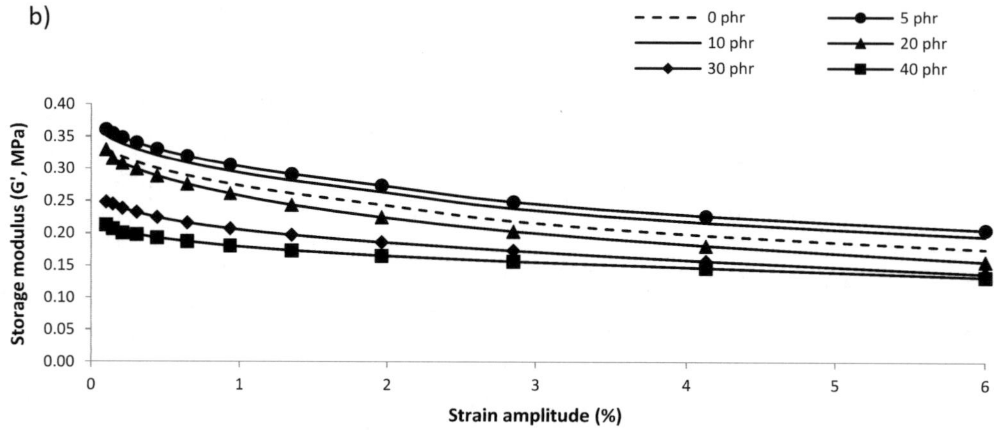

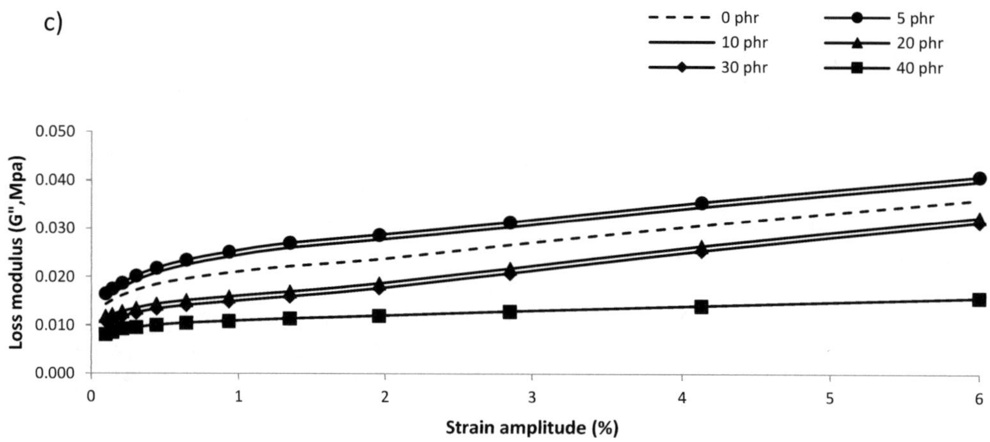

Fig. 4. Tan δ (a) storage modulus (b) and loss modulus (c) of MREs with different content of plasticizer with strain amplitude range of 0.1-6 %

4. Conclusion

Effect of plasticizer on the performance of MREs based on natural rubber and waste nickel zinc ferrite has been experimentally investigated. The addition of paraffin oil decreased the viscosity of polymer matrix due to reduction in cohesive attraction forces between polymer chains. SEM micrograph revealed that the waste nickel zinc ferrite particles organized into chain like columnar structure as a consequence of curing the material under an applied magnetic field at elevated temperature. Tan δ was found to be highest for the 5 phr plasticizer over the whole frequency and strain amplitude explored.

Acknowledgements

The authors gratefully acknowledge support from the Ministry of Education Malaysia under FRGS Grant 6071349

References

[1] Wang Y *et al* 2006 Effects of rubber/magnetic particle interactions on the performance of magnetorheological elastomers *Polym. test.*, **25**(2) 262-267
[2] Yu M *et al* 2017 Understanding the reinforcing behaviors of polyaniline-modified carbonyl iron particles in magnetorheological elastomer based on polyurethane/epoxy resin IPNs matrix *Compos. Sci. Technol.*, **139** 36-46
[3] Hu Y *et al* 2005 Magnetorheological elastomers based on polyurethane/Si-rubber hybrid *Int. J. Mod. Phys. B*, **19** 1114-1120
[4] Perales-Martínez I A *et al* 2017 Enhancement of a magnetorheological PDMS elastomer with carbonyl iron particles *Polymer Testing*, **57** 78-86
[5] Fan Y *et al* 2012 Effect of cross-link density of the matrix on the damping properties of magnetorheological elastomers *Ind. Eng. Chem. Fundam.*, **52**(2)771-778
[6] Oh J *et al* 2014 Effect of plasticizer and curing system on freezing resistance of rubbers *J. Appl. Polym. Sci.* **131**(2)
[7] Boczkowska A *et al* 2009 Image analysis of the microstructure of magnetorheological elastomers *J. Mater. Sci.*, **44**(12)3135-3140
[8] Boczkowska A and Awietjan S 2012 Microstructure and properties of magnetorheological elastomers, in Advanced Elastomers-Technology, Properties and Applications : InTech
[9] Jung, H S *et al* 2016 Magnetic carbonyl iron/natural rubber composite elastomer and its magnetorheology. *Compos. Struct.* **136** 106-112
[10] Stephen R *et al* 2007 Flow properties of unvulcanised natural rubber/carboxylated styrene butadiene rubber latices and their blends *J. Appl. Polym. Sci.*, **104**(4)2528-2535
[11] Wu J *et al* 2012 Improving the magnetorheological properties of polyurethane magnetorheological elastomer through plasticization *J. Appl. Polym. Sci.*, **123**(4)2476-2484
[12] Khairi M H A *et al* 2018 Performance of magnetorheological elastomer based green epoxidized natural rubber/sucrose acetate isobutyrate hybrid matrix *IOP Conference Series: Materials Science and Engineering* **342** 012034
[13] Richaud E *et al* 2015 Plasticizer effect on network structure and hydrolytic degradation. *Eur. Polym. J.* **69** 232-246.
[14] Vieira M G A *et al* 2011 Natural-based plasticizers and biopolymer films: A review. *Eur. Polym. J.*, **47**(3) 254-263.

IOP Conf. Series: Materials Science and Engineering **548** (2019) 012008 doi:10.1088/1757-899X/548/1/012008

Evaluation of Impact Absorbing Properties of Thermoplastic Elastomers

N Yajima[1], **S Kawahara**[2] and **A Matsuda**[3]

[1] Graduate Student of Systems and Information Engineering, University of Tsukuba, Ibaraki, Japan

[2] Kuraray Plastics Co., Ltd., Gifu, Japan

[3] Faculty of Engineering of Information and Systems, University of Tsukuba, Ibaraki, Japan

E-mail: a_matsuda@kz.tsukuba.ac.jp

Abstract. The thermoplastic elastomers (TPEs) are physical mixtures of polymers. Impact absorbing materials in sports equipment are one of the expected applications of TPEs due to good process ability and less-smell. For examples, face guards, head-gears and protectors are expected to be applied. In this paper, static and dynamic mechanical properties of TPEs were investigated to expand the applicability of TPEs to the impact absorbing material for the sports equipment. In this study, TPEs developed by the Kuraray Plastics Co., Ltd. were applied for static and dynamic tests. To investigate the static mechanical properties of TPEs, the uniaxial tensile loading tests were conducted. The nonlinear viscoelasticity considering damage were applied to representation of the experimental results. The FEM analysis showed good applicability to static mechanical properties including damage which was excited by the maximum deformation. TPE. To investigate the dynamic mechanical properties of TPEs, a drop weight impact testing machine was developed. Impact absorbing properties of TPEs were evaluated by the impact force and impact time with baseball weight.

1. Introduction

The thermoplastic elastomers (TPEs) are physical mixtures of polymers. The TPEs are manufactured by pressing or injection process without vulcanization agents. The mechanical properties of the thermoplastic elastomers are getting similar to rubber by technology developments. The TPEs show good process ability and flexibilities at low temperature, even if TPEs were made by extrusion or injection molding[1]. Now, the TPEs are widely used for numerous products such as automobile parts, lifestyle products and electric appliances. The strong points of TPEs are light weight, rubber-like mechanical behavior and less-smell, so they are expected to apply to various sports equipment, for the purpose of absorbing impact[1],[2],[3]. The purpose of this study is to expand the applicability of thermoplastic elastomers to impact absorbing material for sports equipment. Tensile loading test and cyclic tensile loading test were conducted to obtain the static mechanical properties of TPEs. An original drop weight impact testing machine for TPEs with real baseball weight was developed and impact tests were conducted to obtain the dynamic mechanical properties of TPEs for shock absorbing material.

2. Tensile loading test

Tensile loading tests and cyclic tensile loading tests were conducted to evaluate the static mechanical properties of TPEs. Materials of the specimens were styrene elastomers, polyolefin and process oil. They were processed by injection molding after heat melting. First, tensile loading tests on specimens were conducted using a uniaxial tensile loading machine. The deformation of the specimens was measured by using a displacement camera system. The strain rates were set to 0.1%/s, 1.0%/s and 10%/s. Fig.1 shows the relationship between stress and strain given by the

IOP Conf. Series: Materials Science and Engineering **548** (2019) 012008 doi:10.1088/1757-899X/548/1/012008

tensile loading test. From the tensile loading test results, the TPEs was possible to deform over 400% of strain.

Also, the cyclic tensile loading tests were conducted of which stretch for cyclic deformations were 150%, 200% and 250%. Four cycle of deformation were applied to the tensile specimen continuously and maximum stretch decrease from 150% to 250%. Fig. 2 (a), (b) and (c) show the cyclic loading test results for the first cycle and the third cycle.

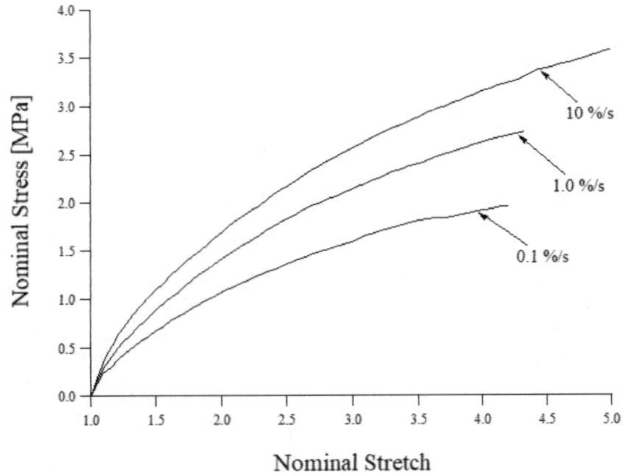

Fig. 1. Tensile loading test results of TPEs

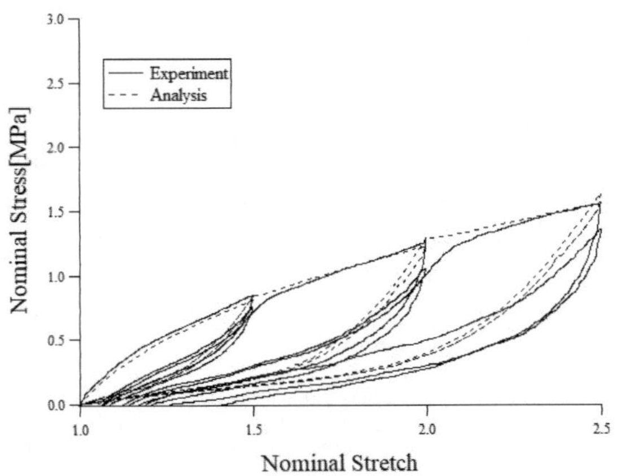

(a) Strain rate= 0.1%/sec

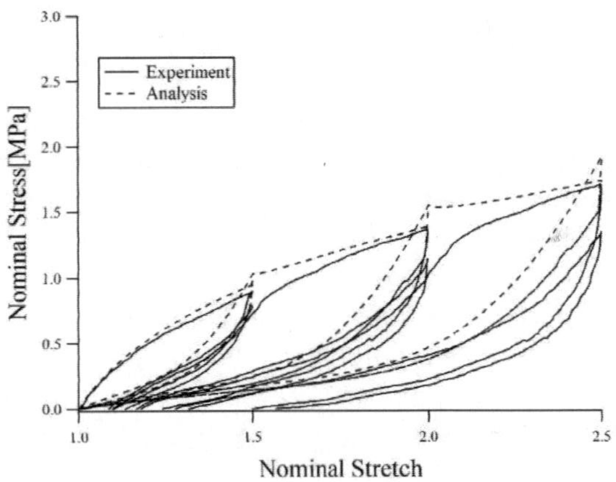

(b) Strain rate= 1.0%/sec

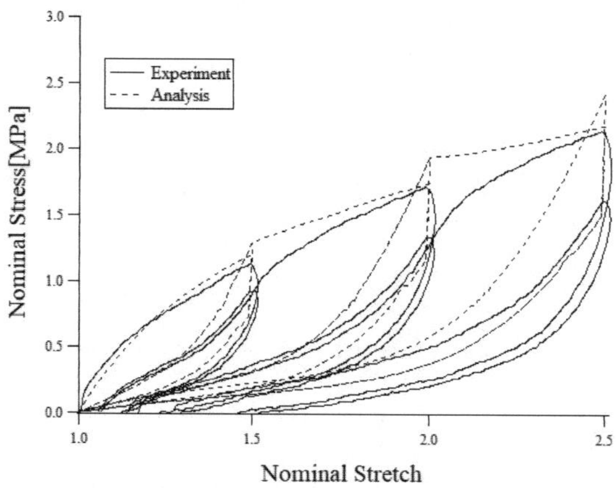

(c) Strain rate= 10%/sec

Fig. 2. Cyclic tensile loading test results of TPEs

3. Nonlinear viscoelastic model

Stiffness softening were obtained by the results. In this study, generalized relaxation model and softening model were applied, in order to reproduce the mechanical properties of TPEs. The following elastic energy function capable of considering compressibility by complement strain energy has been applied to TPEs in this study[4].

$$W(\overline{\boldsymbol{C}}, p, \boldsymbol{Q}_i) = \overline{W}^0(\overline{\boldsymbol{C}},\ \Xi_t^m) - \sum_{i=1}^{N_d} \frac{1}{2} \overline{\boldsymbol{C}} : \boldsymbol{Q}_i + p(J-1) - U_C^0 \tag{1}$$

where W is the elastic energy function, $\overline{\boldsymbol{C}}$ is the volume-preserved right Cauchy-Green tensor, the symbol$(\ \ )^0$ mean the elastic component, $\overline{W}^0$ is the elastic energy function of deviation component of hyperelastic constitutive equation, Ξ_t^m is the internal variable for damage like the Mullins effect, p is the Lagrange multiplier which is equivalent to hydrostatic pressure, U_C^0 is the complement strain energy representing relationship between volumetric strain and pressure, $\boldsymbol{Q}_i$ is viscous stress and N_d is number of dashpots. Viscoelastic model consists of springs and dashpots[4], as shown in Fig. 3. The 2nd Piola-Kirchhoff stress is given by

$$\boldsymbol{S} = 2\frac{\partial W}{\partial \boldsymbol{C}}. \tag{2}$$

The evolutionary equation of viscous stress $\boldsymbol{Q}_i$ under constant deformation is given as follows:

$$\dot{\boldsymbol{Q}}_i(t) + \frac{1}{\tau_i}\boldsymbol{Q}_i(t) = \frac{\gamma_i}{\tau_i} DEV\left(2\frac{\partial \overline{W}^0}{\partial \overline{\boldsymbol{C}}}\right) \tag{3}$$

where the notation DEV is correspond to $DEV(\bullet) = (\bullet) - \frac{1}{3}[(\bullet)\cdot \boldsymbol{C}]\boldsymbol{C}^{-1}$, τ_i is the relaxation time, γ_i is the stiffness ratio. From cyclic loading test results, the material parameters of viscosity were determined as $N_d = 2$, $\gamma_1 = 0.34, \tau_1 = 12s$, $\gamma_2 = 0.12, \tau_2 = 350s$, respectively. From the cyclic tensile loading test results, the stress-strain relationships of the TPEs were confirmed to depend on the maximum stretch[5]. To represent this behavior, following internal variable Ξ_s was applied to introduce the damage function.

$$\Xi_s = \sqrt{2\overline{W}(\overline{\boldsymbol{C}}(s))} \tag{4}$$

Now, let Ξ_t^m be the maximum value of Ξ_s over the past history up to current time t .The damage function $\bar{g}(\Xi_t^m)$ was adopted exponential form as follows:

$$\bar{g}(\Xi_t^m) = \beta + (1-\beta)\frac{1-\exp(-\Xi_t^m/\alpha)}{\Xi_t^m/\alpha} \tag{5}$$

Here, α and β are material parameters of the damage model. The elastic potential function $\bar{W}^0(\bar{\boldsymbol{C}},\ \Xi_t^m)$ was replaced by the following function.

$$\bar{W}^0(\bar{\boldsymbol{C}},\ \Xi_t^m) = \bar{g}(\Xi_t^m)\{C_{11}(\bar{I}_1-3) + C_{12}(\bar{I}_1-3)^2 + C_{21}(\bar{I}_2-3)\} \tag{6}$$

From the cyclic tensile loading test results, material parameters C_{11}, C_{12} and C_{21} were determined as 0.555MPa, -0.003MPa, 0.22MPa, respectively. Also, the parameters of damage model α and β were determined as 0.735 and 0.0243. In Fig. 2, calculated stress-strain relationships are shown.

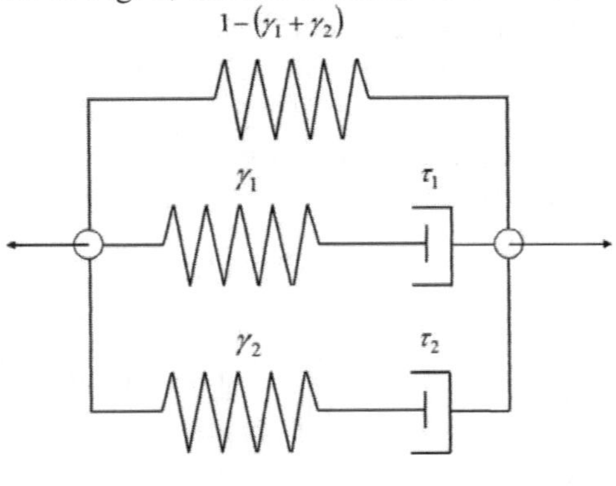

Fig. 3. Generalized Maxwell model (N_d=2)

The nonlinear viscoelastic model which is applied to rubber like material show high applicability to the TPE's mechanical behavior considering the effect of the strain rate and the damage phenomena.

4. Drop weight impact test

The drop weight impact testing method is one of the major method to investigate the dynamic properties of any materials. Drop weight impact testing of materials is a method to measure the ability of a specimens to absorb energy before breaking[6]. Also, the drop weight impact test is applicable to characterize the mechanical properties of thermoplastic material at relatively high strain rates[7]. In order to evaluate impact absorbing properties of materials such as polymers, concrete and carbon fiber reinforced plastics, many drop weight impact tests with a steel impactor have been conducted[6],[8],[9]. It is hard to measure the accurate energy absorbing performance of the TPEs with the normal weight impact testing method because TPEs show fracture at the contact surface with steel impactor.

In this study, original drop weight impact test machine for TPEs was developed to evaluate the applicability to sports equipment. For investigating of applicability to sports fields, some balls in baseball were introduced as the impact weight. The drop weight impact test machine is shown in Fig. 5. The test machine consists of a base plate, an impact plate, four supports and a top plate. Three load-cells were applied between the impact plate and the base plate. The size of TPE specimen which was placed on the impact plate were 50 mm ×50 mm ×10 mm in width, length and depth, respectively. The baseball which was used as the drop weight is shown in Fig. 4. The weight of the baseball (manufactured by the Rawlings Co., Ltd.) was 147g (0.147kg) and the diameter was 74mm. The ball was dropped from a height of 0.5m, 1.0m and 1.5m on to the impact plate 8-times (n=8) under the same conditions. The impact velocity v_{in} was estimated by $v_{in} = \sqrt{2gH}$. Here, notation g is the gravitational acceleration and H is the height releasing the ball. A high-speed camera was applied to record the digital image during impact tests. The motion tracking software (TEMA 3D) was used to measure the impact velocity v_{in}. The impact velocity v_{in} was confirmed that was appropriate by comparing with predicted velocity. The coefficient of restitution (COR) between the impact plate and the ball was calculated as the ratio of initial speed v_{in} and rebound speed v_{re} as follows:

$$\mathrm{COR} = v_{re}/v_{in} \tag{7}$$

where v_{re} was speed of rebounded ball. The COR was calculated on the assumption that the impact plate did not move during the tests. The impulse of impact test was calculated as $I = \int_0^t F(t)dt$. $F(t)$ is the total load measured by three load-cells at time t. The relationships between impact load and time given by the drop weight impact tests are shown in Fig. 6 (a), (b) and (c). These results show the effects of TPEs which are extension of contact time and reduction of peak load. The COR was decreased as the ball speed increased. Input momentum was calculated as $M_{in} = mv_{in}$.The experimental results are summarized in Table1.

Fig. 4 Baseball for impact test

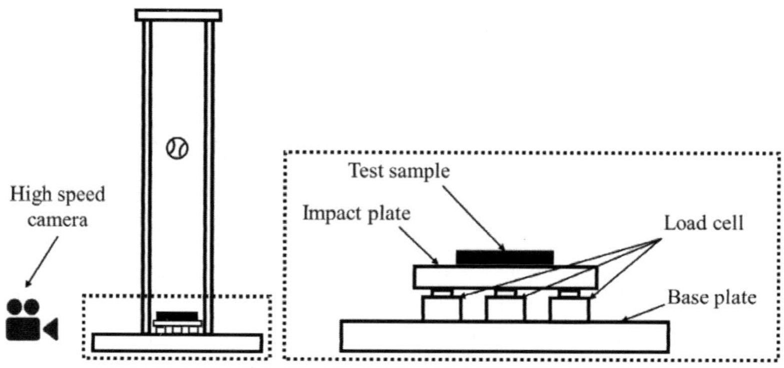

Fig. 5 Drop weight impact test machine

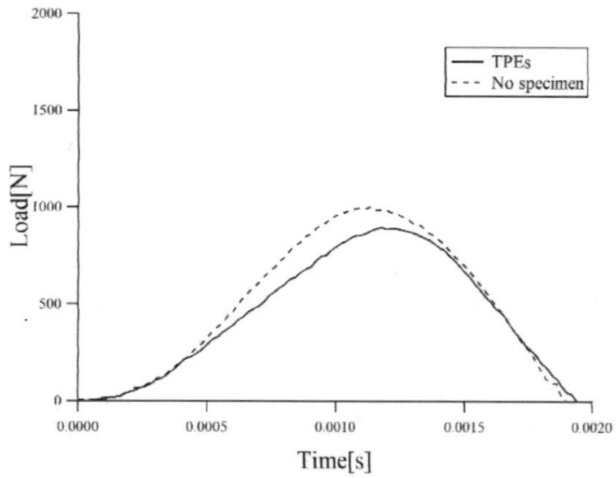

(a) H=0.5m

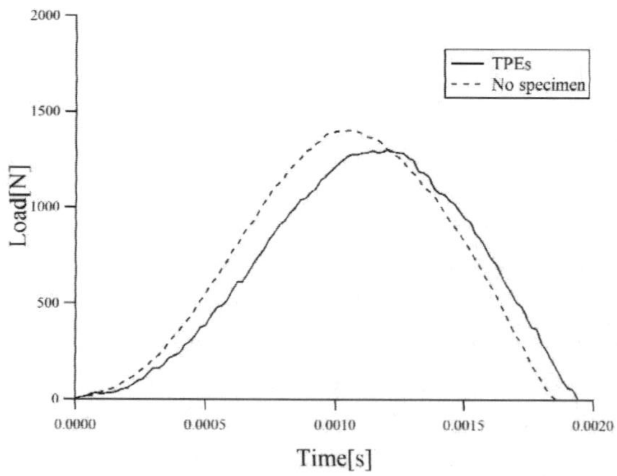

(b) H=1.0m

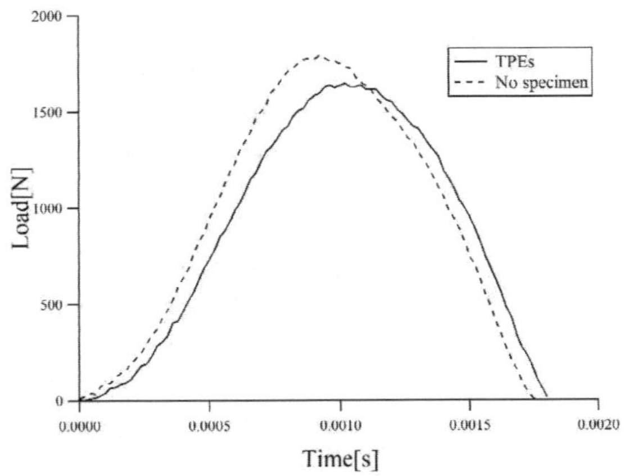

(c) H=1.5m

Fig. 6 Drop weight impact test results

Table 1 Relationships between drop weight speed and peak load

Drop height(m)	Specimen	Peak force(N)	Contact time(s)	Reduction of peak force(%)	Extension of contact time(%)	COR
0.5	No specimen	990	0.00188	6.58	1.06	0.640
	TPE	925	0.00190			0.550
1.0	No specimen	1,410	0.00184	6.97	1.08	0.625
	TPE	1,312	0.00186			0.542
1.5	No specimen	1,778	0.00177	10.5	2.26	0.613
	TPE	1,609	0.00181			0.533

5. Results and discussion

From the tensile loading test results, TPEs show viscoelastic properties and deformation over 400% of strain. The nonlinear viscoelastic model which is applied to rubber like material show high applicability to the TPE's mechanical behavior considering the effect of the strain rate and damage phenomena. TPEs are considered to have the same processability as rubber.

From the drop weight impact test results, the shock absorbing properties of TPEs were conducted with the ball weight to evaluate the TPE's applicability to sports equipment. Developed dropped weight impact test machine worked correctly. Impact load in 0.003s were able to be measured. From the drop weight impact tests results, the peak load was from 1500N to 1800N when the ball was dropped from a height of 1.5m ($v_{in} = 20km/h$). Reduction of the peak load by 6% to 10% were confirmed regardless of ball speed v_{in}. The contact time of impact tests were slightly extended when the TPEs were applied to the shock absorbing material.

6. Conclusion

In this study, static tensile loading tests and drop weight impact tests were carried out to expand the applicability of TPEs to impact absorbing materials for sports. Tensile loading test and cyclic loading test were conducted under three different strain rates. These results showed that TPEs was possible to deform over 400% and TPEs showed the viscosity and the stiffness softening like rubber material. An original drop weight impact test machine for TPEs with real baseball weight was developed and drop weight impact tests were conducted. The results show that the TPEs reduce the peak load by 6% to 10% and extend the contact time slightly.

References

[1] Holden G 2017 Applied plastics engineering handbook (Second Edition) *William Andrew* pp 91-107.

[2] Fraser M A, Grooms D R, Guskiewicz K M and Kerr Z Y 2017 Ball-contact injuries in 11 national collegiate athletic association sports: the injury surveillance program 2009-2010 through 2014-2015 *J. Athletic Training* **52** pp 698-707

[3] Maron B J, Ploliac L C, Kaplan J A and Mueller F O 1995 Blunt impact to the chest leading to sudden death from cardiac arrest during sports activities *The New England Journal of Medicine* **333** pp 337-342

[4] Simo J C and Huges T J R 1998 Computational Inelasticity *Springer* pp 336-373

[5] Simo J C 1987 On a fully three-dimensional finite-strain viscoelastic damage model: formulation and computational aspects *Computer Methods in Applied Mechanics and Engineering* **60** pp 153-173

[6] Söver A, Frormann L and Kipscholl R 2009 High impact-testing machine for elastomers investigation under impact loads *Polym. Test* **28** pp 871-874

[7] Velasco J I, Martínez A B, Arencónm D, Rodríguez- Pérezj M A and De Saja J A 1999 Application of instrumented falling dart impact to the mechanical characterization of thermoplastic foams *J. Mater. Sci.* **34** pp 431-438

[8] Zhang W, Chen S and Liu Y 2017 Effect of weight and drop height of hammer on the flexural impact performance of fiber-reinforced concrete *Construct. Building. Mater.* **140** pp 31-35

[9] Farooq U and Myler P 2016 Finite element simulation of damage and failure predictions of relatively thick carbon fiber-reinforced laminated composite panels subjected to flat and round noses low velocity drop-weight impact *Thin-Walled Structures* **104** pp 82-105

IOP Conf. Series: Materials Science and Engineering **548** (2019) 012009 doi:10.1088/1757-899X/548/1/012009

Investigation of Crude Palm Oil as an Alternative Processing Oils in Natural Rubber : Effect of the Unsaturated Fatty Acid

Khaliesah Abbas[1] and S.K. Ong[1]

[1] Universiti Kuala Lumpur, Malaysian Institue of Chemical and Biengineering Technology (UniKL MICET), Lot 1988 Kawasan Perindustrian Bandar Vendor, Taboh Naning, 78000 Alor Gajah, Melaka

E-mail: skong@unikl.edu.my

Abstract. Petroleum based processing oils is the hydrocarbon oils derived from petroleum refining process and commonly used as extender or processing oils to improve the rubber processability. Hydrocarbon oils can be classified as paraffins, aromatics, naphthene and olefins oil based on their molecular structure. The studies on replacement of hydrocarbon oils with natural processing oils had been studied by vast number of researchers due to finite resources and the toxicity of the hydrocarbon oils which enhanced awareness in using natural based oils. These natural processing oils are from vegetable or animal. The vegetable oils contain mainly unsaturated fatty acids. Studies showed that the unsaturated fatty acids affect the cure characteristics and the mechanical properties of the filled vulcanizates as these fatty acids took part during the vulcanization and promote the filler-rubber interaction. In this study, natural rubber (NR) of grade Standard Malaysian Rubber 20 (SMR 20) with the loadings of paraffin oil (PO) and crude palm oil (CPO) between 1 to 5 phr was investigated. Within the loadings of PO and CPO studied, the unsaturated fatty acid in CPOs were reported to results in plasticizing effect which results in lower minimum and maximum torques. Though that the unsaturated fatty acid which took part during the vulcanization prolonged the cure time, it resulted in higher CB-SMR 20 interaction. The tensile properties, tear strength, hardness and abrasion resistance of the CPO recipes were affected by combination of plasticizing effect and better CB-SMR 20 interaction.

1. Introduction

Over the past 20-30 years, the tire and rubber industry has focused on using renewable sources of raw material, recycling materials or used materials for its new application. There is significant potential to use crude palm oil (CPO) as an alternative for processing oils in rubber compound to replace petroleum-based processing oils. CPO is one of the four leading vegetable oils traded on the world market. Malaysia has a rich oil palm industry that generates excess CPO, in vast quantities for consumption [1]. The use of natural processing oils in rubber compounds has already been commercially demonstrated by the incorporation of *d*-limonene, sunflower oil, castor oil, orange oil, and olive oil by three manufacturers in the large, high performance application of passenger car [2].

Hydrocarbon processing oils can be classified into four types which are paraffins, aromatics, naphthene and olefins [3]. The classification is based on the molecular structure. Paraffins are also available with both normal and iso-paraffins which the former is alkanes while the latter is the branched paraffins [3]. The normal paraffins oil (PO) is used in this study.

CPO are obtained from the mesocarp of the oil palm fruit (*Elaeis Guineensis*) and consists of 49.91% saturated fatty acids (0.93 % myristic, 45.48 % palmitic and 3.49 % stearic) and 50.9 % unsaturated fatty acids (40.17 % oleic and 9.92 % linoleic) [4]. Lai *et al.* also reported that in general, palm oil contained approximately 50 % saturated, 40 % monounsaturated fatty acids and

10 % polyunsaturated fatty acids [5]. The fatty acids are a long stearic chains compounds containing an even number of carbon atoms C_{10} to C_{18} [6]. Like others vegetable oils, the main component of palm oil is triacylglycerol. It also contains a minor amount of monoacylglycerols and diacylglycerols. Apart from that, CPO is rich with phytonutrients such as tocopherols, tocotrienols and phytosterols [5].

Some studies evaluated several vegetable oils in comparison with hydrocarbon processing oil. Kukreja *et al.* used castor oil (loading: 1–5 phr) and cashew nut shell liquid (CSNL) (loading: 1 phr) in natural rubber/polybutadiene (NR/BR) blends and found that with an increasing amount of castor oil and CSNL, the modulus at 300 % elongation (M300), tear strength and tensile strength increased up to optimum oil loading [7]. The presence of unsaturated fatty acid in castor oil and CSNL was suggested to assist the interaction between carbon black (CB) and the rubbers which resulted in better reinforcement. Dasgupta *et al.* also studied the effect of various natural oils in comparison to aromatic, naphthenic and paraffinic oils at 8 phr [8]. They observed that the mechanical properties, particularly with the use of soybean oil, exhibited a reduction in M300 and tensile strength but increased in the elongation at break because the elastomer molecular chains have greater mobility when vegetable oils was employed. Fernandez *et al.* studies on incorporation of linseed oils of (loading: 2–10 phr) in Nitrile rubber/Expandable Graphite (NBR/EG) and found that the linseed oil (LO) compound showed higher maximum torque with the LO loading. This is due to esters and double bond presence in LO which enhanced the physico-chemical interaction between EG and polymer matrix. However, decrement of cure time with the increased of LO loading was reported as the unsaturated fatty acid interrupted the vulcanization [9]. Khalaf *et al.* studies the effect of orange and olive oil as multipurpose additives in NBR and concluded that the physico-mechanical results indicated that olive and orange oils (loading: 5 phr) impart good mechanical properties to NBR vulcanizates due to the polar characteristic of the oils which improve compatibility with NBR [10]. Based on the studies carry out, natural oils improved the properties of the filled rubber vulcanizate up to its optimum loading. However, there are limited study on the effect of CPO on CB filled vulcanizate. To date there are no study carried out on the effect of CPO on CB filled NR vulcanizate.

In this study, the effect of vegetable oil; i.e. CPO in comparison with PO was investigated. Since CPO is rich with unsaturated fatty acid, better performance of the filled vulcanizate in comparison with CB filled NR vulcanizate prepared using PO is hypothesized.

2. Materials and Method

2.1. Materials

NR of grade SMR 20 was supplied by Lembaga Getah Malaysia, CPO was obtained from Sime Darby Plantations Sdn.Bhd, PO and CB of grade N330 were purchased from Sah Petroleum Ltd, Daman, India and Cabot Corporation Sdn.Bhd., respectively. Sulphur was supplied by Orice Australia Ptd. Ltd Australia, zinc oxide is manufactured by Sigma-Aldrich (M) Sdn. Bhd, stearic acid is by Fisher Scientific USA, 6PPD (N-cyclohexyl-2-benzothiazyl sulphonamide) is by Flexys America L. P USA and CBS (N-(1,3-dimethylbutyl)-N'-phenyl-p-phenylenediamine) is by Rongcheng Chemical General Factory Co. Ltd.

IOP Conf. Series: Materials Science and Engineering **548** (2019) 012009 doi:10.1088/1757-899X/548/1/012009

2.2. Recipes

The recipes used in this study are show in Table 1. The loading of PO and CPO were varied between 1 to 5 phr while the other compounding ingredients were constant. The recipes used were adopted from the standard tire tread recipe [11].

Table 1 : Rubber compounding recipes

Compounding Ingredients	PO_1 (phr)	PO_3 (phr)	PO_5 (phr)	CPO_1 (phr)	CPO_3 (phr)	CPO_5 (phr)
SMR 20	100.0	100.0	100.0	100.0	100.0	100.0
Zinc Oxide	5.0	5.0	5.0	5.0	5.0	5.0
Stearic Acid	2.0	2.0	2.0	2.0	2.0	2.0
CBS	1.0	1.0	1.0	1.0	1.0	1.0
6PPD	1.5	1.5	1.5	1.5	1.5	1.5
N330	50.0	50.0	50.0	50.0	50.0	50.0
PO	1.0	3.0	5.0	-	-	-
CPO	-	-	-	1.0	3.0	5.0
Sulphur	2.5	2.5	2.5	2.5	2.5	2.5

2.3. Rubber Compounding

SMR 20 compounds were prepared using two roll mills (Model: SLIMC SXK-160A) at room temperature. The compounding ingredients were added to SMR 20 according to the order in Table 1. N330 with either PO or CPO were added into the rubber compound alternately. Finally, the sulphur was incorporated into the rubber compound. The rubber compounds were kept in the freezer for 24 hours.

2.4. Cure Characteristics

Cure characteristics was carried out using Monsanto Rheometer (Model: Monsanto 100) according to ASTM D2084-01. A sample sealed in a temperature-controlled die cavity surrounds a rotor. The test temperature of 160 °C and torque ranges from 0 to 100 lb-in were used. The test was conducted for 30 minutes and data for cure characteristics such as minimum torque (M_L), maximum torque (M_H), scorch time (t_{s2}),cure time (t_{c90}) and cure rate index (CRI) were obtained. For CRI measures the rate at which cross-linking and the development of stiffness of the compound occur after the scorch point. The CRI was calculated using the equation below:

$$\mathrm{CRI} = \frac{100}{t_{c90} - t_{s2}} \quad \text{(ASTM D2084-01)}$$

2.5. Compression Moulding

The test specimens were prepared by compression moulding of model 4129-039 under pressure 10 MPa at 160 °C according to the cure time, t_{c90} reported in cure characteristics result.

IOP Conf. Series: Materials Science and Engineering **548** (2019) 012009 doi:10.1088/1757-899X/548/1/012009

2.6. Mechanical Tests

a) *Tensile Test:* The test was performed using INSTRON universal testing machine according to ASTM D638-14. The test specimen dimension is 165 mm x 19 mm. Cross head speed of 500 mm/min was employed. Average of 5 test specimens was reported.

b) *Tear Test:* The INSTRON universal testing machine was used to carried out the test as outlined in ASTM D1004-13. The test specimens of Die C were extended with the crosshead speed of 500 mm/min. Average of 5 test specimens was reported.

c) *Hardness Test:* The hardness test was conducted by using the Shore A Durometer according to ASTM D2240-15. Average of 5 test specimens was reported.

d) *Abrasion Resistance Test:* The test is evaluated using Akron Abrasion Tester according to ASTM D1630-00. The test is carried out for 5000 revolutions. The average in abrasion resistance index (ARI) for 3 test specimens were reported. The ARI was calculated according to the following equation :

$$ARI = \frac{V_s}{V_t} \times 100\%$$

Where, V_t is the original volume of the test specimen (in mm^3) and V_S is the volume loss of the test specimen (in mm^3) determined under the same test conditions.

3. Results and Discussion

3.1. Cure Characteristics

The processability of rubber compounds can be determined by evaluating the cure characteristics such as minimum torque (M_L), maximum torque (M_H), scorch time (t_{s2}),cure time (t_{c90}) and cure rate index (CRI) as shows in Table 2.

Table 2: Cure characteristic of different loading of PO and CPO in SMR 20

Materials	PO_1 (phr)	PO_3	PO_5	CPO_1	CPO_3	CPO_5
M_L (dNm)	3.70	4.00	4.20	4.90	3.60	2.00
M_H (dNm)	31.22	35.70	36.5	32.3	38.7	34.5
M_H-M_L (dNm)	27.52	31.7	32.3	27.4	34.1	32.5
t_{s2} (min)	4.00	3.21	3.38	4.13	3.52	4.03
t_{c90} (min)	11.23	10.91	11.39	12.47	12.56	12.15
CRI (min^{-1})	13.83	12.98	12.48	11.99	11.06	12.32

The value of M_L measures the compound viscosity before vulcanization [12]. It gives an indication on the ease of rubber flow during processing [13]. M_L results in Table 2 shows that increased amount of PO exhibit an increased amount of M_L. This result shows a lower plasticizing effect happened in rubber compound of NR and PO due to the lack of different polarities in PO resulting in low interactions with NR [10]. For M_L results of CPO shows that the increased of CPO loadings resulted in decrement in M_L. The results indicate that higher plasticizing effect is due to the unsaturated fatty acids presence in CPO which had facilitated the rubber compound processability [14].

The M_H value is related to molecular stiffness of rubber compounding [12]. From the results in Table 2, it shows that the M_H increased with the increased of PO loading. These results are attributed to the cross-linking in NR increased with PO loading [15]. For CPO loading, M_H is highest at 3 phr CPO loading and lowest at 1 phr CPO loading. The results suggested that CPO improved the CB-NR interface at 3 phr loading [16].

t_{s2} indicates the period of time before vulcanization starts. No significant difference on the t_{s2} for both PO and CPO was observed. The results suggested that neither PO nor CPO affects the t_{s2}.

The results for t_{c90} shows that recipes with PO had slightly faster cure time compared to CPO. The results indicated that CPO which contained higher unsaturation fatty acids had caused sulphur dilution effect due to interaction between sulphur and double bond in CPO molecules as well as promoting the CB-rubber interaction [17]. Therefore, lower cure rate and the higher cure time were observed for CPO loadings in NR compared to PO.

From the results, it shows that CPO cured slightly slower compared to PO. The present of fatty acid groups react with ingredients presented in rubber compounds reduced the cure rate at increase loading of CPO [8] [18].

3.2 Mechanical Properties

The results for tensile strength are shown in Fig. 1. Processing oils has plasticizing effect, thus decrement in tensile strength with the increase of CPO loading is expected. Increase of PO loading does lead to plasticizing effect. This might be due to the amount of PO studied is low. The filled vulcanizate with CPO has lower tensile strength as compare to the filled vulcanizate with PO at the same loading. The results are attributed to the capability of CPO to undergo self-vulcanization to form factice which will results in the decreased value of tensile strength [19].

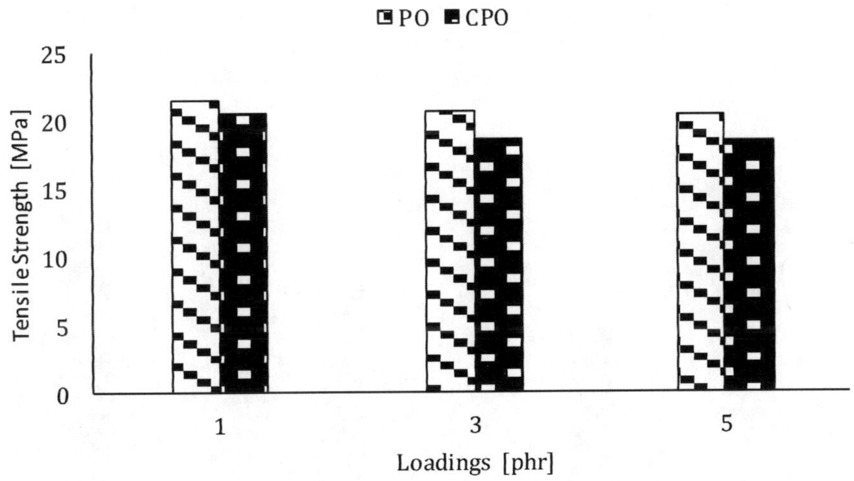

Fig. 1. Effect of PO and CPO loadings on the tensile strength of the CB filled vulcanizates

In Fig. 2, it shows the results for elongation at break for PO and CPO are consistent with the tensile strength trend, the presence of PO does not show plasticizing effect on the PO recipes. The decrement of elongation at break with the increment of CPO loadings was probably due to coupling action, which happened between CB and NR molecular chains. CPO which had unsaturated fatty acid takes part in dual interactions. The polar end of the chain could interact with CB surface while the non-polar unsaturation rich ends could interact with rubber molecules. Although higher stress is applied, the entanglement that happened inside rubber compounds prevents freedom of the molecular chains to move thus, lowered the elongation at break was observed [7] [20].

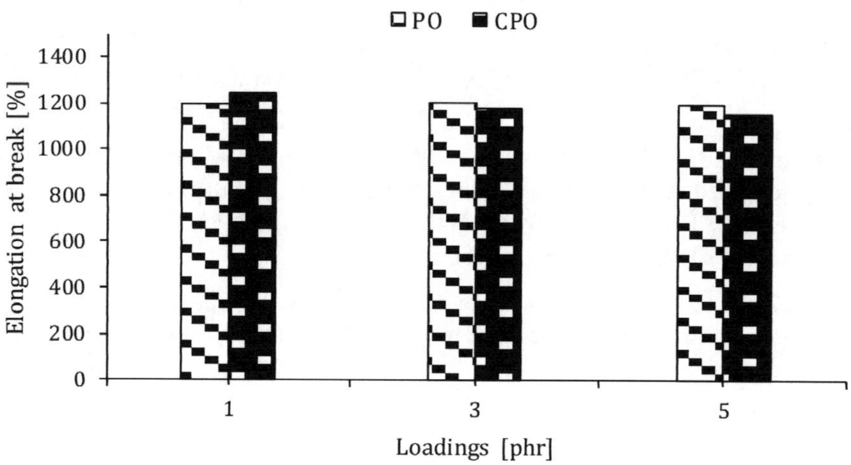

Fig. 2. Effect of PO and CPO loadings on the elongation at break of the CB filled vulcanizates

Tear strength results for PO and CPO filled vulcanizates are show in Fig. 3. The results showed that the incorporation of CPO had resulted in higher tear strength as compared to the recipes with PO. The is because PO had low degree of unsaturation as compared to CPO, thus, lower tear strength was reported [7]. The CPO recipes consistently had higher tear strength as compared to the PO recipes at the same processing oil loading. The highest tear strength was reported at CPO loading of 5 phr. This is due to the high amount of unsaturated fatty acid which actively participated in curing and coupling at the CB-rubber interface resulted in higher tear strength [7].

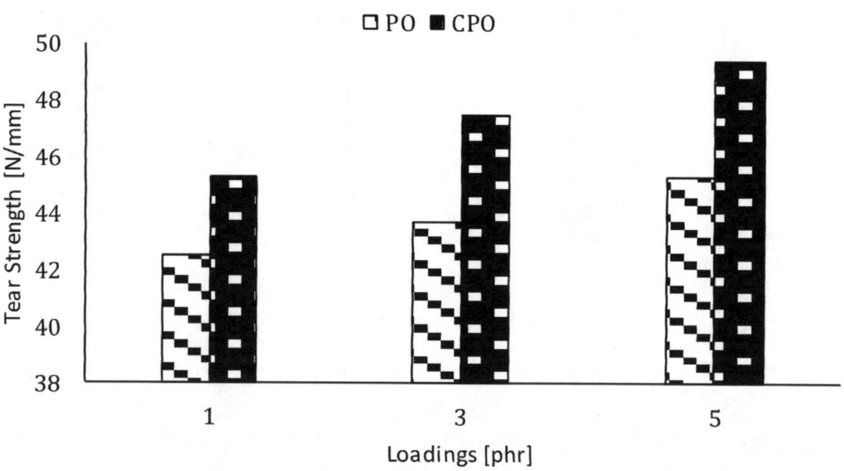

Fig. 3. Effect of PO and CPO loadings on the tear strength of the CB filled vulcanizates

The hardness results for PO and CPO filled vulcanizates were show in Fig. 4. The results do not show the significance differences on the hardness of the filled vulcanizate within the loading of PO studied, and similar results were observed for CPO_1 and CPO_3. A slight increment in the hardness of CPO_5 in comparison with CPO_1 and CPO_3 was reported. Chandrasekara *et al.* reported that longer cure time of CPO recipes implies good CB-rubber interaction and resulted in higher hardness in CPO recipes [21]. Consistently Kukreja *et al.* reported similar trend which is due to the dual interactions of CPO with CB and rubber is more significant at higher loading [7].

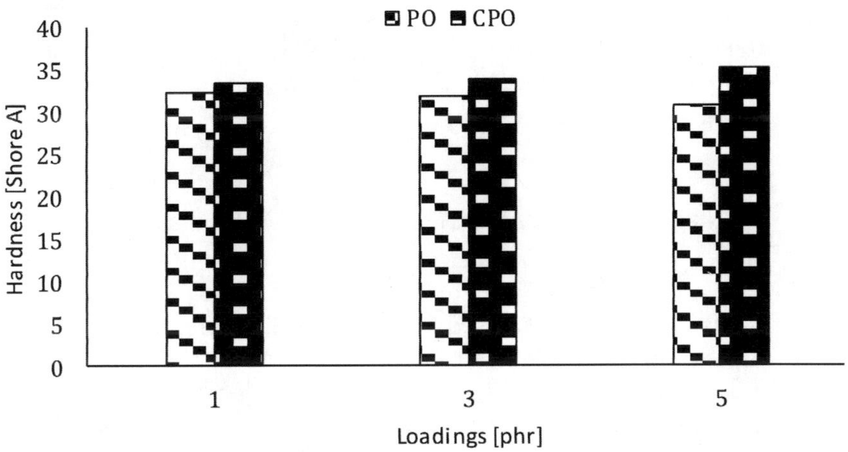

Fig. 4. Effect of PO and CPO loadings on the hardness of the CB filled vulcanizates

Fig. 5 shows the results on the Abrasion Resistance Index (ARI) for PO and CPO recipes studied. From the results, it was observed that the difference of ARI between PO recipes are insignificant. The ARI depends on the softness of the rubber surface, the CB dispersity and the rubber vulcanizates crosslink density [7]. Thus, within the range of PO studied, the CB dispersity and the rubber vulcanizate crosslink density does not differs. The slight decrement in the abrasion resistance was reported on CPO_3 and CPO_5 which due to the high amount unsaturated fatty acid in CPO. The CPO which participated in the vulcanization and the CB-rubber interaction leading to lower ARI [14].

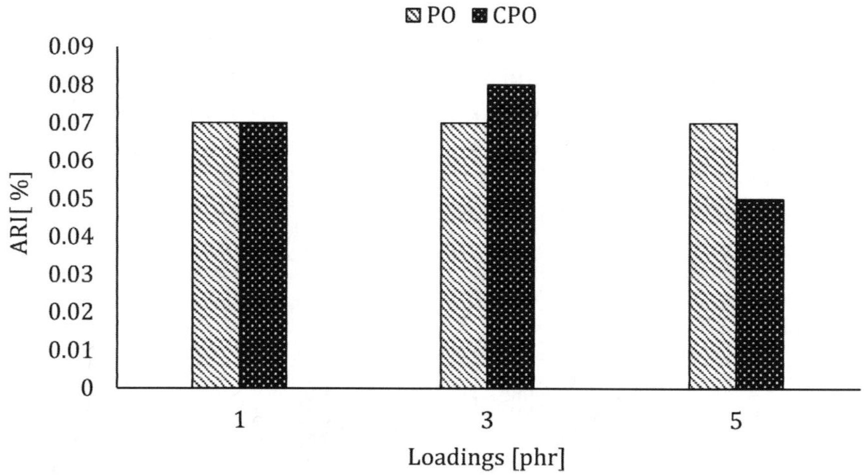

Fig. 5. ARI of the PO and CPO recipes

4. Conclusions

As a conclusion, the presence of unsaturated fatty acid in CPO resulted in plasticizing effect and facilitated the CB-rubber interaction. This resulted in lower M_l and M_H, longer t_{c90} as well as lower tensile strength and elongation at break. Harder and better tear strength of the CB filled vulcanizates were also observed when CPO was employed as the processing oil. Thus, CPO can be a potential and alternative processing oils for CB filled SMR 20 vulcanizates within the CPO loadings studied.

References

[1] Crabbe E , Nolasco-Hipolito C, Kobayashi G , Sonomoto K and Ishizaki A 2001 Biodiesel production from crude palm oil and evaluation of butanol extraction and fuel properties. *Proc. Biochem.*, **37**(1), pp. 65–71

[2] Brentin R and Sarnacke P 2011 Rubber Compound A Market Opportunity Study. *United Soybean Board*

[3] Chaudhuri M 2016 Fundamentals of Petroleum and Petrochemical Engineering *(CRC Press)* chapter 1 p 1

[4] Che Man Y B , Haryati T, Ghazali H M, & Asbi B A 1999 Composition and Thermal Profile of Crude Palm Oil and Its Products. *J. Amer. Oil Chem. Soc.*, **76**(2) pp 237-242

[5] Lai O M, Tan P C and Akoh C 2015 Palm Oil Production Processing Characterization and Uses *(AOCS Press)* chapter 13 pp 383-388

[6] Ismail H and Ruhaizat T A 1997 Effect of Palm Oil Fatty Acid on Curing Characteristics and Mechanical Properties of $CaCO_3$ Filled Natural Rubber Compounds. *Iran. Polymer J.*, **6**(2) pp 97–104

[7] Kukreja T R, Chauhan R C, Choe S, and Kundu P P 2002 Effect Of The Doses And Nature Of Vegetable Oil On Carbon Black/Rubber Interactions: Studies On Castor Oil And Other Vegetable Oils *J. Appl. Polym. Sci.* **87** 10 pp 1574–1578

[8] Dasgupta S, Agrawal S L, Bandyopadhyay S, Chakraborty S, Mukhopadhyay R, Malkani R K and Ameta S C 2008 Characterisation of Eco-Friendly Processing Aids for Rubber Compound *Polym. Test* **26** pp 489-500

[9] Fernandez S S, Kunchandy S and Ghosh S 2015 Linseed Oil Plasticizer Based Natural Rubber/Expandable Graphite Vulcanizates: Synthesis and Characterizations *J. Polym. Environ.* **23**(4) pp 526–533

[10] Khalaf A I, Ward A A, Abd El-Kader A E and El-Sabbagh S H 2015 Effect of Selected Vegetable Oils on the Properties of Acrylonitrile-Butadiene Rubber Vulcanizates *Polimery/Polymers* **60** (1) pp 43–56.

[11] Jayewardhana W G D, Perera G M, Edirisinghe D G and Karunanayake L 2009 Study on Natural Oils as Alternative Processing Aids and Activators in Carbon Black Filled Natural Rubber *J. Nat. Sci. Found. of Sri Lanka* **37**(3) pp 187–193

[12] Moresco S, Giovanela M, Carli L N and Crespo J S 2016 Development of Passenger Tire Treads: Reduction in Zinc Content and Utilization of a Bio-Based Lubricant *J. Clean. Prod.* **117** pp 199–206

[13] Petrović Z S, Ionescu M, Milić, J and Halladay J R 2013 Soybean Oil Plasticizers as Replacement of Petroleum Oil in Rubber *Rubber Chem. Technol.* **86**(2) pp 233–249

[14] Pechurai W, Chiangta W and Tharuen P 2015 Effect of Vegetable Oils as Processing Aids in SBR Compound Macromol. Symp **354** pp 191–196

[15] Jacob M, Thomas S and Varughese K T 2004 Mechanical Properties of Sisal/Oil Palm Hybrid Fibre Reinforced Natural Rubber Composites *Comp. Sci. Technol.* **64**(7–8) pp 955–965

[16] Kuriakose A P and Varghese M 2003 Use of Rice Bran Oil and Epoxidized Rice Bran Oil in Carbon Black-Filled Natural Rubber-Polychloroprene Blends *J. App. Polym.Sci.* **90** (14) pp 4084–4092

[17] Akolisaibemesi J 2007 From Studies in Polymer and Vegetable Oils to Sanitization of The Academic System *19th Inaugural Lecture of the University of Nigeria*, Nnsukka Nigeria.

[18] Boontawee H, Nakason C, Kaesaman A, Thitithammawong A and Chewchanwuttiwong S 2017 Benzyl Esters of Vegetable Oils as Processing Oil in Carbon Black–Filled SBR Compounding: Chemical Modification, Characterization, And Performance *Adv. Polym. Tech.* **36**(3) pp 320–330

[19] Nandanan V, Joseph R and George K E 1999 Rubber Seed Oil: A Multipurpose Additive in NR and SBR Compounds *J. App. Polym. Sci.* **72**(4) pp 487–492

[20] Arayapranee W and Rempel G L 2013 Effects of Expandable Graphite and Modified Ammonium Polyphosphate on The Flame-Retardant and Mechanical Properties of Wood Flour-Polypropylene Composites. *Polym. and Polym. Compos.* **21**(7) pp 449–456

[21] Chandrasekara G, Mahanama M K, Edirisinghe D G and Karunanayake L 2011 Epoxidized Vegetable Oils as Processing Aids and Activators in Carbon-Black Filled Natural Rubber Compounds *J. Nat. Sci. Found. Sri Lanka* **39**(3) pp 243–250

IOP Conf. Series: Materials Science and Engineering **548** (2019) 012010 doi:10.1088/1757-899X/548/1/012010

Study on Voxel Finite Element Analysis of Open-Cell Polyurethane Foam

Y Kimura[1], K Kato[2] and A Matsuda[3]

[1] Graduate School of Systems and Information Engineering, University of Tsukuba, Ibaraki, Japan

[2] NHK Spring Co., Ltd, Kanagawa, Japan

[3] Faculty of Engineering, Information and Systems, University of Tsukuba, Ibaraki, Japan

E-mail: a_matsuda@kz.tsukuba.ac.jp

Abstract. Establishment of a design method for the macroscopic stress-strain relationship of open-cell foam has been investigated because of the foam's complicated microstructures. In this study, a material design method for the mechanical behavior of open-cell polyurethane foam using voxel finite element analysis is proposed to consider the microscopic structure and matrix properties. A finite-strain hyperelastic model was applied to a matrix of polyurethane foam. The Mooney-Rivlin model was adopted for the potential energy function of the hyperelastic model. To determine the mechanical properties of the matrix, original, solid specimens of polyurethane were prepared. The material parameters of the matrix were identified based on the tensile loading test results of the solid specimens. Microstructures of the open-cell polyurethane foam were determined using a CT-scanning system. Finite element models constrained by the original microstructural configurations were applied to numerical simulations. Periodic finite element models based on the modified microstructural configurations, the relative density of which was the same as that of the original models, were applied to the simulation. Based on the numerical results, verification of the design performance of the simulation code for open-cell polyurethane foam was performed. The simulation results were reasonably consistent with compression loading test results up to the plateau region.

1. Introduction

Polymer foam materials have been widely used in a variety of applications such as car seats, shoe soles and packing materials. The strong points of polymer foam are that it is light weight, has good impact energy absorption and has manufactural formability [1]. Polymer foam exhibits a nonlinear stress-strain relationship, which is divided into three regions – the linear elastic region, the plateau region and the high-density region. It is known that the mechanical properties of polymer foam depend on its microstructure and matrix [2]. Prediction methods for the mechanical properties of polymer foam have been proposed by numerous researchers. For example, Gibson and Ashby proposed a formula to describe the relationship between relative density and initial stiffness [2]. Relative density represents the ratio of the matrix in the foam. Studies to analyze the elastic properties of open-cell foams using the Kelvin cell foam model have been done [3, 4]. However, while this method can be applied to the initial stiffness, prediction of the stress-strain relationship with greater strain remains difficult.

Especially in the medical field, the voxel FEM has been used to predict the mechanical properties of materials with complex microstructures, such as human bones [5, 6]. The validity of this method has been confirmed in numerous studies. In this study, the applicability of voxel FEM analysis to predict the stress-strain relationship up to the large-strain region of polymer foam material was verified. CT scan data of polyurethane foam was acquired in voxel form. Parts of the CT scan data were applied to voxel FEM analysis. The material parameters of the polyurethane matrix employed in the simulations were identified using tensile loading test results for solid specimens. In addition, a periodic model which represents the microstructures of polymer foam was also prepared. The simulation results were in near agreement with experimental results up to the plateau region.

2. Tensile loading tests on solid specimens

Preparing a specimen without voids is difficult because numerous voids in polymer foam are generated during the chemical reaction. For this reason, there have been few discussions of the precise mechanical properties of the matrix. In this study, in order to identify the material parameters of the polyurethane matrix, original solid specimens of a polyurethane foam matrix without air voids were prepared. Fig. 1 shows the solid specimens. Tensile loading tests on three solid specimens were conducted using a uniaxial machine (Autograph AG-20kNXplus, Shimadzu Corporation); the strain rate for all tests was set to 0.1%/s. Fig. 2 shows the stress-strain relationships given by the tensile loading tests. The three specimens showed similar stress-strain relationships. The results did not exhibit the complicated mechanical characteristics peculiar to polyurethane foam. Thus, the mechanical properties of the polyurethane matrix which were obtained with the loading tests excluded the influence of the microstructure.

To determine the material parameters of the polyurethane matrix, the incompressible hyperelasticity was applied to the mechanical behavior given by the tensile loading test results. Thus, the second Piola-Kirchhoff stress tensor $\boldsymbol{S}$ of a hyperelastic material is given by the partial differentiation of the strain energy function W with respect to the right Cauchy-Green deformation tensor $\boldsymbol{C}$ as follows:

$$\boldsymbol{S} = 2\frac{\partial W(\bar{\boldsymbol{C}})}{\partial \boldsymbol{C}} \tag{1}$$

where $\bar{\boldsymbol{C}}$ is the modified right Cauchy-Green deformation tensor. The following equation for elastic energy function was applied to represent the stress-strain relationships of the polymer matrix:

$$W(\bar{\boldsymbol{C}}) = W(\bar{I}_1, \bar{I}_2)$$
$$= C_{11}(\bar{I}_1 - 3) + \frac{C_{12}}{2}(\bar{I}_1 - 3)^2 + \frac{C_{13}}{3}(\bar{I}_1 - 3)^3 + \frac{C_{14}}{C_{15}} exp\{C_{15}(\bar{I}_1 - 3)\} \tag{2}$$

where $\bar{I}_1$ is the modified first invariant, $\bar{I}_2$ is the modified second invariant, and C_{11},C_{12},C_{13},C_{14}, and C_{15} are material parameters. The material parameters were identified using the least square method with the tensile loading test results for the solid specimens. Table 1 shows the material parameters of the polyurethane matrix approximated using the above method.

Table1. Material parameters identified from the result of tensile loading test for solid specimens

C_{11}	C_{12}	C_{13}	C_{14}	C_{15}
2.37	-0.0486	0.0312	3.08	-6.75

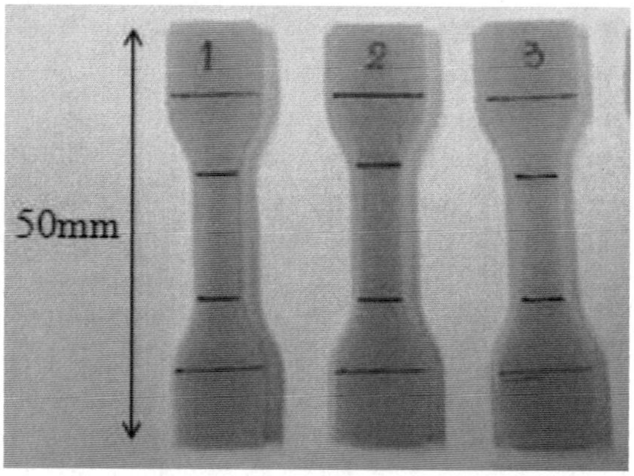

Fig. 1. Polyurethane solid specimens

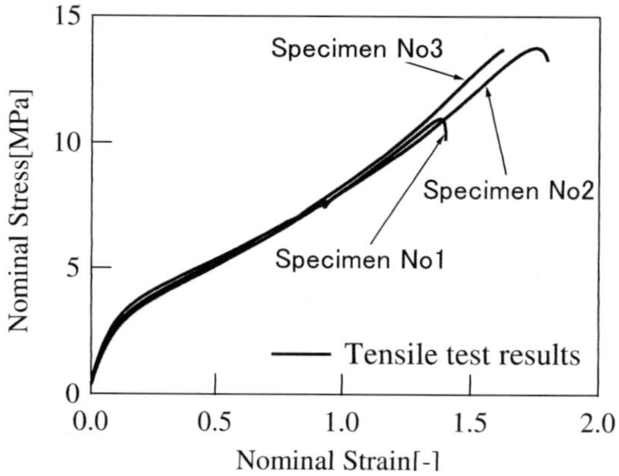

Fig. 2. Stress-strain relationships of polyurethane solid specimen

3. Compression loading tests on polyurethane foam

In order to determine the mechanical properties of polyurethane foam, compression loading tests were conducted. The relative densities of the compression test specimens were 3.73%, 4.82% and 5.94%, respectively. The specimens were compressed up to a nominal strain of 0.6, and unloaded. The strain rate was set to 0.1 %/s for all tests. Fig. 3 shows the stress-strain relationships given by the compression loading test. As confirmed in previous research, initial stiffness increases as relative density increases [6]. All specimens exhibited linear-elastic regions, plateau regions and high-density regions. These are characteristics peculiar to polymer foam material. Initial stiffness was calculated based on the definition that the linear-elastic region is up to a strain of 0.05. Fig. 4 shows the relationship between initial stiffness and relative density. These results confirmed that this polyurethane foam satisfies the theoretical formula for the relationship between initial stiffness and relative density, which was introduced by Gibson and Ashby [2]. The formula is as below:

$$\frac{E_f}{E_m} = C_1 \left(\frac{\rho_f}{\rho_m}\right)^2 \tag{3}$$

Where E_f is the initial stiffness of the foam material, E_m is the initial stiffness of the matrix, C_1 is the constant of proportionality, ρ_f is the density of the foam material and ρ_m is the density of the matrix.

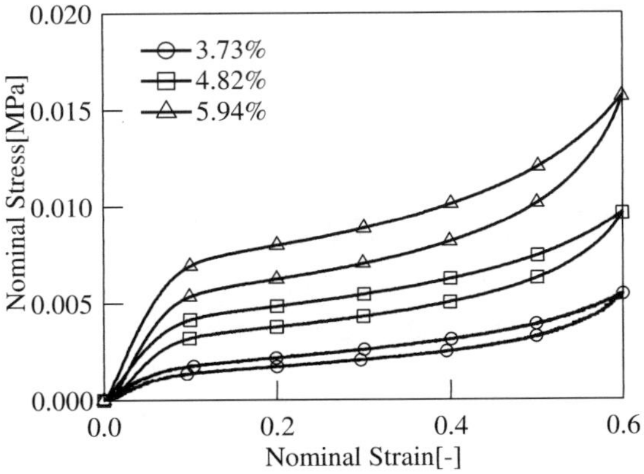

Fig. 3. Stress-strain relationships of compression loading test

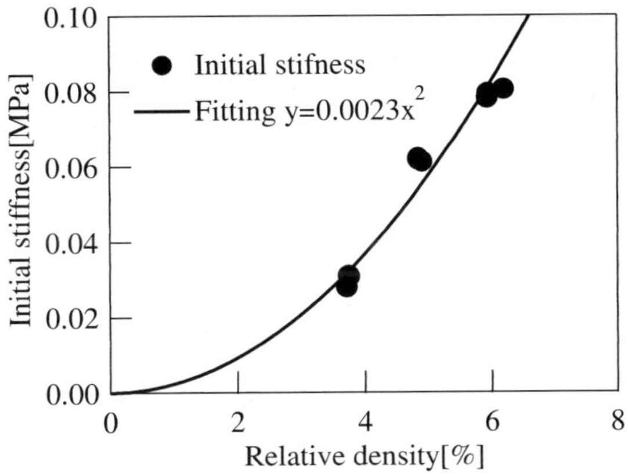

Fig. 4. The relationship between initial stiffness and relative density

4. Voxel FEM analysis based on CT scan data

Voxel FEM analysis with the model based on CT scan data was conducted in order to consider the effect of microscopic structures on the mechanical properties of polymer foam. CT scan data for the polyurethane foam was acquired in voxel form. The polyurethane foam scanned was same as the specimens used for the compression loading tests described in a later section. Rectangular areas were extracted from the CT scan data. The length of one side of each area was 2 mm. Hereinafter this model is referred to as the CT model. Voxel FEM analysis of the CT model was conducted under the following conditions. The top and bottom surfaces were completely constrained in x_1, x_2 and x_3 directions, and the side surfaces were free. Strain was applied to the top surface in increments of 0.015 up to a strain of 0.3 at the 20th increment. The material parameters of the matrix as determined by the tensile loading tests of the solid specimens were applied to this simulation. Fig. 5 is a diagram of the model at strains 0, 0.15 and 0.3. Fig. 6 shows the stress-strain relationships obtained with the simulations. The percentage annotations in the graph indicate the relative density of each model. The simulation results all roughly agree with each other. These results suggest that the sample areas contained representative microstructures of polymer foam. In addition, the results show the bend of stress-strain relationship, which represents the plateau region. Therefore, it is implied that applying the voxel FEM with a CT model can predict the stress-strain relationship of polymer foam up to the plateau region.

IOP Conf. Series: Materials Science and Engineering **548** (2019) 012010 doi:10.1088/1757-899X/548/1/012010

Fig. 5. Stress distribution charts for the simulation with the CT model

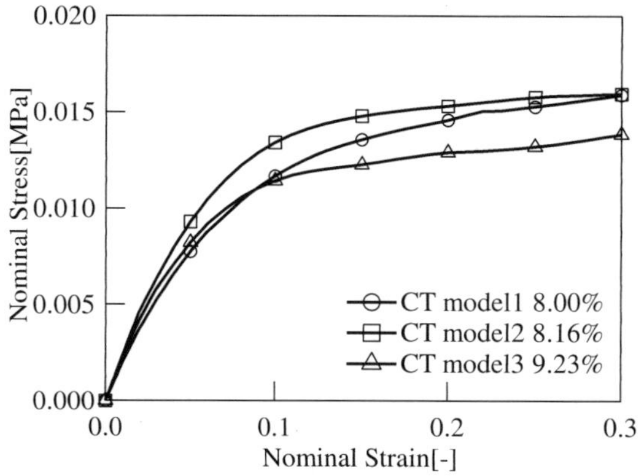

Fig. 6. Stress-strain relationships obtained by voxel FEM of CT model

5. Voxel FEM analysis of the body-centered cubic model

The number of elements of the CT model was so large that the simulation required a high calculation cost. Therefore, a simplified voxel model reproducing the microstructure of the polymer foam was prepared by creating voids in the cubes. The length of one side of each cube was 30 voxel, and the model was a body-centered cubic lattice, which had cavities in the center and corners. The radius of the center void and the corner void were 19 voxel and 5 voxel, respectively. This simplified model is referred to as the BCC model(the Body-Centered Cubic model) because of its structure. Analysis conditions were same as those of the voxel FEM analysis of the CT model. Fig. 7 shows diagrams of the model at strains of 0, 0.15 and 0.3. Fig. 8 shows the stress-strain relationship obtained with the simulation. As with the CT model, the initial stiffness and the bend of the stress-strain relationship was reproduced reasonably well, which indicates good agreement with the simulation results of the CT model.

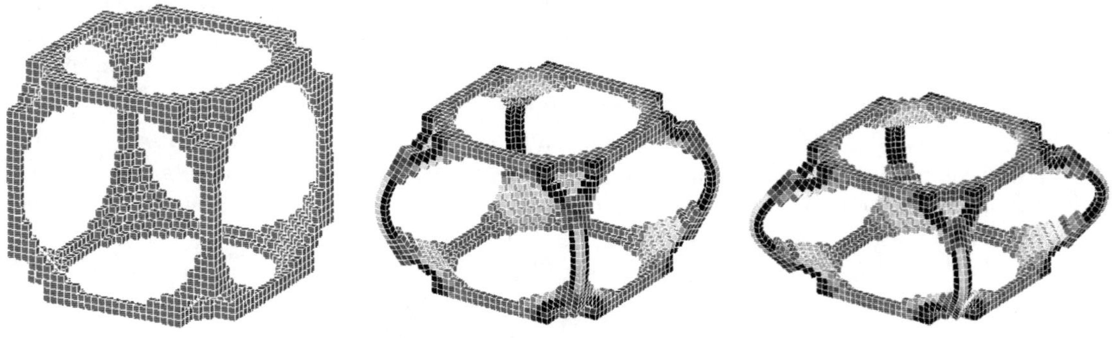

Fig. 7. Stress distribution chart for the simulation of the BCC model

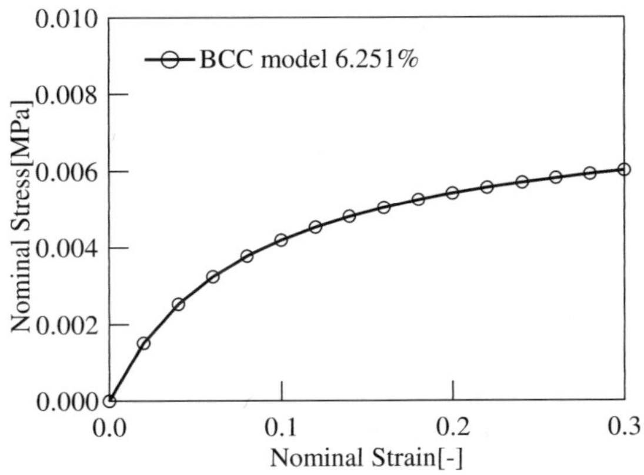

Fig. 8. Stress-strain relationship obtained by voxel FEM of BCC model

6. Result and discussion

6.1. Verification of the applicability of the voxel FEM to polymer foam material

In order to verify the applicability of the voxel FEM to polymer foam materials, comparison of the experimental results with simulations was required. However, since the relative densities of the CT model specimens were higher than those of the compression loading test specimens, simple comparison was not permitted. Thus, an assumption that the equation (3) was possible to apply up to the plateau region was introduced. If it is possible, the formula holds:

$$\frac{\sigma_1}{\sigma_2} = C_1 \left(\frac{\rho_1}{\rho_2}\right)^2 \tag{4}$$

Where, σ_1 and σ_2 are nominal stresses obtained by experiment or simulation, and ρ_1 and ρ_2 are relative densities of test specimens or models. To verify the applicability of this assumption, two experimental results of which relative densities were 4.82% and 5.94% with modification by equation (4), were compared. Thus, for this study, the following values were used in formula (4): σ_2 was the nominal-stress experimental data obtained for specimens with a relative density of 4.82%, ρ_1 was the estimated result's relative density of 5.94%, ρ_2 was the test specimens's relative density of 4.82%, solving for σ_1. This result is referred to as 4.82% modified, and its curve is compared to the 5.94% curve in Fig. 9. The estimated stress-strain relationship shows rough correspondence with experimental results over the entire range of the strain. Therefore, equation (4) was considered to be applied up to the plateau region of experimental results which was conducted in this research.

The results of the voxel FEM analysis of the CT model include randomness affected by the relative density. By applying formula (4) to the results, estimated stress-strain relationships were calculated with the condition that relative density was 5.94%, which is the same as the compression test specimens. Fig. 10 shows a comparison of the estimated stress-strain relationship based on the CT model and the experimental results. The estimated results are consistent with experimental results. Therefore the voxel FEM is reasonably applicable to polymer foam material.

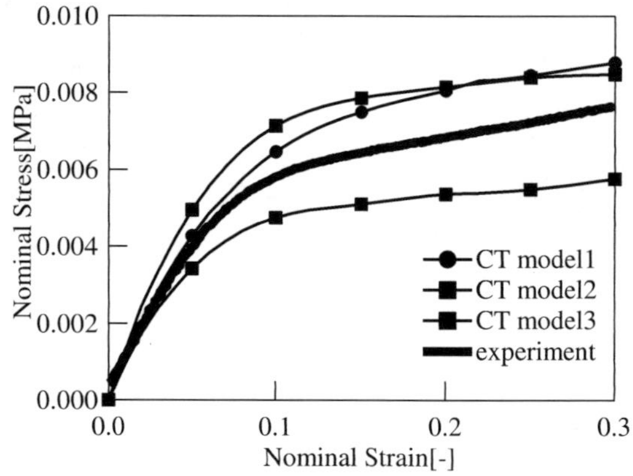

Fig. 9. Stress-strain relationships of compression loading test and result modified with formula (4)

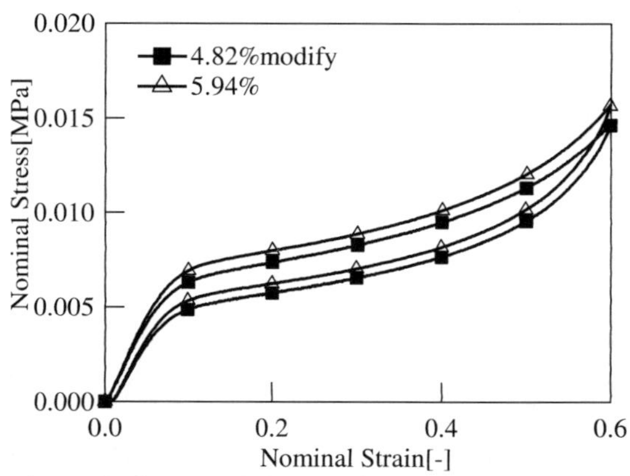

Fig. 10. The results of voxel FEM of CT model modified with formula (4) and experimental result

6.2. Verification of the applicability of the BCC model

The stress-strain relationship of the BCC model with a relative density of 5.94%, was calculated with equation (4). Fig. 11 shows a comparison of the modified stress-strain relationship of the BCC model and the experimental results. From these results, it can be seen that the bending of the stress-strain relationship of the BCC model is consistent with experiment results. In addition, the results obtained with the CT model also reproduced the bending strain. Therefore, voxel FEM can reproduce the stress-strain relationship of polymer foam up to the plateau region using a model that contains some randomness of structure. The simulated initial stiffness was lower than that of the experimental results due to the boundary conditions. The microscopic boundary conditions of the BCC model in the voxel FEM simulation did not correspond exactly to the microscopic boundary conditions of the real specimens in the loading tests. The microscopic deformation of the real specimens was satisfied with the equilibrium of stress with surrounding parts. Conversely, the BCC model was not stressed from the side surfaces by the adjacent cells. For this reason, the differences in the initial stiffness between the voxel FEM simulation and the loading tests are thought to be due to the boundary conditions rather than the microstructure.

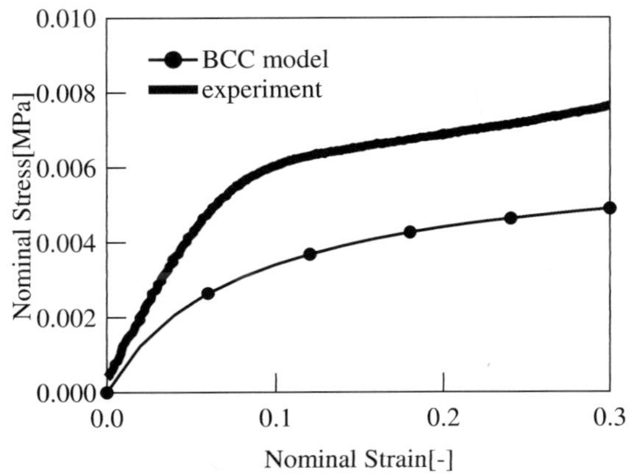

Fig. 11. The results of voxel FEM of BCC model modified with formula (4) and experimental result

7. Conclusion

The applicability of voxel FEM analysis to polymer foam was investigated. Incompressible hyperelasticity was applied to the voxel FEM code. CT models obtained using a CT-scanning system and BCC models that reproduced the microscopic structure of polymer foam were prepared. The simulated results of the CT model employing material parameters of the matrix were consistent with the experimental results. Voxel FEM using the CT model made it possible to design polymer foam that incorporate its macroscopic stress-strain relationship by considering its microscopic structure and matrix properties. Additionally, the results for the BCC models were consistent with the experimental results in terms of bending strain. It was possible to consider the geometrical effects of microscopic voids on the mechanical characteristics of polymer foam using the proposed voxel methods. Voxel FEM simulation using the results of the CT-scanning system were applicable to the mechanical characteristics of polymer foam up to the plateau region.

References

[1] Obi BE 2017 Polymeric foams structure-property-performance *William Andrew*

[2] Gibson L J and Ashby M F 1999 Cellular solids : structure and properties – second edition *Cambridge University Press* pp 175-231

[3] Zhu H X, Knott J Fand Mills N J 1997 Analysis of the elastic properties of open-cell foams with tetrakaidecahedral cells *Int. J Solids Structures* **45** pp 319-343

[4] Gong L, Kyriakides S and Jang W Y 2005 Compressive response of open-cell foams. Part I: morphology and elastic properties *Int.l J. Solids Structures* **42** pp 1355-1379

[5] Keyak J H, Meagher J M, Skinner H B and Mote C D Jr 1990 Automated three-dimensional finite element modeling of bone: a new method *J. Biomed. Eng.* **12** pp 389-397

[6] Keyak J H, Rossi S A, Jones K A and Skinner H B 1997Prediction of femoral fracture load using automated finite element modeling *J. Biomechanics* **31** pp 125-133

IOP Conf. Series: Materials Science and Engineering **548** (2019) 012011 doi:10.1088/1757-899X/548/1/012011

Rubber Particles Coating Layer on Macroballoons and Its Physical and Compression Properties in Syntactic Foam

L Anusha[1], Z Zakaria[1] and KK Chew[1]
[1]School of Materials Engineering, Universiti Malaysia Perlis, Kompleks Pusat Pengajian Jejawi 2, Taman Muhibah, 02600 Jejawi, Arau, Perlis, Malaysia.

E-mail: zunaida@unimap.edu.my

Abstract. The effect of epoxy and acrylonitrile butadiene rubber (NBR) layers on physical and compression properties of macroballoons in syntactic foams was investigated. Epoxy mixture was used to coat on the surfaces of expended polystyrene (EPS) beads template, and the beads were cured to produced epoxy macroballoon (EM). The cured EM coat with NBR powders in order to form epoxy rubber macroballoon (ERM). Subsequently, the cured ERM was coated with epoxy mixture as the outermost layer on macroballoon to form epoxy rubber epoxy macroballoon (EREM). The EM, ERM, EREM were then filled into syntactic foam and post cured to ensure the shrinkage of EPS beads to produce hollow structures. Three syntactic foams with EM, ERM and EREM were characterized in terms of physical and compression properties, respectively. It was found that EREM exhibited the highest density and average wall thickness in syntactic foam. Overall, the inclusion of NBR powders as rubber coating of macroballoon with epoxy as the outermost layer resulted in improved compressive strength and toughness by absorbing energy with enhanced interfacial adhesion.

1. Introduction

Syntactic foams have gained considerable attention due to their unique properties such as lightweight, good impact behaviour and high specific compressive properties [1]. Syntactic foams are composite materials consisting of macroballoon embedding in a polymeric matrix [2]. Macroballoon acts as reinforcing particulate dispersed in the polymeric resin of syntactic foam. The addition of macroballoons enhances the properties of syntactic foam and reduces the material costs simultaneously. The common macroballoons used are glass macroballoon, polymer macroballoon and metal macroballoon [3,4,5]. Syntactic foams are classified as closed-cell foam as the porosity in syntactic foams exists in the form of discrete hollow particles. The closed-cell configuration of syntactic foam provides the benefits of superior strength, low moisture absorption and low density [6].

On the other hand, epoxy is widely used as based syntactic foam due to its chemical and physical cross-linked rigid 3D network of polymer matrix [7,8]. Syntactic foams have been applied in various applications, such as marine structures, aircraft, construction and transportation [9,10]. Previous study reported that syntactic foam has been used in marine structural application because of its low density of buoyancy materials which are used to balance gravity and buoyancy of the oilfield drilling pipe. Alternatively, high pressure resistance materials are needed to withstand the high hydrostatic pressure in deep water [11]. In order to overcome this problem, the rubberized syntactic foam has been introduced for absorbing impact energy. Li and Jones found that syntactic foam filled with microballoons coated with rubber latex aid increased the toughness and impact tolerance [12].

In the present study, rubberized syntactic foam by coating a layer of recycled acrylonitrile butadiene rubber (NBR) gloves powder on macroballoon was prepared. The effect of a layer of

IOP Conf. Series: Materials Science and Engineering **548** (2019) 012011 doi:10.1088/1757-899X/548/1/012011

rubber particles content on macroballoon on physical (i.e density and morphology) and compression properties of syntactic foam were investigated.

2. Materials and Methods

2.1. Materials

Clear epoxy D.E.R 331 or diglycidyl ether of bisphenol A (DGEBA) and isophorone diamine (IPD) or hardener 8161 were supplied from Euro Chemo-Pharma Sdn. Bhd. (Malaysia). Epoxy resin D.E.R 331 was a clear liquid of epoxide equivalent weight of 182–192, and density and viscosity at 25 °C are 1.16 g/mol and 11000–14000 mPa·s, respectively. IPD has viscosity and amine value of 300–600 mPa·s and 260–284 mg KOH/mg, respectively at 25 °C. Potassium hydroxide (KOH) pellets with content more than 85% was supplied from Sigma-Aldrich (M) Sdn. Bhd. KOH acted as diluent to lower down the viscosity of resin mix. Calcium carbonate ($CaCO_3$) was obtained from the Sun Minerals Sdn. Bhd and used to prevent expended polystyrene (EPS) beads from sticking to each other. The EPS beads with particle size in the range of 3-6 mm was supplied from San Yong Enterprise Sdn. Bhd. Rejected acrylonitrile butadiene rubber (NBR) glove was supplied from Top Gloves Sdn. Bhd.

2.2. Preparation for Micron-Sized NBR Powders

The rejected NBR gloves were milled into powders with ambient milling method using two rolls mill machine model X(S)K-160X320. The gloves were added into the mill between two nips, as a continuous sheet onto the front roll. The milling of NBR gloves was repeated until micron-sized of rubber powders was obtained. Dino-Lite Digital Microscope was then used to characterize the rubber powders. The rubber powders were sieved at the size of 250 μm after the characterization of the rubber particles.

2.3. Preparation of Epoxy Macroballoons (EM)

The weight ratio of epoxy resin and IPD was 2E:1IPD. Total weight of epoxy resin and IPD was considered as one hundred part resin. Epoxy was mixed with IPD using overhead mechanical stirrer at 300 rpm for 2 min to produce epoxy system. EPS beads were added into the prepared epoxy system until half-filled the polypropylene (PP) cup to ensure that the EPS beads were fully-coated with the epoxy system. The epoxy-coated EPS beads were transferred to the PP tray that was covered with adequate amount of $CaCO_3$. A wooden chopstick was used to roll over the uncured epoxy-coated beads with $CaCO_3$ powder until their surfaces were fully-coated to hinder them from clumping one another. The $CaCO_3$-coated EPS beads were cured in the air-circulating oven at 80 °C for 15 min. Excess $CaCO_3$ powder on the surfaces of macroballoons was washed with distilled water and dried in the air-circulating oven at 100 °C for 1 h.

2.4. Preparation of Epoxy Rubber Macroballoons (ERM)

EM was added into the prepared epoxy system until half-filled the PP cup to ensure that all the macroballoons were fully-coated with the epoxy system. They were then transferred to the PP tray with sufficient amount of NBR powders. The macroballoons were rolled on the NBR powders with wooden chopstick to produce rubber layer on macroballoons. The macroballoons were rolled until the surfaces were fully-coated with the rubber particles. ERM were cured in the air-circulating oven at 80 °C for 15 min.

2.5. Preparation of Epoxy Rubber Epoxy Macroballoons (EREM)

ERM was used as template and was added into the prepared epoxy system until half-filled the PP cup. The macroballoons in the epoxy system were stirred gently until they were completely coated with epoxy to produce epoxy rubber epoxy macroballoons. The epoxy-coated macroballoons were transferred to the PP tray that was covered with adequate amount of $CaCO_3$. A wooden chopstick was used to roll over the uncured EREM with $CaCO_3$ powder until their surfaces were fully-coated to prevent them from clumping to each other. The calcium carbonate-coated macroballoons were cured in the air-circulating oven at 80 °C for 15 min. The excess $CaCO_3$ powder on the macroballoon surfaces was washed with distilled water and dried in the air-circulating oven at 100 °C for 1 h.

2.6. Preparation of Syntactic Foam with Coated Macroballoon

EM was poured and filled in the PP mold, and the stacking volume of 100% was ascertained. EM was then removed from the mold and left aside after measuring the amount of macroballoons for 100% of stacking volume. The mixture of epoxy resin and hardener was prepared by mixing the epoxy and hardener together with concentration of 3% KOH aqueous solution (10% by weight with respect to the resin mix) at 300 rpm for 5 min. EM was added into the epoxy mixture at regular interval subsequently within 15 min. Uncured matrix compound was then casted into the PP mold with the dimension of 113×113×25 mm. A constant load of 5 kg was placed on the top of the lid to maintain the well-dispersed state. The mixture was left to cure in room temperature for 24 h and post-cured at 120 °C for 90 min to shrink all the EPS beads inside the epoxy-coated macroballoons to produce hollow structure within the sphere. The cured syntactic foam was demoulded and cut into standard dimension of 50×50×25 mm for further testing and characterization. The steps for preparing syntactic foam were repeated for ERM and EREM.

2.7. Characterization of Samples

The density of syntactic foam was measured according to ASTM D3574. The syntactic foam specimens with the dimension of 50×50×25 mm [13] were measured by Mitutuyo digital vernier calliper. The mass of syntactic foam was measured within ±1%. Samples were weighed to an accuracy of 0.001 g. The densities of the syntactic foams were obtained by calculating based on Equation 1. The average value of 5 samples was recorded for syntactic foam.

$$\rho = \frac{m}{V} \qquad (1)$$

where ρ, *m*, and *V* are density, mass, and volume of syntactic foam, respectively.

The morphology of syntactic foams was captured using the Dino-Lite digital optical microscope (AM521ZT) DinoLite Capture 2.0 software. Image J software was used to analyze microscope images; the average wall thickness of the EM, ERM and EREM were measured by taken 50 readings to represent a better results as according the Equation 2.

$$\text{Wall thickness} = r_o - r_i \qquad (2)$$

where r_i and r_o are inner radius and outer radius of macroballoon.

The compression load deflection of syntactic foams was determined using Shimadzu Universal Tensile Machine fitted with a compression jig based on ASTM D3575. The specimens dimension of 50×50×25 mm [13] were compressed with the crosshead speed of 2.5 mm/min up to 60 % of strain under load of 250 kN, and the compressive value was recorded.

3. Results and Discussion

3.1. Density of Syntactic Foams

The density of syntactic foams with different type of macroballoon is illustrated in Fig. 1. It can be seen that, the syntactic foams with EREM marked the highest density of 1.2269 g/cm^3, followed by syntactic foam with ERM (0.8686 g/cm^3). Meanwhile, syntactic foam with EM marked the lowest density of 0.5736 g/cm^3. This significantly proved that the density of syntactic foam increased when the number of coating layer on macroballoons increased. In other words, the density of ERM in syntactic foams has increased by 51.4% by adding a layer of rubber powder on EM and its density inclined by 41.3% when another layer of epoxy mixture was coated on ERM. Moreover, EREM had the three layers coated of macroballoon contribute to the increasing density due to attribute the thicker shell of macroballoon compared to EM and ERM. Thus, the number of coating layer on macroballoon could alter the final density of syntactic foam.

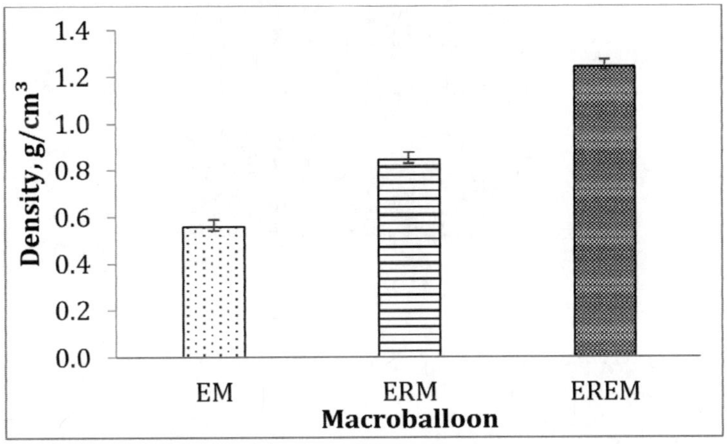

Fig. 1: Density of different type of macroballoon in syntactic foams.

3.2. Morphology Analysis

The microscopy images of syntactic foam comprising different types of macroballoon are displayed in Fig 2. It can be seen that EM possessed a layer coated on the surface of macroballoon as shown in Fig. 1(2). Besides, ERM showed two layers coated on the macroballoon surface (Fig. 2(b)), while EREM had three layers coated on the macroballoon surface (Fig. 2(c). In addition, it shows that by increased the number of coating layer will increase the diameter of macroballoon. This observed morphology is stated that the number of coating layer on macroballoons attributed the bigger the diameter of macroballoon, the thicker the wall thickness of the macroballoon thus affected the final density of syntactic foam as discussed in Section 3.1. Similar finding were reported by Wu et al, by increase the layer of glass fiber on macrospheres will increase the wall thickness and diameter of macrospheres [14].

Fig. 3 shows the relationship between the number of coating layer on macroballoons and their average wall thickness. The average wall thicknesses of EM, ERM and EREM were 0.1673 mm, 0.4850 mm and 0.7775 mm, respectively. As the number of coating layer on macroballoon increased, the average macroballoon wall thickness increased [14]. It was found that wall of EREM was thicker than that of ERM and EM as seen the trend in Fig. 2. Furthermore, the average wall thickness of EM increased at 65.5% when a layer of rubber powder was added onto the surfaces of EM, but it increased by 37.62% when another layer of epoxy mixture was coated on the outermost surface of ERM. This proven that, by increase the number of coating layer will increase the volume fraction of rubber powder and thus increased the wall thickness of the macroballoon.

IOP Conf. Series: Materials Science and Engineering **548** (2019) 012011 doi:10.1088/1757-899X/548/1/012011

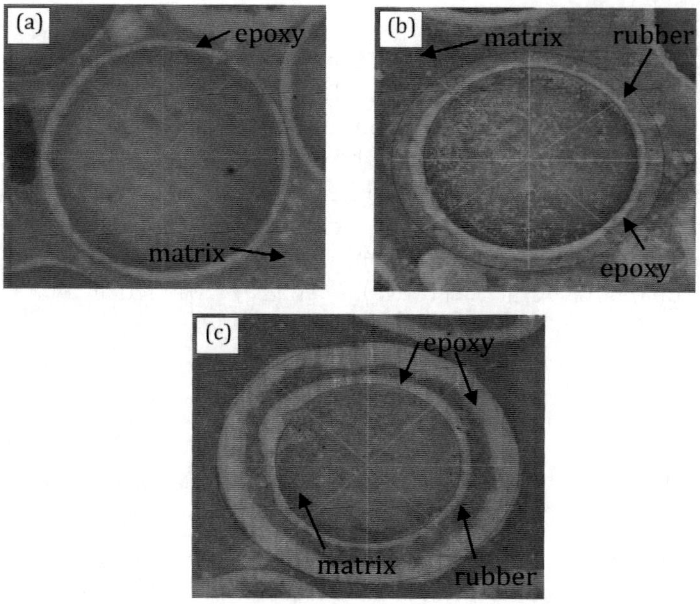

Fig. 2: Microscopic images of different types of macroballoon in syntactic foam at magnification of 20X: (a) EM (b) ERM and (c) EREM.

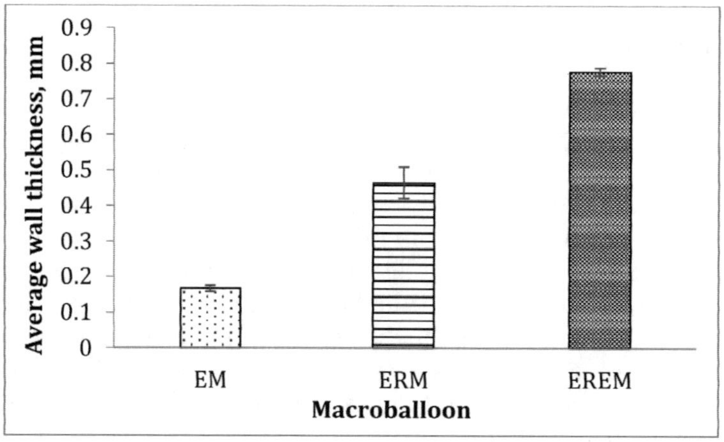

Fig. 3: The effect of number coating layer on macroballoons on its wall thickness.

3.3. Compression stress of Syntactic Foams

Fig. 4 shows the compressive strength of different wall thickness of macroballoon in syntactic foam. The compressive strength of EM, ERM, and EREM in syntactic foam were 7.9139, 13.2105, and 24.7986 MPa, respectively. ERM showed higher compressive strength compared to EM as depicted in Fig. 4. It can be claimed that the presence of rubber layer could absorb deformation energy and delays the crack propagation in syntactic foam during compression process [15]. Interestingly, the results indicated that EREM had the highest compressive strength compared to others. This could be due to outstanding interfacial area between the rubber layer and outermost layer of epoxy resin [16]. Therefore, EREM exhibited resistant to compression when load was applied which resulted in enhanced compressive strength of syntactic foam.

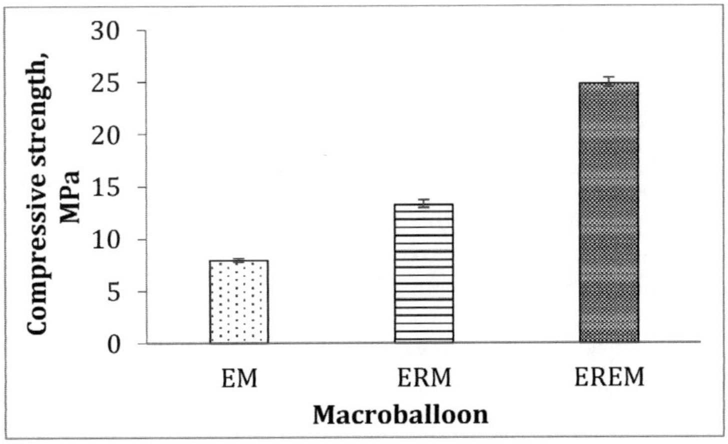

Fig. 4: Compression strength of syntactic foams with different type of macroballoon.

Compressive toughness is defined as the ability of material to absorb energy and deform plastically before failure by its density [17]. Material with high toughness tends to have high ductility and strength. Fig. 5 depicts the compressive toughness syntactic foam with EM, ERM and EREM. It can be observed that the compressive toughness was increased as the increase of number of coating layer on macroballoon surface. The compressive toughness of syntactic foam of ERM was improved by 34.74% through addition of a layer of rubber powder on the macroballoon surfaces. The crack deflection at the rubber particles and plasticizing effect of rubber reduced the brittleness, as well as increased the ductility of syntactic foam. Besides, rubber layer on the macroballoons surfaces delays crack initiation by absorbing higher amount of energy during compression test which results in higher compressive toughness [18]. Thus, syntactic foam with ERM has higher specific compressive toughness compared to syntactic foam with EM. Surprisingly, the compressive toughness of syntactic foam with EREM was dramatically improved by 69.19%. It can be explained that when another layer of epoxy coated on the outermost surfaces of EREM in syntactic foam (cleary seen in Fig. 1) that the compressive toughness of syntactic foam indicated substantial increment.

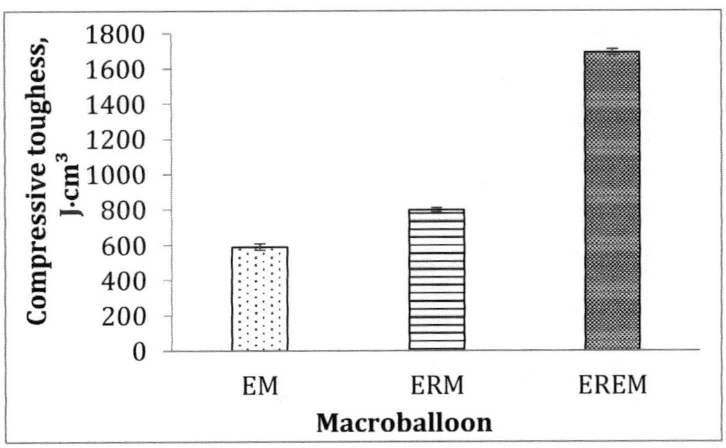

Fig. 5: Compression toughness of syntactic foams with different type of macroballoon.

4. Conclusion

The addition of NBR powders as coating layer on macroballoons in syntactic foam was developed to improve the compression properties of syntactic foam. Besides, another layer of epoxy mixture coated on the outermost surfaces of macroballoons was also developed to improve the surface adhesions between the matrix and macroballoons. The macroballoon wall thickness increased by increasing the number of coating layer on the macroballoon. Thicker macroballoon wall thickness of EREM resulted in higher density (1.2269 g/cm^3) of syntactic foams. The compressive strength and compressive toughness of syntactic foam with EREM were the highest compared to the other foams. Syntactic foam with EREM had achieved excellent compressive properties due to the ability of NBR powders as coating material in absorbing high energy before failure as well as epoxy outermost layer on macroballoon surface which improved the interfacial adhesion to syntactic foam matrix.

References

[1] Wang L, Zhang, J, Yang, X, Zhang C, Gong W, and Yu J 2014 Fiber *Mater. Design* **55** 929–936
[2] Dando K R *et al* 2018 *J. Cell. Plast.* **54(3)** 499–514
[3] Altenaiji M *et al* 2014 *Mater. Design* **59** 296–302
[4] Vogiatzis C A *et al* 2015 *Mater. Design* **85** 444–454
[5] Wu X *et al* 2018 *Fibers and Polym.* **19(1)** 170–187
[6] Salleh Z, Islam M M and Ku H 2014 *J. Multifunct. Composite.* **2(1)** 21-27
[7] Li Y and Kessler M R 2014 *Polym.* **55(8)** 2021–2027
[8] Wang L *et al* 2014 *Composites Part B Engineering* **56** 724–732
[9] Yazici M *et al* 2014 *Acta Physica Polonica A* **125** 526– 528
[10] Salleh Z, Islam M M and Ku H 2014 *J. Multifunct. Composite.* **3** 71–77
[11] Wu X *et al* 2014 *Polym. Composite.* **37** 497-502
[12] Li G and Jones N 2007 *Compos Part A Appl Sci Manuf.* **38(6)** 1483–1492
[13] Samsudin S S *et al* 2011 *Express Polym Lett* **5(7)** 653–660
[14] Wu X F *et al* 2015 *RSC Adv* **5** 61204–61217
[15] Basuli U *et al* 2012 *Elastomers Compos.* **47(4)** 297–309
[16] Sanjay O S 2012 Effect of Crumb-Rubber Particle Size on Mechanical Response of Polyurethane Foam *Composites* Oklahoma State University
[17] Lakshminarayanan A K, Balasubramanian V and Salahuddin M 2010 *J. Iron Steel Res. Int.* **17(10)** 68–74
[18] Zhang L and Ma J 2010 *Compos. Sci. Technol.***70(8)** 1265–1271

IOP Conf. Series: Materials Science and Engineering **548** (2019) 012012 doi:10.1088/1757-899X/548/1/012012

3-Dimensional Homogenization Finite Element Analysis of Foamed Rubber

A Matsuda[1] and S Oketani[2]

[1] Faculty of Engineering, Information and Systems, University of Tsukuba, Ibaraki, Japan

[2] Graduate School of Systems and Information Engineering, University of Tsukuba, Ibaraki, Japan

E-mail: a_matsuda@kz.tsukuba.ac.jp

Abstract. In this study, 3-dimensional homogeneous finite element analysis of a truncated octahedral unit cell which consisted of hyperelastic beams was conducted to predict macroscopic mechanical characteristics of the foam rubbers considering its microscopic structure. The homogenization theory was applied to finite element analysis code and the unit cells for simulation were assumed to have the periodic boundary condition. An original finite-strain homogenization FEM code for hyperelastic material and truncated octahedral unit cell were developed. The relative density of the homogenization analysis was adjusted by width of the beams in unit cell model. The rubber matrix was assumed to be represented by the nearly-incompressible hyperelasticity. The Mooney-Rivlin model and nearly-incompressible condition were applied to the hyperelasticity. The material parameters of the Mooney-Rivlin model for foamed rubber matrix were identified by the tensile loading test results. The developed code showed enough stability to predict the mechanical property of foamed rubber in the large strain region.

1. Introduction

Foam rubbers such as polyurethane foam have been used widely in transport equipment, sports equipment, living ware, and so on. The advantage reasons of which the foam rubber are used, are large deformable performance, soft stiffness, durability for cyclic deformations and manufacturing formability. The foam rubber includes numerous microscopic cavities inside of the solid body. So, the mechanical characteristics of the foam rubber are strongly depending on mechanical properties of rubber matrix and its microstructure[1,2,3]. The relative density is important material parameter which defined mechanical calculated by the volume and density of foam rubber. The relative density which is calculated by the ratio of rubber matrix in foam rubber are used as parameter which present material characteristics. Material design using numerical simulation considering the microstructure would be required to improve quality, size and productivity of foam rubber products. In this study, 3-dimensional homogeneous finite element analysis of a truncated octahedral unit cell which consisted of hyperelastic beams was conducted to predict macroscopic mechanical characteristics of foam rubbers[4,5] (shown in Fig. 1). The homogenization theory was applied to original finite element analysis code of hyperelasticity and the unit cells for simulation were assumed to have the periodic boundary condition. An original finite-strain homogenization FEM code for hyperelastic material and truncated octahedral unit cell models for the homogenization analysis were newly developed. The relative density of the homogenization analysis was adjusted by width of the beams in the unit cell models. The rubber matrix was assumed to be represented by the nearly-incompressible hyperelasticity. The Mooney-Rivlin model and the nearly-incompressible condition were applied to the hyperelasticity. The material parameters of the Mooney-Rivlin model for foam rubber matrix were identified by the tensile loading test results. Homogenization FE-analysis were conducted to verify the applicability of developed code.

2. Homogenization finite element analysis

In this study, the microstructure of the foam rubber like polyurethane foam was assumed to have the periodic structure for analysis. The formulation of homogenization analysis for large deformation was described with two different coordinates $\boldsymbol{X}$ and $\boldsymbol{Y}$. The macroscopic behavior was defined in $\boldsymbol{X}$-coordinate, and the microscopic behavior was defined in $\boldsymbol{Y}$-coordinate. The two coordinates were related by the scale ratio ε as follows[6]:

$$\boldsymbol{Y} = \boldsymbol{X}/\varepsilon \tag{1}$$

When the scale of the microstructure was much smaller than the scale of whole structure, the scale ratio ε was a very small value. Therefore, macroscopic characteristics, such as stiffness, stress and strain were calculated from the average of microscopic characteristics using the homogenization theory. The all microscopic structure deformed uniformly, and microscopic periodicity was kept under finite deformation. The total position of microscopic structure $\boldsymbol{y}$ was divided into macroscopic displacement $\boldsymbol{Y}$ and microscopic periodical displacement $\boldsymbol{w}$ as follows:

$$\boldsymbol{y} = \widetilde{\boldsymbol{F}}\boldsymbol{Y} + \boldsymbol{w} \tag{2}$$

where, $\widetilde{\boldsymbol{F}}$ is the macroscopic deformation gradient tensor. The deformation gradient tensor $\widetilde{\boldsymbol{F}}$ was given as the boundary condition of numerical simulation. The microscopic deformation gradient tensor $\boldsymbol{F}$ was calculated from Equation 2 as follows:

$$\boldsymbol{F} = \nabla\boldsymbol{y} = \widetilde{\boldsymbol{F}} + \frac{\partial \boldsymbol{w}}{\partial \boldsymbol{Y}} \tag{3}$$

The macroscopic deformation gradient tensor $\widetilde{\boldsymbol{F}}$ was supposed to be constant and calculated as average of volume integration of microscopic deformation gradient tensor $\boldsymbol{F}$, so $\widetilde{\boldsymbol{F}}$ was calculated as follows:

$$\widetilde{\boldsymbol{F}} = \frac{\boldsymbol{1}}{\boldsymbol{V}}\int_{Y_0} \boldsymbol{F}dY \tag{4}$$

Also, the average of volume integration of microscopic deformation gradient $\boldsymbol{F}$ was calculated using Equation 4 as follows:

$$\frac{\boldsymbol{1}}{\boldsymbol{V}}\int_{Y_0} \boldsymbol{F}dY = \frac{\boldsymbol{1}}{\boldsymbol{V}}\int_{Y_0} \left(\widetilde{\boldsymbol{F}} + \frac{\partial \boldsymbol{w}}{\partial \boldsymbol{Y}}\right)dY = \widetilde{\boldsymbol{F}} + \frac{\boldsymbol{1}}{\boldsymbol{V}}\int_{Y_0} \left(\frac{\partial \boldsymbol{w}}{\partial \boldsymbol{Y}}\right)dY \tag{5}$$

From Equation 5, it is necessary to satisfy the following equation. By giving $\boldsymbol{w}$ a periodic boundary condition, the following equation was satisfied automatically.

$$\frac{\boldsymbol{1}}{\boldsymbol{V}}\int_{Y_0} \left(\frac{\partial \boldsymbol{w}}{\partial \boldsymbol{Y}}\right)dY = 0 \tag{6}$$

In order to accurate evaluation in consideration of nearly-incompressibility of the hyperelastic material, the displacement/pressure mixing method with the Langrangian multiplier p was applied to the FEM formulation. Therefore, the strain energy function of nearly incompressible hyperelastic material was defined as follows[7]:

$$W(\overline{\boldsymbol{C}}, p) = \overline{W}(\overline{\boldsymbol{C}}) + p(J-1) - U_C(p) \tag{7}$$

Here, p is the Lagrangian multiplier which is correspond to the hydrostatic pressure, $\overline{W}(\overline{\boldsymbol{C}})$ is the elastic potential function, $\boldsymbol{C}$ is the right Cauchy-Green deformation tensor, $\overline{\boldsymbol{C}}$ is the modified right Cauchy-Green deformation tensor removing the volume component, $U_C(p)$ is the complementary strain energy expressing the relation between volumetric strain and pressure. $U_C(p)$ was expressed using the bulk modulus k as follows:

$$U_C(p) = {p^2}/{2k} \tag{8}$$

Also, $\boldsymbol{C}$ and J were expressed as follows:

$$\boldsymbol{C} = \boldsymbol{F}^T \boldsymbol{F} \tag{9}$$

$$J = det(\boldsymbol{F}) \tag{10}$$

where, J is the determinant of the deformation gradient tensor and represented the volumetric deformation. The elastic potential $\overline{W}(\overline{\boldsymbol{C}})$ was expressed using the modified first principal invariant $\bar{I}_1$ and the modified second principal invariant $\bar{I}_2$ of $\overline{\boldsymbol{C}}$ as follows:

$$\overline{W}(\overline{\boldsymbol{C}}) = \overline{W}(\bar{I}_1\ ,\bar{I}_2) \tag{11}$$

Here, the modified deformation gradient tensor $\overline{\boldsymbol{F}}$ that doesn't contain volume components, $\overline{\boldsymbol{C}}$, $\bar{I}_1$ and $\bar{I}_2$ were expressed as follows:

$$\overline{\boldsymbol{C}} = \overline{\boldsymbol{F}}^T \overline{\boldsymbol{F}} = J^{-2/3}\, \boldsymbol{F}^T \boldsymbol{F} \tag{12}$$

The second Piola-Kirchhoff stress tensor $\boldsymbol{S}$ of hyperelastic material is given by the partial differentiation of the strain energy function W with respect to the right Cauchy-Green deformation tensor $\boldsymbol{C}$ as follows:

$$\boldsymbol{S} = 2\frac{\partial W}{\partial \boldsymbol{C}} \tag{13}$$

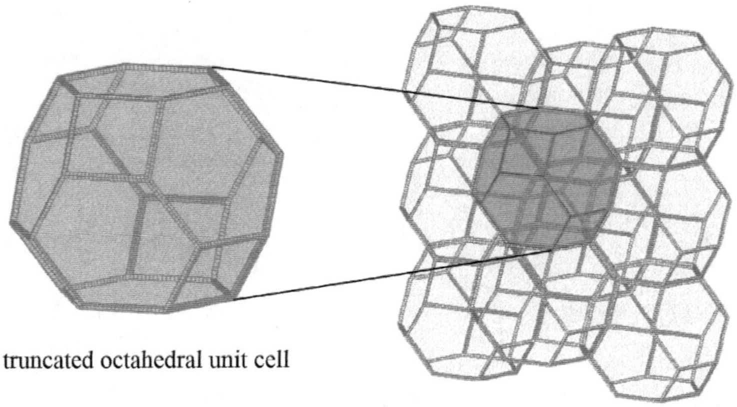

Fig. 1. FEM model based on a truncated octahedron

The periodic boundary condition was set on the boundary surface of FEM model. The macroscopic deformation gradient tensor $\widetilde{\boldsymbol{F}}$ was given by multiplying the deformation gradient tensor $\widetilde{\boldsymbol{F}}_0$ by the rotation tensor $\boldsymbol{R}$ as follows:

$$\widetilde{\boldsymbol{F}} = \boldsymbol{R} \cdot \widetilde{\boldsymbol{F}}_0 \cdot \boldsymbol{R}^T \tag{14}$$

3. Material modeling and simulated results

To verify the applicability of the homogenization analysis to foam rubber, compression tests of polyurethane foam and tensile loading tests of polyurethane matrix were conducted.

In Fig. 2, compression stress-strain behavior of polyurethane foam specimens were shown. The relative density of compression specimens were 3.73%, 4.82% and 5.94%, respectively.

In order to identify the material parameters of polyurethane matrix, tensile loading tests on 3-solid specimens were conducted. The solid specimens were filled with the polyurethane foam matrix. Material parameters of the hyperelasticity were identified by using the stress-strain relationship of the solid specimens shown in Fig. 3.

The truncated octahedral unit-cell for foam rubber like polyurethane were constructed. As shown in Fig. 4, it is three-dimensional models based on a truncated octahedron. Relative densities for the homogenization analysis were adjusted based on the width of the beams in unit cell. Uniform macroscopic deformation $\widetilde{\boldsymbol{F}}$ was given as analysis condition and periodic microscopic displacement $\boldsymbol{w}$ was calculated by homogenization FEM analysis. Homogenization analysis of foam rubbers were conducted to evaluate applicability of developed code.

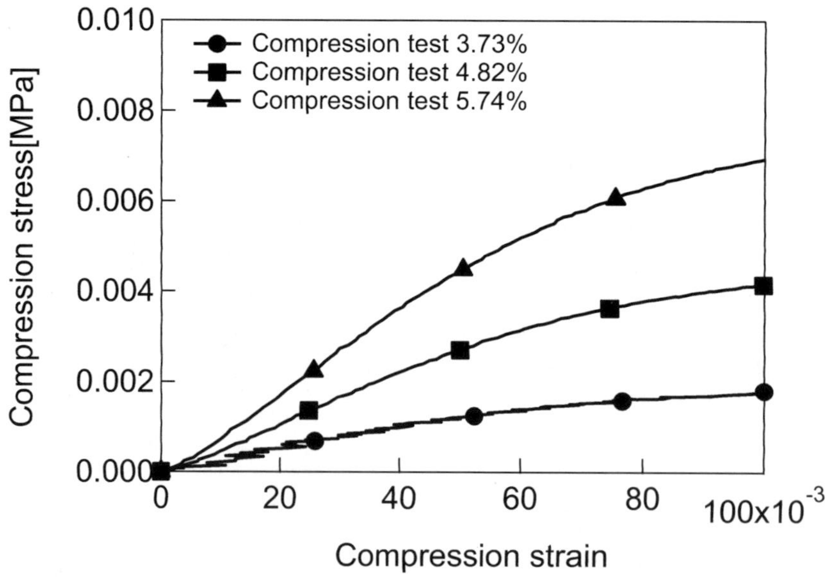

Fig. 2. Stress-strain relationships of compression tests of the polyurethane foam

The relative densities of the homogenization analysis were adjusted to 3.7%, 4.8% and 5.9%. The stress-strain relationships are shown in Fig. 5. The stress-strain relationships calculated by the homogenization analysis shows good agreement with the experimental results shown in Fig. 3.

The diagrams of the compression deformation of the 5.94%-relative density unit cell are shown in Fig. 6. Vertical deformation was applied to the truncated octahedron unit cell under the periodic boundary condition which is shown in Fig. 1. The lateral stress was constrained to be zero by the penalty method. The simulated results had good stability with the compression deformation to a strain of 0.5.

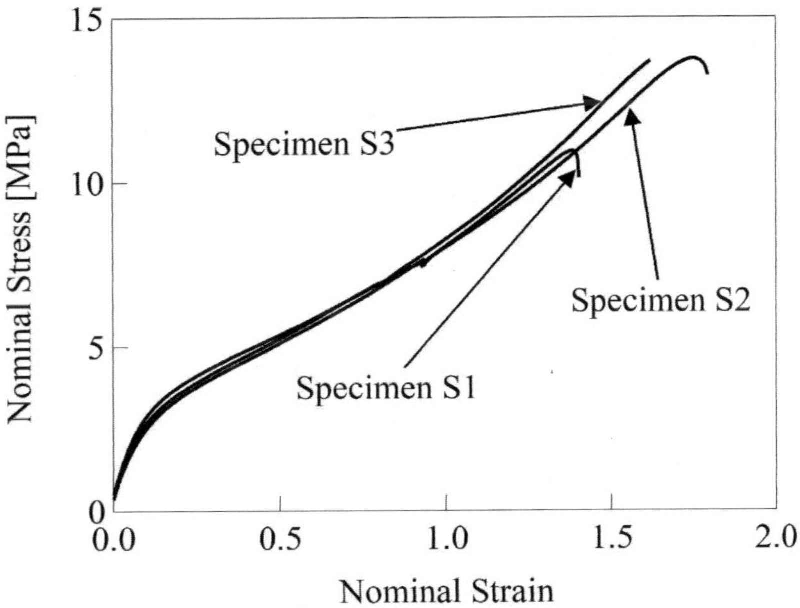

Fig. 3. Stress-strain relationships of tensile test of solid specimens which was filled with polyurethane matrix

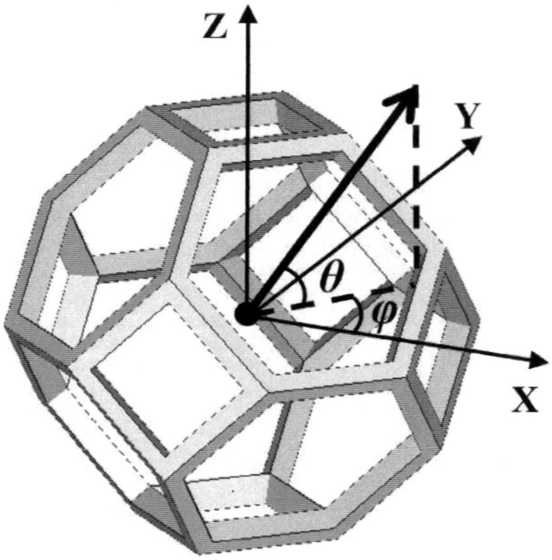

Fig. 4. Definition of load direction for analysis, φ is rotation angle around Z axis, θ is rotation angle around Y axis.

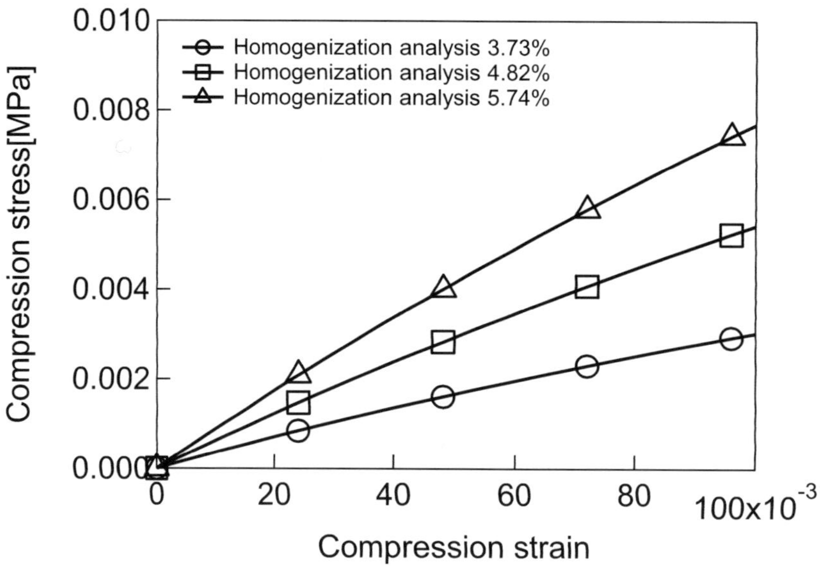

Fig. 5. The stress-strain relationships calculated by the homogenization method

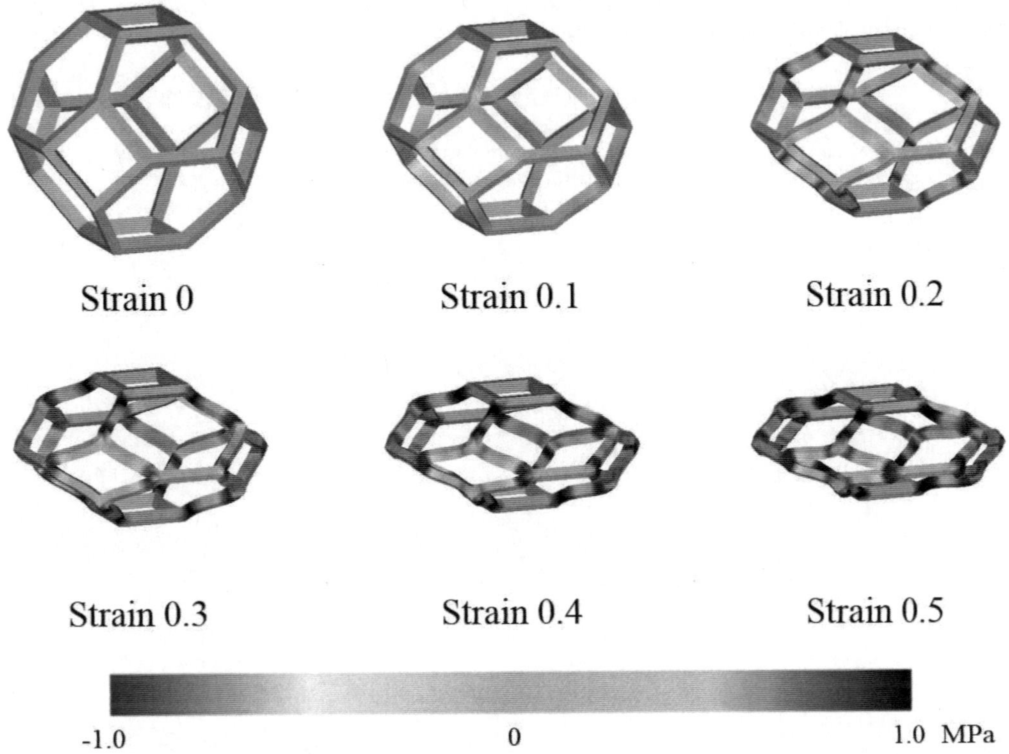

Fig. 6. The stress distribution for a unit cell with a relative density of 5.94%

4. Conclusion

An evaluation method for foam rubber that considers its microstructure was investigated in this study. In order to reproduce the foam rubber's microstructure, three-dimensional homogeneous FEM analysis of truncated octahedral unit cells were conducted. The microstructure of foam rubber like polyurethane was assumed to have the periodic structure. The polyurethane matrix was assumed to be represented by the nearly-incompressible hyperelasticity. The developed FEM code with homogenization theory shows enough stability to represent the mechanical characteristics of foam rubbers.

Acknowledgment

This work was supported by JSPS KAKENHI Grant Number 15K05671.

References

[1] Zhang J, Kikuchi N, Li V, Yee A and Nusholtz G 1998 *Constitutive Modeling of Polymer Foam Material Subjected to Dynamic Crash Loading* International *J. Impact Engine.* **21** **5** pp 369-386

[2] Youssef S Marie E and Gaertner R 2005 *Finite Element Modelling of the Actual Structure of Cellular Materials Determined by X-ray Tomography Acta Materials* **53** pp 719-730

[3] Gibson L J and Ashby M F 1997 *Cellular Solids: Structure and Properties* (Cambridge Solid State Science Series) (Cambridge: Cambridge University Press) Chapter 3 pp 175-231

[4] Shimazu R, Yautaka H, Nomoto A and Matsuda A 2015 *3-dimensional Homogenization FEM Analysis of Hyperelastic Low Density Polymer Foams* European Conference on Constitutive Models for Rubbers IX pp 675-681

[5] Shinoda Y and Matsuda A 2013 *Homogenization Analysis of Porous Polymer Considering Microscopic Structure Procedia Engine.* **60** pp 343 – 348

[6] Terada K, Hori M and Kyoya T and Kikuchi N 2000 Simulation of the Multi-scale Convergence in Computational Homogenization *Approaches Int. J. Solids Structures* **37** pp 2285-2311

[7] Treloar L R G 2005 *The Physics of Rubber Elasticity* (Oxford Classic Series)(Oxford: Oxford University Press) Chapter 10 pp 211-229

IOP Conf. Series: Materials Science and Engineering **548** (2019) 012013 doi:10.1088/1757-899X/548/1/012013

Ultra High Molecular Weight NORDEL™ EPDM for TPV and Low Hardness Molded Goods Enabled by Advanced Molecular Catalyst

S. Wu[1], T. Han[2], V. Thakur[3], R. Weeks[1], C. Li Pi Shan[1], and J. Tuberquia[1]

[1] The Dow Chemical Company, Lake Jackson, Texas, USA

[2] Dow Chemical (China) Investment Co. Ltd., Shanghai, China

[3] Dow Europe GmbH, Horgen, Switzerland

E-mail: [1]xswu@dow.com, [2]than2@dow.com

Abstract. Ethylene propylene diene synthetic rubber (EPDM) is the material of choice to manufacture various automotive parts due to its natural stability and resistance to heat, UV, and ozone. In addition, its polyolefinic nature results in consistent processing and good filler and oil acceptance for use in compounds ranging from automotive weather sealing, under-the hood belts and coolant hoses, roofing membranes, gaskets and other general rubber articles. Building on a fifty-year history of performance, Dow's Advanced Molecular Catalyst (AMC) and process technology advancements enable highly tailored design of EPDM molecular architecture including, composition, molecular weight and long chain branching (LCB). This is not only resulting in capacity efficiency and sustainable production but also increasing design options of the molecular features to tailor the performance for high in-use efficiency. This paper summarizes a new ultra high molecular weight (UHMW) NORDEL™ EPDM product and its potential applications. Case studies will be presented to showcase its use in thermoplastic vulcanizate (TPV), black color low hardness compounds, and lighter color (gray) washing machine gasket application. The EPDM polymer structure, compound formulation, mixing performance, compound rheology, vulcanization characteristics, and mechanical property improvements are discussed.

1. Introduction

1.1 EPDM Catalyst Technology

There are six key variables that define the properties of EPDM rubbers: (a) polymer molecular weight, (b) molecular weight distribution, (c) comonomer (propylene) content, (d) comonomer distribution, (e) termonomer type, content and distribution and (f) long chain branching level. The most important factor in controlling these variables is catalyst technology. Commercial olefin polymerization catalysts not only have to produce EPDM products with the desired architectures and properties outlined above, but they also must exhibit high catalytic activity under the desired polymerization conditions. Additionally, precatalysts (neutral, unactivated complexes) need to be thermally stable at and above ambient temperature, as they are often stored for weeks before they are used.

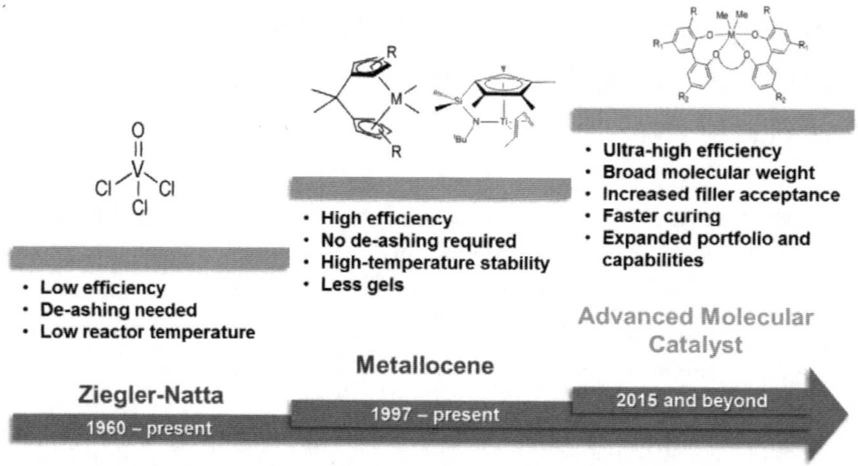

Fig. 1. Different catalyst technologies to produce EPDM Rubber

Fig. 1shows the differences between the different catalyst technologies that exist today to produce EPDM rubbers. The heritage Ziegler-Natta/Vanadium (ZN) catalysts developed in 1960s are still in use. However, this process technology is now antiquated due to its low production efficiency and high energy and water usage. In terms of catalyst and process, the Z-N catalyst suffers from low catalyst efficiency that operates at ambient and sub-ambient reaction temperatures and requires the removal of catalyst residues by steam stripping followed by drying of the polymer crumb [1].A recent life cycle analysis reports that a Ziegler-Natta EPDM production process can use up to 50% more energy and has 30% more impact on the environment than Dow's state-of-the-art solution process [2].

Dow first revolutionized the production of EPDM in 1997 with the introduction of its INSITE™ technology and use of metallocene catalysts in a high temperature solution process [3]. The direct advantages included catalyst efficiency, ten times higher than that of a Z-N catalyst and operated at significantly higher reaction temperatures and does not require the removal of catalyst residues. Due to the plants high temperature operation and special plant design, resulted in EPDM grades with almost no gels and improved product consistency that elevated the quality of EPDM and enabled Dow's industrial partners to extrude products with Class A surface finish and reduced scrap.

Since the introduction of metallocene catalysts, Dow has continued with its generational development of polyolefin catalysts. Using high-throughput (HTR) catalyst screening technology, new catalysts were screened against targeted product-process improvements [4]. In the high-throughput screening and molecular modelling process, thousands of catalyst combinations are screened and optimized. As an output of the high-throughput screening process, Dow has introduced its new Advanced Molecular Catalyst technology (AMC) that leads the EPDM industry by enabling even higher process sustainability and catalyst efficiency. For the end-user of EPDM rubber, Dow's AMC technology unlocks new molecular capabilities for the development of new products. For instance, the new AMC technology can efficiently produce EPDM with ultra-high molecular weight, higher diene content, and increased levels of long chain branching. In addition to process upgrades, oil-extension capability further extends the possibility to design & produce EPDMs with even higher molecular weights than conventionally used.

1.2 Thermodynamic Vulcanizate

Thermoplastic vulcanizate (TPV) is a plastic elastomer alloy which consists of a high volume content fine particle size cross-linked rubber particles dispersed in a low volume content thermoplastic matrix. TPV is produced via dynamic vulcanization of immiscible blends of a rubber and a thermoplastic, i.e., the selective crosslinking of the rubber while melt mixing with the thermoplastic [5-8]. Today, the major commercial available TPVs are based on blends of ethylene propylene diene rubber (EPDM) with polypropylene (PP), and crosslinked with phenolic resin or peroxide curing agents [9].

TPV are considered as a high performance thermoplastic elastomers since they have the excellent elasticity and mechanical properties of thermoset rubbers and robust processability and recyclability of thermoplastics. This unique feature enable TPV to penetrate into many applications previously dominated by thermoset rubber.

TPV has been enjoying fast growth for it offers a number of advantages over thermoset rubbers. First of all, it addresses the light weight trend. TPV weighs 30% less than thermoset EPDM, Another advantage is that any scrap/flash generated, or even the end product itself, can be recycled back into the production. Therefore there's is less waste. TPVs also require little or no compounding by the converter nor curing. For example, injection molding cycle times can be reduced to seconds rather than the minutes generally needed for molding and cure of thermoset rubbers.

Molecular weight and composition have been identified as the top design parameters for TPV grade EPDM. MW determines viscosity, which in turn has significant impact on phase inversion and morphology development. EPDM Crystallinity of EPDM strongly affects compression set at room temperature and at low temperature, low temperature oil retaining and tensile properties. For TPV applications, an ultra-high MW and tailored composition is desired.

1.3Molded Rubber Goods

Molded goods are a diverse array of products using a mold to produce the required geometry such as seals, diaphragms, bushings, and mountings [10]. Compression molding, transfer molding, and injection molding are the major molding methods. Natural rubber and variety of synthetic rubbers are being used to produce molded rubber parts for various applications. EPDM has been the material choice to make grommets, boots, bellows, muffler hanger, engine mounts, and washing machine gasket where the weathering and heat resistance are important. In this case study, a low hardness soft compound designed for grommets with excellent elasticity and compression set resistant will be shared. Design a low hardness formulation with excellent mechanical properties is always a challenge. The ultra high molecular weight is required to have large number of chain entanglement therefore to provide the vulcanizate excellent mechanical properties after plasticized with large amount of mineral oil. Without pre-extended with oil, this ultra high molecular weight EPDM will have an extremely high Mooney viscosity and rubber elasticity making the processing a very challenging task therefore 75 phr hydrogenated paraffinic oil has been added to this EPDM rubber to lower the Mooney viscosity to 55 MU (ML1+4@125C) providing it excellent processablity.

IOP Conf. Series: Materials Science and Engineering **548** (2019) 012013 doi:10.1088/1757-899X/548/1/012013

2 Experiment

2.1 Materials and Formulations

The product specification of the newly developed ultra high molecular weight NORDEL™ EPDM is listed in the Table 1.

Table 1. Characteristics of the new Dow UHMW NORDEL™ EPDM

Polymer	Mooney Viscosity*	Ethylene content (wt%)	Diene content (wt%)	Oil (phr)
UHMW EPDM	55	65	4.5	75

*Note: Rubber Mooney viscosity is measured at 125 °C following ASTM D1646-15.

The phenolic resin cured TPV formulation is listed in Table 2.

Table 2. TPV formulation with UHMW NORDEL™ EPDM

Ingredient	PHR
UHMW EPDM	175
Polypropylene (2 MFR)	58
Clay	42
Paraffinic oil	55
Phenolic resin curatives	6.26
Zinc oxide	2
Total	338.26

The low hardness molded compound formulations are listed in Table 3.

Table 3. Low hardness molded part formulations with UHMW NORDEL™ EPDM

Ingredient	*Black*	*Gray*
UHMW EPDM	175	175
Zinc oxide	5	5
Stearic acid	1	1
Carbon black N550	70	0.35
TiO2	0	10
Mineral fillers	0	170
Coupling agent	0	2.25
Paraffinic Oil	25	50
Processing Aid	3	5
Stabilizer	1	1
Curatives	6.75	6.75
Total (PHR)	286.75	428.35

2.2 *Sample Preparation*

2.2.1 TPV Sample Preparation

A single step compounding process was used for making UHMW EPDM based TPVs. The following components were blended (dry mixed): EPDM, polypropylene, clay filler, stannous chloride, and zinc oxide, to form a pre-blend. The pre-blend was added to a 26-mm twin-screw extruder with a total length-to-diameter ratio (L/D) of 60. Oil (50 PHR) and phenolic resin (5 PHR) were added through an injection port at barrel 3 or barrel 5. Additional oil (40 PHR) was injected at a barrel 5 or barrel 10. Before injection, phenolic resin was dissolved in heated process oil, at 120°C, and the solution was then added to the extruder using a Leistritz Gear Pump cart, with heat traced liquid feed systems. The crosslinked (compounded) composition (TPV) exiting the extruder was then pelletized using an underwater pelletization unit.

The TPV molded sample were injection molded using a Krauss Maffei KM 110-390/390 CL Injection Molder. Each test plaques were 4 inch x 6 inch x 0.125 inches in dimension. About twelve test plaques were prepared for each TPV composition.

The TPV tape sample was extruded using a Haake Rheocord System 9000 torque rheometer with a Rheomix 254 extruder. The Rheomix 254 is a 25 L/D single screw extruder equipped with ¾ inch screw and four zones. The extrusion process parameters were optimized for each material based on its rheological properties. A 0.25 mm thick tape sample with smooth surface was produced with each TPV material.

2.2.2 Low Hardness Molded Rubber Goods Mixing, Molding, and Curing

For the molded parts formulations, all compounds were mixed using an "up-side-down" mixing procedure. All filler, polymer, 2/3 of oil and all chemicals excluding sulfur were mixed for 90 seconds, after which the left over 1/3 oil were added. Mixing was continued for another 120 seconds, after which sulfur was added. The compounds were mixed for another 70 seconds then dropped. The mixer rotor speed was 35 RPM and the fill factor was 0.86. Mixing was completed on a 6" 2 roll mill at ambient conditions, a blanket was sheeted out to use for rheology and mechanical testing.

Compounds were cured on a hot press for t90 + 5 minutes at 180 °C for tensile and tear testing following ASTM D3182-15. Testing specimens were cut from the cured sheets. Compression set specimen were cured for t90+12 minutes at180 °C. Cured specimens were conditioned for at least 18 hours before testing.

2.3 Characterization

2.3.1TPV Sample Characterization

Atomic Force Microscopy (AFM)

Samples were polished under cryogenic conditions using a Leica UCT/FCS microtome operated at 120°C. Thin sections (about 160 nm) were cut from sample and placed on a mica surface for AFM analysis. Topography and phase images were captured at ambient temperature by using a Digital Instruments (now Veeco) Multi-Mode AFM equipped with a NanoScope IV controller.

Shore A Hardness

Hardness was measured with a Shore A type durometer (ASTM D2240). The durometer was placed onto a plaque of ~0.125 inch thickness, prepared by injection molding procedure described earlier.

IOP Conf. Series: Materials Science and Engineering **548** (2019) 012013 doi:10.1088/1757-899X/548/1/012013

Compression set

Compression set was measured according to ASTM D 395 at 70 °C and 120 °C. Pucks with 29.0 mm (±0.5mm) diameter were extracted from the injection molded plaques of ~3 mm thickness. For each sample, four pucks were inspected for notches, uneven thickness and inhomogeneity, and were stacked such that the total height was 12.5 mm (± 0.5 mm), equating to compressive strain of 25% when a 9.5 mm spacer was used.

Stress-strain Properties

Tensile properties were measured at room temperature following the method of ASTM D-412, on micro-tensile specimens that were die cut from the injection molded plaques.

Oil resistance

Sample specimens were die cut from injection molded plaques, which were made as described in the Injection Molding section below, into tensile bars. Oil resistance was measured in accordance with ASTM D471-98 in which the sample was immersed in oil, specifically IRM-903, at 125 °C.

Extrusion Surface Quality

The surface quality of the extruded tape made from the TPV compound was evaluated using a visual rating of the tape appearance and smoothness, including the edge uniformity and surface smoothness. Additionally, the surface roughness can be measured by confocal laser scanning microscopy (CSLM). A coarse surface texture is indicative of poor dispersion of the rubber phase, and may also represent an incomplete phase inversion (when thermoplastic polymer (e.g., PP) becomes the continuous phase).

2.3.2 Low Hardness Molded Rubber Goods Sample Characterization

Mooney Viscosity

Mooney viscosity of gum and compounds were measured using an Alpha Technologies Mooney Viscometer following ASTM D1646-15.

Moving Die Rheometer (MDR)

Cure kinetic profiles at 180 °C were measured using an Alpha Moving Die Rheometer (MDR-2000) following ASTM D5289-15.

Tensile strength

The tensile properties of the solid vulcanizate were acquired following ASTM D1708.

Die-C Tear strength

Die-C tear strengths of the solid vulcanizate were obtained using ASTM D624.

Compression set

The compression set has been measured at three conditions (72 hours @ 100 °C, 150 °C) according to ASTM D 395 Method B.

3 Results and Discussion

3.1 Ultra High Molecular Weight NORDEL™ EPDM based TPV

A new ultra-high MW NORDEL™ EPDM was developed based on the understanding how molecular architecture affects TPV properties. The features of the new polymer include ultra-high molecular weight for better morphology development and mechanical properties, tailored composition for better balance of properties and 75 phr (43 wt%) oil extended for balanced throughput and formulation flexibility. The product specification was listed in Table 1.

Table **4**4 summarized physical properties of a TPV product based on the UHMW EPDM, compounded on a commercial-scale twin screw extruder with a single step process. TPV sample showed excellent properties such as compression set, tensile strength, tear strength and oil swell. The extruded TPV also demonstrated excellent extrudates surface quality. Surface roughness as characterized by Sq of 25 micron was measured with confocal laser scanner as shown in Fig. 2.

Table 4: Typical physical properties of TPV based on UHMW NORDEL™ EPDM

TPV Mechanical Properties	Value
Hardness (Shore A)	62
Compression Set %(22 hours@70°C)	27
Compassion Set %(70 hours@120°C)	48
Die-C Tear Strength kN/m (MD/CD*)	25.7/28.5
Elongation % (MD/CD*)	222/436
Tensile Strength psi (MD/CD*)	5/6.2
Oil Swell % (168hr@120°C)	103

*Note: MD is machine direction; CD is cross direction.

Fig. 2. Surface Roughness (Sq) measured on Extruded TPV tape by Confocal Laser Scanning (a) left image: surface roughness (b) right image: surface morphology

IOP Conf. Series: Materials Science and Engineering **548** (2019) 012013 doi:10.1088/1757-899X/548/1/012013

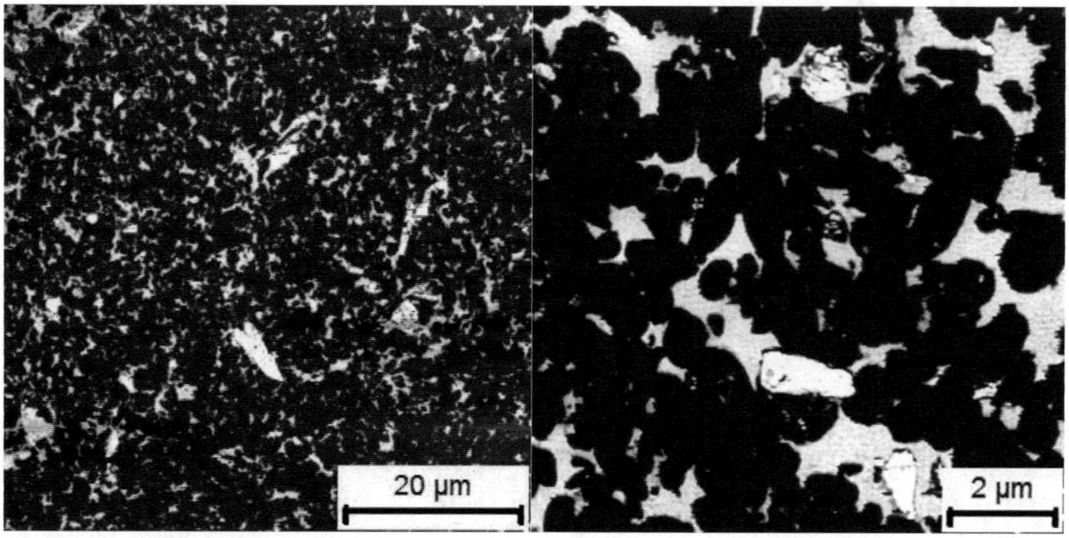

Fig. 3. Morphology of Dow UHMW EPDM based TPV. Left image: AFM image with scale bar of 20 micrometer (b) right image: AFM image with scale bar of 2 micrometer

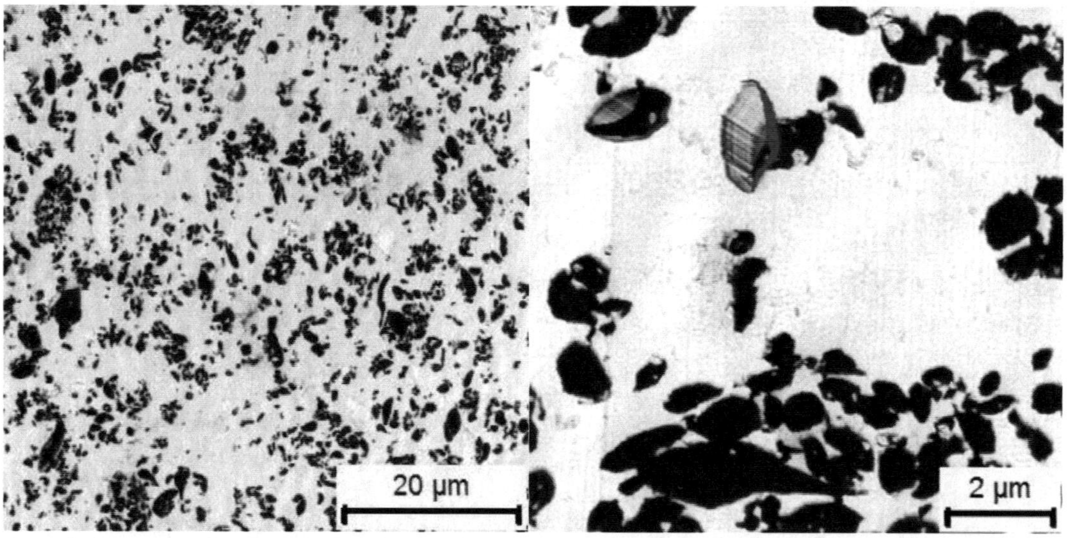

Fig. 4. Morphology of Dow UHMW EPDM based TPV sample diluted with PP at 50:50 volume ratio. Left image: AFM image with scale bar of 20 micrometer (b) right image: AFM image with scale bar of 2 micrometer

All TPV morphologies were characterized by agglomerated rubber particles dispersed in the polypropylene matrix. The agglomeration was due to the high volume ratio of the oil-swelled rubber phase. The diluted phase morphologies clearly showed ultra fine EPDM particles at 1-3 micron size depicted by Fig. 4.

Thermoplastic Vulcanizate (TPV) is a unique member of thermoplastic elastomer family as it could be melt processed as thermoplastic and has thermoset rubber like elastic properties. The EPDM/PP type TPV is the most common one, it contains micron-sized fully crosslinked EPDM rubber particles dispersed homogenously in a continuous PP matrix. By tuning the molecular architecture of EPDM produced by Dow's Advanced Molecular Catalysis and process technology advancements, a new Ultra-high MW NORDEL™ EPDM was developed to enable TPV with excellent physical properties and superior surface smoothness.

3.2 Low Hardness Molded Rubber Goods (Black and Gray Color)

The compound Mooney is a simply indicator of the rubber compound processability, the different processing techniques require different compound Mooney viscosity to provide proper performance. Usually the suitable compound Mooney range for injection molding is around 25 to 65 MU (tested at 100 °C) and up to 100 MU for compression molding. The compounds Mooney of the two low hardness formulations were listed in Table 5, both compounds showed proper Mooney viscosities suitable for molding process. The curing density, scorch time, and curing time were tested using moving die rheometer (MDR) at 180 °C and the data were summarized in Table 6.

Table 5. Compounds Mooney Viscosity of low hardness molded parts

Compound Mooney Viscosity	***Black***	***Gray***
ML (1+4) @ 100°C (MU)	51	42

Table 6. Compounds Curing Characteristics measured by MDR at 180 °C

Curing Characteristics	***Black***	***Gray***
ML(dNm)	1.35	1.34
MH(dNm)	8.63	7.9
MH-ML(dNm)	7.28	6.56
ts1 (min)	0.11	1.12
ts2 (min)	0.76	1.47
tc10(min)	0.95	0.97
tc50(min)	0.69	2.04

As shown by Both formulations were soft with hardness around 40 Shore A, had more than 10 MPa tensile strength with over 600% ultimate elongation which are highly required for many automotive and appliance applications such as grommet, bellows, boots, and washing machine gaskets.

IOP Conf. Series: Materials Science and Engineering **548** (2019) 012013 doi:10.1088/1757-899X/548/1/012013

Table 7. Mechanical Properties of Black and Gray Color Low Hardness Molded Goods

Mechanical Properties	***Black***	***Gray***
Density (g/cc)	1.02	1.23
Hardness (Sh.A)	44	38.6
Young's Modulus (MPa)	3.14	2.34
10% modulus(MPa)	1.75	1.39
50% modulus(MPa)	1.03	0.72
100% modulus(MPa)	0.87	0.53
Tensile strength (MPa)	11.77	11.2
Elongation at break (%)	698.71	818.1
Die-C tear (kN/m)	28.83	20.31
Compression set		
72 hours @ 100 °C (%)	48	NA
72 hours @ 150 °C (%)	57	NA

From these two 40 Shore A formulation study, we could conclude that the newly developed oil extended ultra high molecular weight NORDEL™ product is an ideal EPDM for the manufacturing of low hardness molded rubber parts which require high mechanical strength and ultimate elongation at break.

4 Conclusion

The rubber industry uses EPDM materials from three different generations of catalyst and process technologies; traditional ZN, metallocene, and now Dow's new Advanced Molecular Catalyst technology. The new AMC technology developed by Dow Chemical leads the EPDM industry in process sustainability and production efficiency; saving up to 50% less energy and having 30% less impact on the environment in comparison to the traditional ZN process.

By tuning the molecular architecture of EPDM produced with Dow's Advanced Molecular Catalysis and process technology advancements, a new Ultra-high MW NORDEL™ EPDM was successfully developed and proven to be an ideal EPDM to produce premium TPV with excellent physical properties, excellent surface smoothness and also suitable to manufacture low hardness molded rubber parts where high mechanical strength and ultimate elongation at break are required. Please add some general conclusion on molded parts

5 Acknowledgement

The authors would acknowledge the lab support provided by Jun Wu, Lin Shen and Vivian Lu on polymer characterization, compound mixing, curing and testing.

IOP Conf. Series: Materials Science and Engineering **548** (2019) 012013 doi:10.1088/1757-899X/548/1/012013

References

[1] Ravishankar P S 2012 Treatise on EPDM *Rubber Chem. Technol.* 85 327-349

[2] Helling R 2017 Life Cycle Assessment of NORDEL™ EPDM using Advanced Molecular Catalyst technology *The Dow Chemical Company*

[3] Stevens J C 1996 Constrained Geometry and Other Single Site Metallocene Polyolefin Catalyst: A Revolution in Olefin Polymerization *Stud. Surf. Sci. & Catal.* 10111-20

[4] Boussie T et al 2003 A Fully Integrated High-Throughput Screening Methodology for the Discovery of New Polyolefin Catalysts: Discovery of a New Class of High Temperature Single-Site Group (IV) Copolymerization Catalysts. *J. Am Chem Soc.* 125(14) 4306-4317

[5] Coran A, Das B, and Patel R 1978 *US Pat.4130435*

[6] Abdou-Sabet S and Fath M, 1982 *US Pat.4311628*

[7] Walker B M and Rader C P 1988 *Handbook of Thermoplastic Elastomers* (Van Nostrand Reinhold) 85-140

[8] Legge N, Holden G, and Schroeder H 1987 *Thermoplastic Elastomers: a Comprehensive Review* (Hanser Publishers)

[9] Chung O and Coran A 1997 The Morphology of Rubber/Plastic Blends *Rubber Chem. Technol.* 70 781.

[10] Walder J 1990 Molded Goods *The Vanderbilt Rubber Handbook* (R T Vanderbilt Company)13th edition 637-646

Oven Ageing versus UV Ageing Properties of Natural Rubber Cup Lump /EPDM Rubber Blend with Mangosteen Powder (MPP) as Natural Antioxidant

M Rahmah[1], A R Ahmad Khusyairi[2], H Nur Khairunnisya[2] and M Siti Nurbalqis[2]

[1] Green Polymer Research Group, Faculty of Applied Science, UiTM Shah Alam, Selangor, Malaysia

[2] Faculty of Applied Sciences, UiTM Shah Alam, Selangor, Malaysia.

E-mail: rahmahgepsb@gmail.com

Abstract. Natural Rubber (NR) are susceptible to oxidation due to the presence of diene double bond. Improvement of NR properties can be enhanced by presence of antioxidant. Natural antioxidant can be a new source of additives to prevent and delay oxidation process. In this work, mangosteen powder (MPP) is exploited to be good potential additive in improving the aging properties of NR cup lump (NRCL) blended with EPDM rubber. Blended NRCL with EPDM rubber as a sealant had been studied with and without MPP. Aging effects upon thermal and weather exposure after two weeks had been analyzed by FTIR. Increasing amount of MPP had retained the diene bonds upon further rubber degradation. The broad band absorbance peak at 3410 cm^{-1} corresponds to O-H stretching with H-bond which indicates the existence of phenols in the MPP. Carboxylate groups and amine presence in MPP also found to exhibit absorbance peak at 1028 and 1608 cm^{-1} respectively. MPP contained polyphenol proven to be effective antioxidant agent in reducing degradation upon thermal and sunlight exposure. Oven and weathering tests performed for two weeks showed different aging behavior with weathering exposure found to have greater degradation effect due to radical attack by free radical from UV sunlight.

Keywords: Natural antioxidant, Cup lump, Mangosteen powder peel, EPDM, Tackiness

1. Introduction

During the eighteenth century, rubber was known throughout Europe as rubber had greater properties which can be benefitted. Systematic tapping process cause the latex to be extracted regularly and which name it Natural Rubber Cup Lump (NRCL) without process [1].

Initially, there are four types of field coagula that can be derived from Natural Rubber (NR) which are cup lump, tree lace, smallholders' lump and earth scrap. Each of them has their own significant into converting as final products. Cup lump is widely available and preferable to be used as raw material because it is cheap in cost and versatile to be converted into higher grade or Technically specified rubber (TSR) such as SMR20 and TSR20 [2]. Ethylene-propylene-diene monomer (EPDM) rubber is the type of synthetic rubber that are based on 5-ethylidene 2-norbornene (ENB). It is more stable than conventional elastomers-(butadiene, isoprene rubber) as it has sensitive that are very important for subsequent crosslinking by peroxide or sulfur. Applications of EPDM include blending with NR for automotive tires, side walls, cover strips, wires, cables, hoses, belting, roofing barriers and sporting goods [3].

Hevea Brasiliensis is a major commercial source of NR that has been extensively used to manufacture high quality rubber products for instances, tires, gloves, liners, catheters and the tube. NR is a type of renewable source where a lot of ongoing research had been done to refine its processing offering varieties of rubber products. The rubber plantation has been expanding rapidly as there were many high-quality rubber products can be made from NR. The cup lumps are typically collected daily and sorting this raw material to produce various rubber grades is largely based on its moisture content (MC)[4]. The STR 20 is mainly produced from field coagulum or cup lump and has good processing characteristics and physical properties. The raw material for STR 20 is formed either by the coagulation of field latex with acid, or by the evaporation of water from latex in a collection cup. EPDM rubber is used as it has weather resistant properties. Previous studies had shown the poor compatibility of blending NR and EPDM rubber, but the problem can be overcome by incorporating compatibilizer into the rubber blends. Study by El-Sabbagh, he found that grafting the maleic anhydride (MAH) with EPDM /NR introduce greater compatibility and greatly improved the rheological properties of rubber blends [5]. Then, another study was made to add the compatibilizer agent for NR/EPDM blends which uses maleic anhydride grafted with ethylene propylene monomer (EPM). This time the result shows improvement for mechanical properties by adding 5 pphr of MAH-EPM [6]. The tack ability for vulcanized EPDM rubber having hydrocarbon resin found to be compatible at higher concentration (32% wt), better dispersion, lower softening point and better peel strength [7].

In blends, inherent properties of polymer components are retain and most dominant polymer and its properties will stand out. NR/EPDM blend show that TBBS as accelerator is good choice for vulcanizing the blends since it the result reveal both processability and mechanical properties [8]. Benefits of blending are, they can be used to lower the compound cost, for ease of fabrication and to improve the performance of the industrial rubber. NR and EPDM rubber have been blended for a long time for these reasons [9].

Degradation of unstabilized organic polymers occur rapidly upon exposure to UV and weathering the function of this antioxidant is to retard the auto-oxidation which can prolong the useful life of the host substrate. Hindered phenols is the example of the stabilizers which accept the unpaired electron of free radicals. Use of Natural antioxidant are important addtives that can confer environmental friendly additive as compared to other antioxidant available in the market. Phenolics and amines are scrutinized as carcinogenic. Hindered phenols, hindered amines and quinolones are mostly used in rubber industries. [10] Examples of antioxidants used in rubbers include

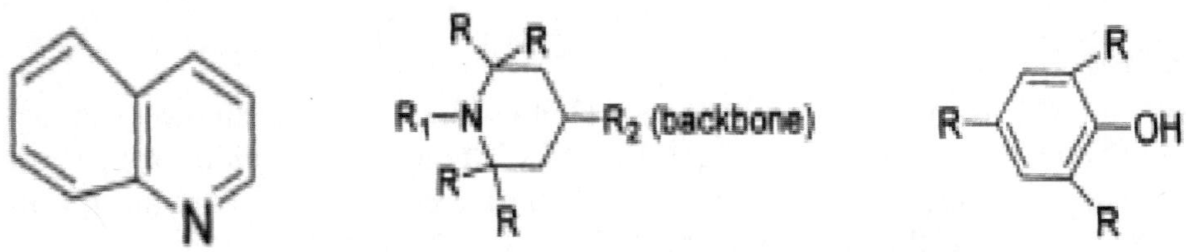

A. Quinoline B. Hindered Amines (HALS) C. Hindered Phenols

Fig. 1. Examples of Antioxidants for rubbers (a) quinolone (b) HALS (c) Hindered Phenols

Upon reacting with free radicals, phenolic are considered as primary antioxidants. Typical loading sare 0.25 to 2.0phr and they are effective in most polymer and sulfur cure systems. Quinoline are particularly useful in rubber goods requiring resistance to high temperatures. They are also used as a copper and manganese inhibitors and provide anti flex cracking properties.

Accelerated ageing of styrene-butadiene rubber nanocomposites stabilized by phenolic antioxidant hed been investigated by Bellas et.al [11]. Synthesis of a new nanosilica-based antioxidant and its influence on the anti-oxidation performance of natural rubber had been perfomed by Lei H [12]. Cibulkova Z, et al had also found the stabilization effect of potential antioxidants on the thermooxidative stability of styrene butadiene rubber (SBR). [13] Oxidation Degradation Study and Use of Phenol and Amina Antioxidant Compounds in Natural Rubber had been investigated and been extensively use [14]

Mangosteen from the fruit carp grown in Thailand , Malaysia and ASEAN countreis, can be extracted .The extract have the phenolic natural xanthones called α-mangostin and γ- mangostin structure . It is a hindered phenol which are suitable to replace the use of synthetic hindered phenols or amine. Fig. 2 shows the structure of mangostin xanthone; type of hindered phenols. It had been investigated to have antibacterial properties too [15]

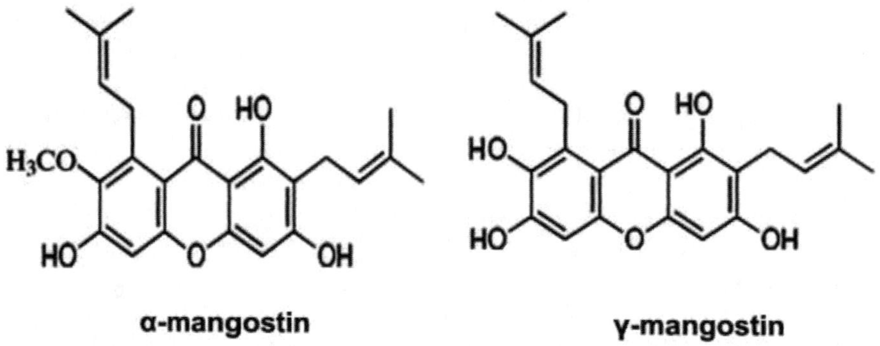

Fig. 2. Structure of Structure of α-mangostin and γ- mangostin xanthone [15]

This research involve the blending of NR cup lump with EPDM rubber and added with natural antioxidant. In this study, NRCL will be blended with EPDM rubber to produce a new environment friendly sealant. Natural antioxidant is added to enhance the ageing properties of elastomer blend. Rahmah et al, [16, 17] had previously use natural additives as antioxidant and can also act as Prodegradant for the polyolefin degradant masterbatch

2. Materials and Methods

In this study, NR cup lump supplied by small holder rubber tapper, EPDM rubber grade VS 3666 supplied by Exxon Mobil Chemical Ltd., natural antioxidant, Mangosteen powder from carp sieved to 50 micron, Phenolic resin tackifier from Malaysian Adhesive & Chemicals Sdn Bhd, Toluene and methyl ethyl ether peroxide (MEKP) from Sigma Aldrich Inc. The formulation mixture of NR cup lump, EPDM rubber and other chemicals are being made as shown in Table 1.

2.1 Preparation of NR cup lump into solution

Raw NRCL were washed, sheared and cleaned thoroughly to remove dirt, sand, twigs and tree bark. The cleaned cup lump was cut into small pieces or as strips and put into oven to dry. Dried cup lump was removed from the oven and they were masticated for about 10-15 minutes to reduce viscosity and molecular weight. Masticated rubber was cut into smaller pieces for easy dissolution with toluene for 2 days. EPDM rubber was cut into smaller size for dissolution into toluene forming EPDM rubber solution.

Table 1. The formulation of NRCL rubber sealant

Ingredient (part by weight)	Control sample	Sample 1 (S1)	Sample 2 (S2)	Sample 3 (S3)
NR cup lump (g)	10	10	10	10
EPDM rubber (g)	2	2	2	2
Natural antioxidant (g)	0	2	4	6
Phenolic resin (g)	0	1.5	1.5	1.5
MEKP (%)	2	2	2	2
Toluene (ml)	100	100	100	100

NRCL solution was mixed with EPDM rubber solution and stirred thoroughly to get homogenous mixture. Then, a tackifier which was add into mixture to give a better tackifying properties. After that, natural antioxidant was added into this mixture at different composition as in Table 1 to improve the ageing properties of rubber blends. MEKP acted as a curing agent that was responsible to cure the rubber blends also added into this mixture. Polymer blend were poured onto the glass and will be left for 1 day to cure the mixture. Then, the sample undergone aging by thermal exposure and weather for 14 days. Upon completion of exposure the samples undergone characterization and testing.

2.2 Characterization and testing

The sample prepared went under characterization and testing to test the properties of sample made. The sample undergoes FTIR characterization test to identify the presence of certain functional groups in a molecule. Standard test for characterization test method is ASTM D3403. The samples which were known as control sample, S1, S2 and S3 were tested after they undergone oven exposure for 0, 2, 7 and 14 days aging at 60°C as according to ASTM D573-(04). Standard Test Method for Rubber—Deterioration in an Air Oven.

The weight loss test is to determine the antioxidative ability of antioxidant affecting properties of sample made. Samples were weighed by using analytical balance before and after aging test for 0, 2, 7 and 14 days. Samples were exposed to weathering process under Malaysian weather having high humidity and temperature between April and May 2017. Recorded humidity ranges between 64% to 94% and min temperature about 25°C and max temperature about 32°C were found for the week mentioned. (Data collected from meteorological website).

Retention and percent were calculated based on percent weight loss upon exposure to sunlight and oven for each duration of 0, 2, 7 and 14 days.

$$Retention, \% = 100\% - percentage\ weight\ loss$$

$$Percent\ weight\ loss = \frac{Initial\ weight\ , g - Final\ weight, g}{Final\ weight, g}\ x\ 100\%$$

3. Result and Discussion

Table 2 below shows the tabulated data for percentage of weight loss and percentage of retention weight experienced by the sample for 0, 2, 7 and 14 days during ageing process for both conditions' oven and weathering using weight loss data from Fig. 3 and 4 .

Table 2. Tabulated data of percentage weight loss and retention percentage for each sample under oven and UV weathering aging up to 14 days

	Oven								Weathering							
	Weight loss (%)				Weight Retention (%)				Weight loss (%)				Retention (%)			
Duration (days)	0	2	7	14	0	2	7	14	0	2	7	14	0	2	7	14
Control sample	0	3.75	25	50	100	96.3	75	50	0	6.3	38.8	87.5	100	93.7	61.2	12.5
S1	0	0	13.8	18.8	100	100	86.2	81.2	0	0	16.3	25	100	100	83.7	75
S2	0	0	6.3	15	100	100	93.7	85	0	0	8.8	21.3	100	100	91.2	78.7
S3	0	0	0	0	100	100	100	100	0	0	0	0	100	100	100	100

3.1 Weight Loss Test

Fig. 1 shows weight loss of sample undergone by oven ageing process at different loading of MPP. Initial weight for control sample before aging was 0.8 g. Weight loss was reduced to 0.4 g with 50 % weight retention after 14 days. With S1 and S2 of 5%MPP loading and below, the % weight retention are high ~80% and highest loading of 6% MPP for S3 had retained its original weight.

Upon exposure to weathering, for S1, the weight before aging was 0.8 g exhibiting least % of about 12.4% after 14 days. For 2% MPP (S2) the percent retention is 75% which improved the weight loss while for S3, no change in weight loss was observed. Comparing the two conditions of exposure, MPP is significantly effective in improving the weight retention especially for weathering exposure which can retain the weight by more than 4x compared to heat aging exposure. Ultra violet light which is the agent of degradation of rubber samples which can cause the breaking down of rubber chain can be impeded by incorporating MPP. Heat exposure had caused less change of structural crosslink integrity than weathering; however both samples for oven and weathering retain their original weight with no signs of weight loss.

Li et al [16] had investigated similar work on the thermal ageing of nitrile-butadiene rubber (NBR) reinforced with different graphene (GE)/The mechanical properties, chemical changes, and thermal stability of ageing samples and neat samples were evaluated and found integrity crosslink were affected by variation in GE used at temperature fixed at100ºC,. Oven or thermal aging for this work do not affect the integrity of crosslink for high MPP loading but significantly affect the control and with less than 5% loading of MPP.

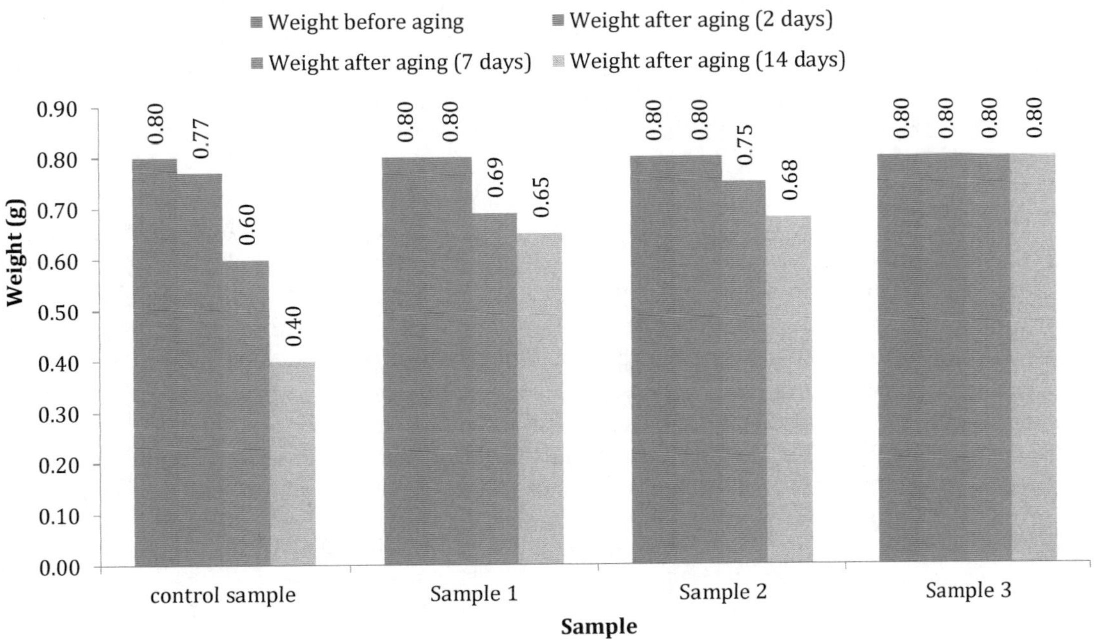

Fig. 3. Graph of weight loss versus percentage loading of MPP by Oven Aging

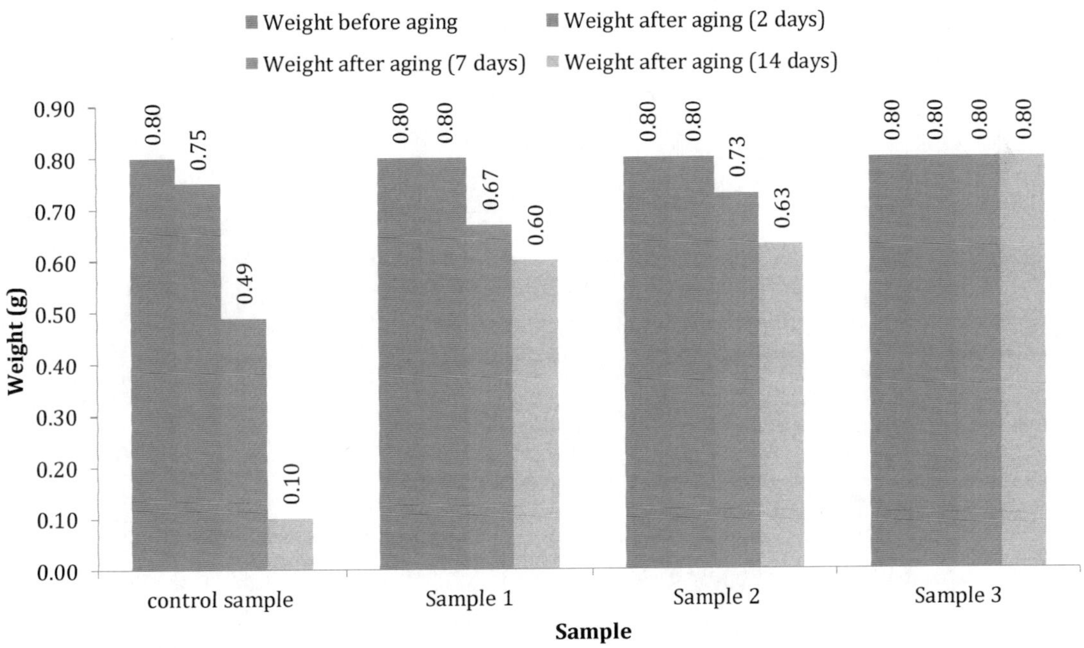

Fig. 4. Graph of weight loss versus percentage loading of MPP by Weathering.

Li et al [16] had investigated similar work on the thermal ageing of nitrile-butadiene rubber (NBR) reinforced with different graphene (GE) / The mechanical properties, chemical changes, and thermal stability of ageing samples and neat samples were evaluated and found integrity crosslink were affected by variation in GE used at temperature fixed at100ºC,. Oven or thermal aging for this work do not affect the integrity of crosslink for high MPP loading but significantly affect the control and with less than 5% loading of MPP.

Control NRCL sample lose weight easily because there is no antioxidant (AO) presence. Without AO, structural integrity of rubber upon expose can be attacked as the diene groups of *cis*-polyisoprene could not withstand the high radiation under UV light or high accelerated oven temperature to form free radical. This free radical can attack the double bond of rubber blend causing oxidation, which leads to degradation. With higher amount MPP added into sample, improved rubber aging properties were attained after undergoing weathering for 14 days. This happened because natural AO which was MPP having high rich polyphenols acted as uv interceptor causing effective resistance to oxidation.. MPP had prevented free radical attack and prevented degradation. According to Li et al, oxidative degradation prevention for NR is by addition of AO. Natural rubber with AO exhibited 12 times more stability towards oxidative degradation. The action of MPP xanthone structure is postulated to follow hindered phenol mechanism.

3.2 FTIR Analysis

FTIR spectra of Natural Rubber (NR) and EPDM rubber are depicted in Fig. 5 and 6 respectively. For the NR graph based on Fig. 5, the functional group present were CH, C=C.CH_2 and CH_3. The functional group of NR were detected at the absorption peaks of 290 cm^{-1}, 1662 cm^{-1}, 1444 cm^{-1}, 1374 cm^{-1} and 839 cm^{-1}. The best region to identify presence of *cis*-Polyisoprene was the absorption at 2960 cm^{-1} (C-H_3 group) and 1444 cm^{-1} (C-H deformation).

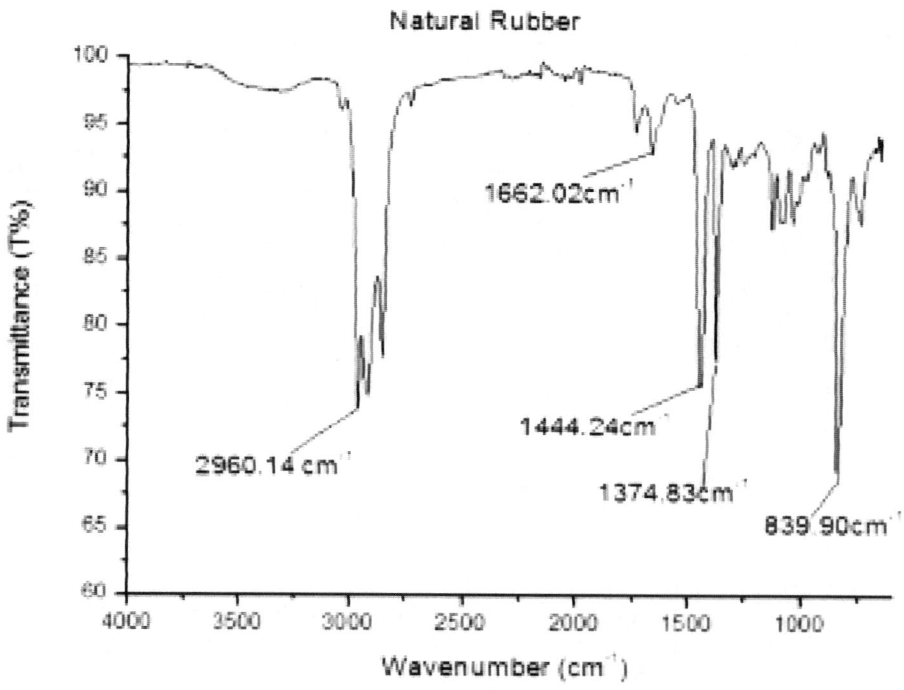

Fig. 5. FTIR spectra for Natural Rubber (NR)

IOP Conf. Series: Materials Science and Engineering **548** (2019) 012014 doi:10.1088/1757-899X/548/1/012014

Table 3. Analysis of NR FTIR spectra

Band intensity wavenumber	Wavenumber cm^{-1}	Absorption	Assignment
strong	2960	CH	Stretching vibration
medium	1662	C=C	Stretching vibration
medium	1444,1374	CH_2	CH_2 and CH_3 deformation
medium	839	CH_3	Out of plane bending

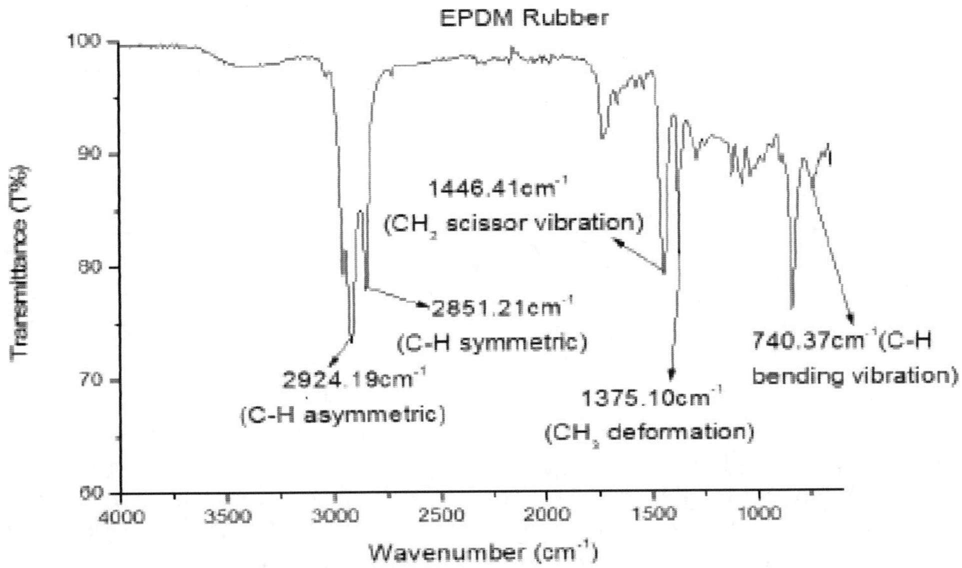

Fig. 6. FTIR spectra for EPDM rubber

Stretching vibration of CH are visible at wavenumber 2960-2924 cm^{-1} for NR and EPDM rubber while rubber has medium to strong peak at ~1662 cm^{-1} representing C=C diene bond which is inherent in NR rubber. For EPDM rubber, strong to medium peak are exhibited around 2850 and significant peaks around CH deformation with bending peak at 740cm^{-1}.

Table 4. Analysis of EPDM FTIR spectra

Band intensity wavenumber	Wavenumber cm^{-1}	Absorption	Assignment
strong	2924	CH	Stretching vibration
medium	2851	CH	Stretching vibration
medium	1446,1375	CH_2	CH_2 and CH_3 deformation
medium	740	CH_3	Bending vibration

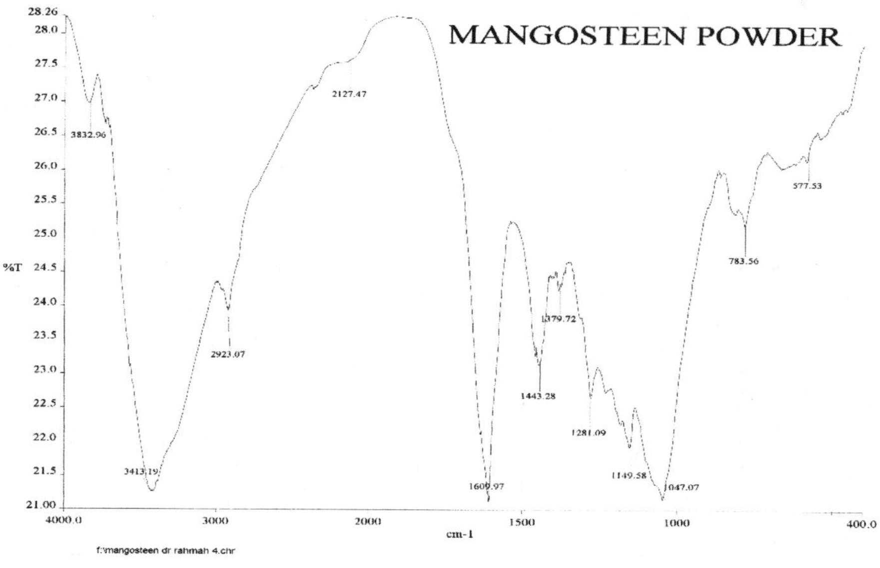

Fig. 7. FTIR spectrum of mangosteen powder

Fig. 7 shows the graph of mangosteen powder with high intensity of OH, C-O and C=O peak of xanthones which shows strong α-mangostin and γ- mangostin peaks. The MPP blended NRCL before aging are shown in Fig. 8. Fig. 9(a) showed MPP absorption peaks which have intensity absorption of strong to medium at 3410, 2918 representing OH phenolic groups and CH bending while 1608, 1337 and 1048 cm^{-1} represents the carbonyl deformation and C-O ether of mangosteen functional groups. The broad peak of OH bonding for all samples was diminished after undergo aging for 2 days because of the sample exposed to sunlight which cause the samples dried but OH bonding formed again after the samples undergone for 7 days aging probably because of the samples exposed to the rain and dew which cause the samples absorb the water.

Referring to Fig. 9(c) to 9(d), the peaks at 3210 cm−1 and 967.2 cm−1 increased in intensity remarkably upon weathering, possibly because most of the antioxidants in the NRCL samples had been consumed. The peak at 3210 cm−1 was mostly due to O–H symmetric stretching vibration and C–H deformation vibration indicating that hydroxyl groups were produced for weathering samples when the NCRL was exposed to weathering which exhibit increased in degradation compared to oven ageing environment. This similar pattern were exhibited by Li et al [17] who investigated with Nitrile rubber exposed to thermal aging condition.

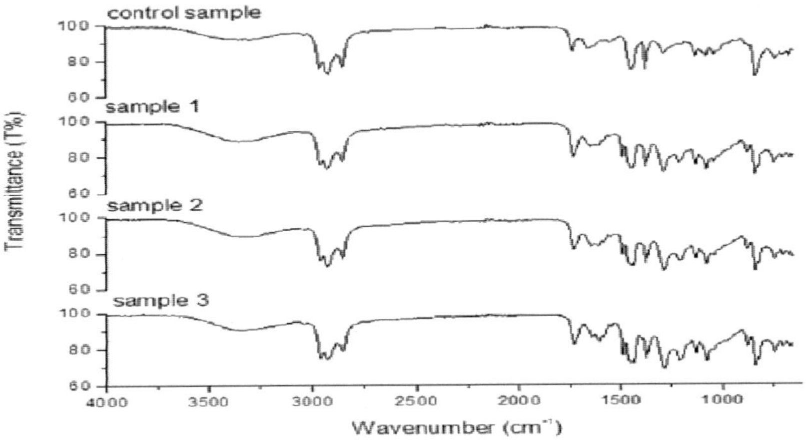

Fig. 8. FTIR spectra of rubber blend for 0 days aging

At 1400 cm^{-1} to 1200 cm^{-1}, more peaks were observed, having more CH deformation. For aging 14 days in Fig.. 9(e), the result showed many structural vibration differences with unexposed samples. For samples aged by oven in Fig. 9(b), the functional group presenting in sample coupled with natural antioxidant was diminished. Based on Fig.. 9(c), the broad peak which represent for OH bonding at 3200 cm^{-1} was diminished when the samples undergo aging since the first 2 days. This indicated that the samples became dried after exposed to thermal exposure. For the samples undergone aging for 14 days, the result showed that only broad peak which were represent OH bonding only diminished. This showed that the sealant is not easily degrade if they only exposed to the thermal exposure compared to the samples that undergo aging by weather where they were exposed to the UV light, rain and dew.

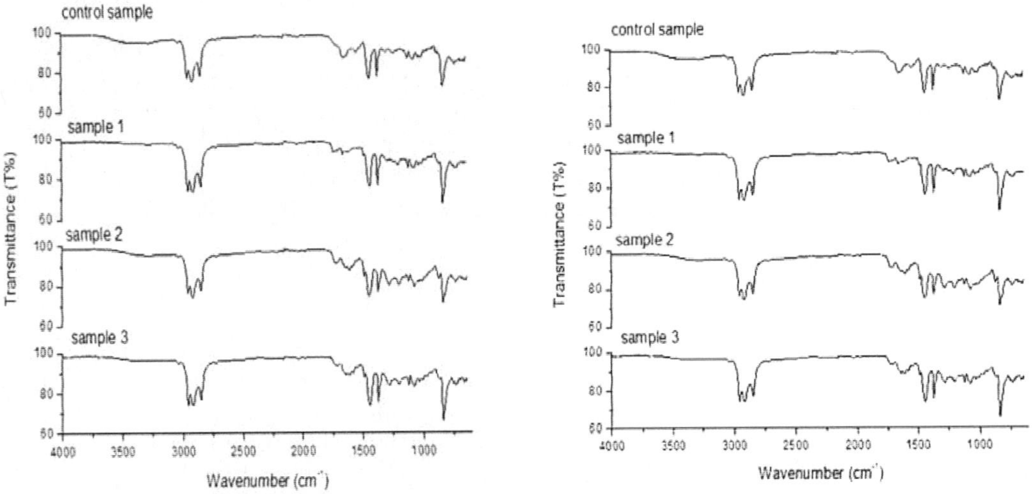

(a) (b)

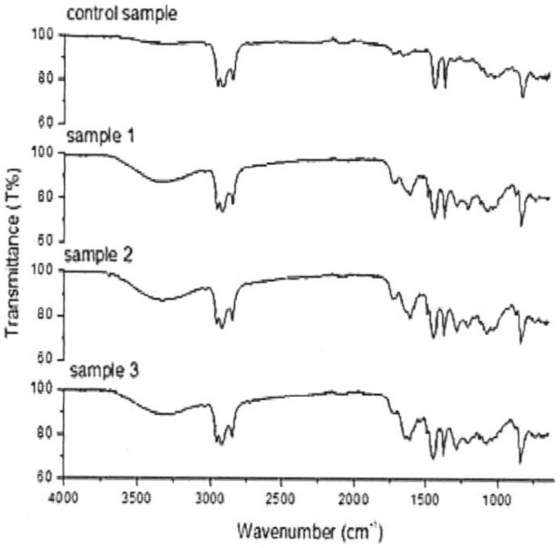

(c)

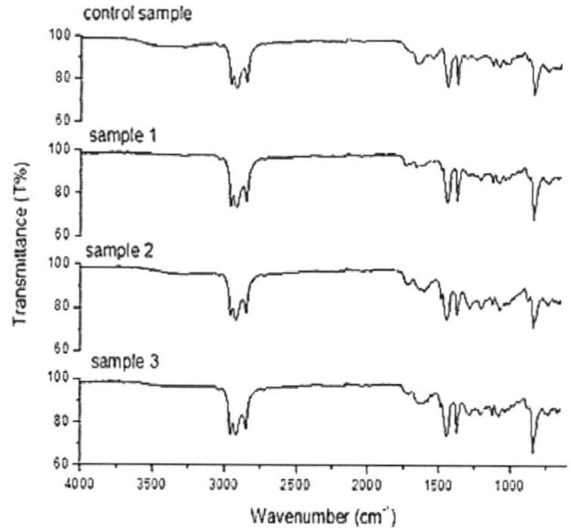

(d)

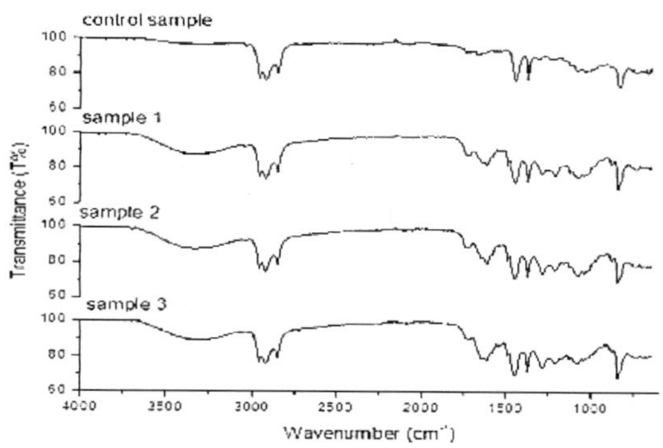

(e)

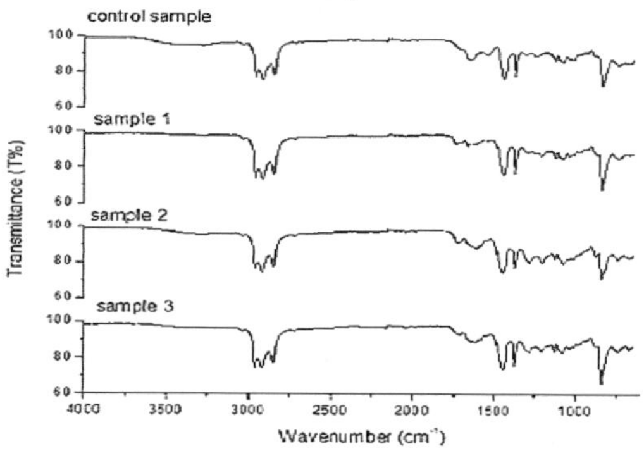

(f)

Fig. 9. FTIR spectras of rubber blend at (a) 2 days by weathering (b) 2 days by oven (c) 7 days by weathering (d) 7 days by oven (e) 14 days aging by weather (f) 14 days aging by oven

4. Conclusion

In conclusion, the addition of MPP as natural antioxidant in the sealant improved the aging resistance as these sealants can reduced their weight loss and maintained their appearance. Its ability as a good sealant was not likely affected by the aging through weather or thermal exposure as high amount of MPP can act as its antioxidant and prevent degradation. High amount MPP added into the sealant prevent degradation to rubber compounds as from structural analyzed by FTIR. Oven and weathering tests showed different behavior on aging. Weathering had greater degradation effect as compared to thermal aging.

5. Acknowledgements

The authors are thankful to the editors and reviewers for their insightful comments/suggestions which helped to bring the manuscript in a better shape. The first author expresses special gratitude to and the Research Management Institute (RMI) UITM to partly finance the project under the Bestari Fund 600-IRMI/PERDANA 5/3 BESTARI (106/2018) and Polymer lab UiTM at Shah Alam, Selangor, Malaysia.

IOP Conf. Series: Materials Science and Engineering **548** (2019) 012014 doi:10.1088/1757-899X/548/1/012014

References

[1] Fainleib A, Pires R V, Lucas E F and Soares B G, 2013 *Polímeros* 23 441–450
[2] Yahaya N Z, Abbas Z, Ibrahim N M, Hafizi M H M and Yahaya M Z 2014 *J. Agri. Bio. Engine.* 7 (5) 48–54
[3] Morlat-Therias S, et. al. 2006 *Polym. Degrad. Stab.* 91(12) 3033-3039
[4] Suchat S 2015 *Industr. Crops Prod.* 76 772-780
[5] El-Sabbagh S 2003 *J. Appl. Polym. Sci.* 90(1) 1-11
[6] Heinen W 1996 13C *Macromolecules* 29(4) 1151-1157
[7] Sasaki M 2008 *International J. Adhesion and Adhesives* 28(7) 372-381
[8] Sae-oui P 2007 *Polym. Test.* 26(8) 1062-1067
[9] Zaharescu T 2000 70(3) 341-345
[10] Jacoby P 2002 http://www.pstc.org/files/public/ Jacoby.pdf [2 April 2014]
[11] Bellas R, Diez J, Rico M, Barral L, Ramirez C and Montero B 2014 *Polym. Compos.* 35 334-343
[12] Lei H, Huang G and Weng G 2013 *J. Macromolecular Sci. Part B Physics* 52 84-94
[13] Cibulkova Z, Cerna A, Simon P, Uhlar J, Kosar K and Lehocky P 2011 *J. Therm. Anal. Calorim.* 105 607-6
[14] Arofah Megasari Siregar, Eddyanto, Basuki Wirjosentono, Ameilia Zuliyanti Siregar 2016 *Int. J Sci Technol. Research* 5, (5) 297-299
[15] Srinivasan Narasimhan1 , Shanmugam Maheshwaran, Imad A. Abu-Yousef, Amin F. Majdalawieh, Janarthanam Rethavathi, Prince Edwin Das and Palmiro Poltronieri 2016 *Molecules.* 1-18
[16] Johnson JJ, Petiwala SM, Syed DN, Rasmussen JT, Adhami VM, Siddiqui IA, Kohl AM, Mukhtar H 2012, *Carcinogenesis* 33 (2) 413–419
[17] Rahmah Mohamed, Amir Fakhrul Islam Abdullah, W A Zalikha A S, Mohd A F 2011 *J. Mater. Sci. Eng.* 5, 555-560
[18] Yusak N A M. Rahmah Mohamed, Mohaimi Mohamed, Siti Sarah A Talib 2015 *Adv. Mater. Res.* 1113, 571-577. Trans TecPublications
[19] Fei-Zhou Li, Mu-Rong Gao and Bian Guo 2018 *J. Chemists and Chemical Engineers* 67(1-2) 29-37

IOP Conf. Series: Materials Science and Engineering **548** (2019) 012015 doi:10.1088/1757-899X/548/1/012015

THE Effect of Azido-Starch as Filler on Tensile and Tear Properties of Natural Rubber Latex Films

S Daud[1], A R Azura [2] and M F Yhaya [3]

[1] School of Materials and Mineral Resources Engineering, USM Engineering Campus, 14300 Nibong Tebal, Penang.

[2] School of Materials and Mineral Resources Engineering, USM Engineering Campus, 14300 Nibong Tebal, Penang

[3] School of Dental Sciences, Health Campus, Universiti Sains Malaysia, 16150 Kubang Kerian, Kelantan.

E-mail: srazura@usm.my

Abstract. The azido starch/ Natural Rubber (azo-starch/NR) latex films were prepared with control (0), 0.5, 1.0, 1.5 and 2.0 phr of azido- starch loadings. The azido starch functional groups were characterized by Fourier Transform Infra- Red (FTIR) analysis to observe the presence of significant peaks of azide (N_3).The control and azido starch-filled NR films were subjected to tensile test and tear test in order to investigate the effect of the addition of azido- starch on the mechanical properties of azo-starch/NR latex films. The tensile strength of azo-starch/NR latex films was increased with increasing azido-starch loading and slightly decreased after 1.0 phr loading. The optimum tensile strength of azo-starch/NR latex films is obtained at 1.0 phr loading. The moduli (M100 and M300) and tear strength also increased with increasing azido- starch loading. However, the elongation at break (EB) was almost similar with increasing azido- starch loading. The swelling test was done to determine the swelling index, crosslink density and rubber-filler interaction of azo-starch/NR films. The swelling index was decreased but the crosslink density increased with increasing azido-starch loading indicating the formation of crosslinks between azido- starch and rubber macromolecules.

1. Introduction

One of the most preferred types of polysaccharides is starch where the research has been carried on the utilization of starch as filler in the production of latex films [1][2]. The ultimate purpose of using lignocellulosic and polysaccharides materials into the rubber and polymer matrices are to induce biodegradation of the products. The incorporation of starches into rubber matrix not only cheapen the cost but also increased the rate of degradation and fragmentation of the rubber products upon disposal [3]. However, the utilization of starches on their own can deteriorate the vital properties of rubber products such as its mechanical properties. Starches are mostly unsuitable for applications without physical or chemical modification. The most common method of modification of starches normally acid hydrolysis, oxidation, esterification, etherification, substitution and crosslinking of the available hydroxyl groups on the α–D-glucopyranosyl units that make up the starch [4]. The starch granule consists of two constituents; amylose and amylopectin. The ratio of amylose and amylopectin varies depend solely on starch species [5]. The chemical reactivity of starch derives from the reactivity of hydroxyl groups of the anhydroglucose units (AGU). Each AGU contains 3 three OH groups, which react similarly to alcohols [6]. Cross-linkable functional group can be directly synthesized onto the starch. The crosslinking approached can be applied to any primary hydroxyl-containing starch. The extent of the modification reactions is generally expressed as the degree of substitution (DS), which corresponds to the average amount of OH groups per AGU that have been substituted by another functional group [7].

Azido saccharides are usually synthesized for the amino group where the chemistry of azides allows it to be used for crosslinking purposes (8). Starch was chemically modified to improve interconnection between starch particles which increased the strength of the product. Azidated starch can be prepared by substituting the free hydroxyl groups with azide groups. Substituted starch with DS 0.2 and above increased the hydrophobicity of the starches, thus, increased the miscibility with a hydrophobic polymer that used as matrix [9]. Compared to native starches, crosslinked starch has superior properties in mechanical, thermal and chemical resistance [10]. The purpose of this research work is to investigate the potential use of azido- starch (azidated starch) as a reinforcing filler in natural rubber latex to increase the mechanical properties of the films produced. Furthermore, it also to develop a new potential cross-linker as an alternative to conventional sulphur/accelerator in natural rubber latex films using modified starch via click chemistry route.

2. Material and Methods

2.1 Materials

High ammonia NR latex with 60% total solid content (TSC) was purchased from Zarm Scientific and Supplies (Malaysia) Sdn Bhd .Sago starch was manufactured by Sago Link Sdn Bhd. Potassium hydroxide was supplied by Merck Sdn Bhd. Other compounding ingredients including anchoid, antioxidant, zinc oxide, zinc diethyldithiocarbamate (ZDEC) and suphur in dispersions form were supplied by HH Saintifik Sdn Bhd and used as received. The tosyl chloride and sodium azide were manufactured Sigma Aldrich.

2.2 Preparation of azido-starch-filled NR latex films

The preparation of azido sago starch (azidated sago starch) involved the tosylation process. The synthesis of tosylated starch was carried out according to Elchinger [11] with a few modifications. First, a 50 g sago starch was mixed with 300 ml distilled water, 58.91 g of tosyl chloride, and 64.70 ml triethylamine under stirring for 24 hours at room temperature. The mixture was then filtered off and washed with distilled water followed by ethanol. Then the filtered product was dried at 40oC for 24 hours. Second, the toslylated starch was carried out by mixing the 50 g of tosylated starch powder with 9.1 g of sodium azide in water/ethanol (50/50 v/v) for 48 hours at 75°C. After that, the reaction was cooled down to room temperature, followed by the filtration process. The product obtained was subjected to centrifugation process in order to remove water and by-products using distilled water and finalized with ethanol. The starch cake obtained was dried at 40°C for 24 hours to obtain azido- starch powder. The 10% total solid content of azido starch dispersion was prepared by mixing the starch with anchoid, potassium hydroxide and distilled water using ultrasonic homogenizer at 50 Hz for 1 hour. The 0, 0.5 1.0, 1.5 and 2.0 phr of azido-starch loading mixtures were constantly stirred for 2 hours with stirring speed at 270 rpm. The latex mixtures were transferred into a pre-heated water bath at temperature 70°C and stirred at 270 rpm for another 20 minutes for the pre-vulcanization process to occur. The compound obtained was sealed and left for 12 hours at room temperature for the maturation process. The azido starch filled NR (azo-starch/NR) dipped latex films were prepared using the steel plates which were dipped in coagulant for 10 seconds and dried in an oven at 100 °C before cooled at room temperature. The steel plates were then dipped into NR latex compounds for 15 seconds and cured in an air oven for 20 minutes at 100 °C before cooled at room temperature for 24 hours.

2.3 Fourier Transform Infra –Red (FTIR)

Azido starch was characterized by Fourier Transform Infra- Red (FTIR), Perkin Elmer Spectrum One Spectrometer to identify the presence of azide group in the azido starch and compared with control starch. The scan was performed in the range of 550 cm^{-1} to 4000 cm^{-1}.

2.4 Tensile and Tear Tests

Tensile test was conducted by using Instron 3366 according to ASTM D 412-06.The specimens were cut into dumbell shape and the thickness of each test piece was recorded. The crossheads speed was set at 500 mm/min. Five samples were prepared and the average values were calculated and reported. Tear test was conducted by using Instron 3366 according to ASTM D 624-00. The specimens were cut into a crescent shape (Die C) and the thickness of each test piece was recorded. The crosshead speed was set at 200 mm/min. Five samples were prepared and the average values were calculated.

2.5 Swelling Test

For the swelling test, small samples of the dried azo-starch/NR latex films were cut, weighed and placed in glass vials which were filled with toluene. The vials were stored at room temperature for 48 hours. The samples were taken out of the solvent, dabbed gently with a tissue to remove the solvent and quickly weighed. The swelling ratio, network density, rubber-filler interaction was calculated using Equation 1, 2, 3 and 4, respectively. The swelling ratio (Q) is calculated using the equation (1)

$$\text{Swelling ratio (Q)} = \frac{W1-Wo}{Wo} \quad (1)$$

Where W_1 is the swollen weight and W_o is the initial weight.

The crosslink density (v) can be calculated using the equation (2)

$$V= \frac{1}{2Mc} \quad (2)$$

Mc is the molecular weight of polymer between crosslinks calculated using equation (2.1),

$$Mc = \frac{-\rho r V s V r f \frac{1}{3}}{In\,[1-Vrf]+Vrf+\chi Vrf2} \quad (2.1)$$

ρ_r is the density of polymer, V_s is the molar volume of the solvent absorbed, and χ is the interaction paramater, and V_{rf} is the volume fraction of rubber in the solvent swollen filled sample. The rubber-filler interactions are calculated using Cunneen-Russell [12] and Lorentz-Park [13] equations; The Lorentz-Park equation is given by equation (3). Where Q_f and Q_g is the amount of solvent absorbed for filled and unfilled latex films, respectively. Z is the weight fraction of filler, a and b are constant. While the Cunneen-Russell equation is derived from Lorentz-Parks equation (4)

$$\frac{Qf}{Qg} = ae^{-z}+b \quad (3)$$

$$\frac{Vro}{Vrf} = ae^{-z} + b \tag{4}$$

3. Results and Discussion

3.1 Fourier Transform Infra- Red (FTIR)

The analysis of the FTIR spectra of the tosylated starch powder and azido starch powder were given in Fig.1 (a). The azido starch showed an increment of OH groups at 3420-3250 cm^{-1} , 2831 cm^{-1}, 1640 cm^{-1} 1150 cm^{-1}, 1076 cm^{-1} and 1000 cm^{-1} due to the removal of amorphous region on the surface of the starch after tosylation, thus, more C-OH, C-O-C and C-C bonds have been exposed [14]. The peak at 1000 cm^{-1} indicates the presence of C-O bonds. The strong transmittance peak at 1000 cm^{-1} indicates the increase in crystallinity of starch [15]. The FTIR spectrum of tosylated and azidated starch contained the following characteristics peaks: a broad band between 3600 cm^{-1} to 3200 cm^{-1} due to stretching vibrations of O-H groups, C-H aliphatic stretching at peak 2927 cm^{-1}, O-H bending due to water absorbed in an amorphous region of starch at peak 1650 cm^{-1}, 1160, 1081, 1015 cm^{-1} attributed to stretching of C-O bond from the AGU ring [16]. The characteristics peaks of the AGU ring can be seen at 1154, 1081, 1025 and 930 cm^{-1} [17]. The IR spectrum showed a strong absorption peak of the C≡C corresponding to the azide group at 2118 cm^{-1} [18]. The absorption peaks at 1360 cm^{-1} and 1176 cm^{-1} (SO_2) confirmed the presence of tosyl chloride [19]. The primary peaks of NR latex spectrum are at 2961 -2852 cm^{-1} for C-H stretching, 1663 cm^{-1} for C=C stretching, and C-H bending vibrations of –CH2 and deformation of CH_3 at peaks 1438 cm^{-1} and 1375 cm^{-1}, respectively [20].

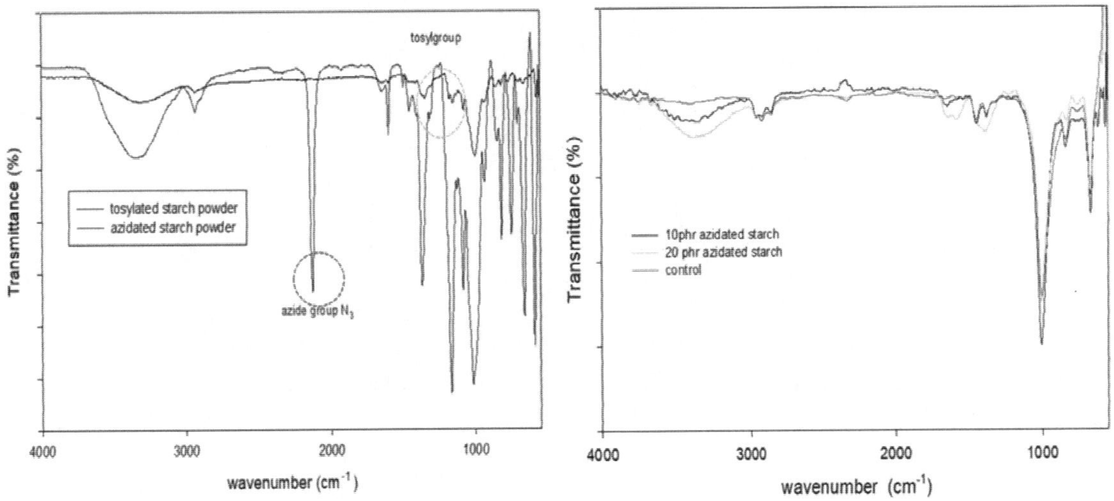

Fig. 1. The spectra of tosylated starch powder and azido-starch powder (b) control NR film and azido-starch- filled- NR films

3.2 Crosslink Density and Swelling

Fig. 2 shows the crosslink density and swelling index of azido starch-filled-NR latex films. The crosslink density is inversely proportional to swelling density which the values provide an idea about the rubber-filler interaction [21]. The results in Fig.2 shows that crosslink density of azido starch- filled NR films are increased significantly as azido starch filler loading increased. Addition of azido starch formed a good chemical crosslink between filler-rubber chains. Swelling index, on

IOP Conf. Series: Materials Science and Engineering **548** (2019) 012015 doi:10.1088/1757-899X/548/1/012015

the other hand, measured the solvent resistance of the specimens which indicates the restricted diffusion of solvent molecules through the voids of the rubber. The higher crosslinks between filler and matrix reduce the available free voids within the matrix, thus, less solvent was absorbed. The interaction between rubber and filler can be assessed using Cunneen-Russell and Lorentz-Parks equations [11]. As the filler loading increases, the Qf /Qg decreases due to the greater interaction between filler- rubber. The Cunneen-Russell and Lorentz- Park plot give positive slopes showed the greater extent of rubber-filler interaction present in the rubber matrix as shown in Fig. 3 [12]. The reinforcing mechanisms might be explained by the catalytic effect of the azido starch on the vulcanization or additional crosslinks are formed by azido starch filler acts as complex crosslinks network [22]. The negative slope of the plot indicates the reinforcing effect by azido starch on the rubber matrix. The ratio Vro / Vrf less than in Fig.3 shows that azido starch has strongly adhered to the surrounding the rubber molecules preventing complete absorption of solvent molecules.

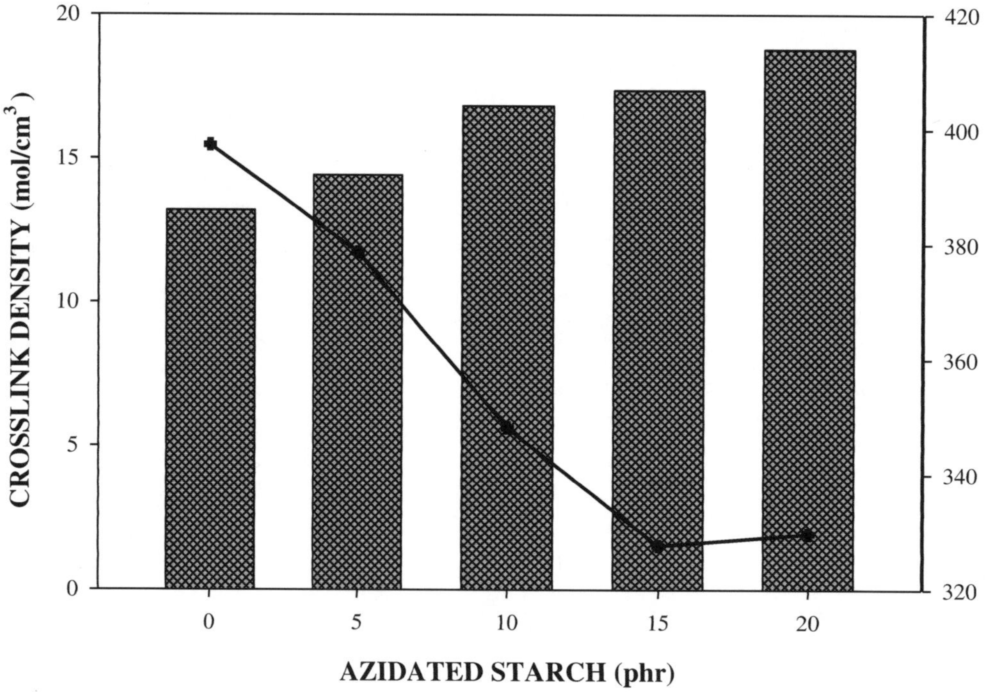

Fig. 2. The crosslink density (Line) and swelling index (bar-chart) of control and azido starch-filled NR latex films.

IOP Conf. Series: Materials Science and Engineering **548** (2019) 012015 doi:10.1088/1757-899X/548/1/012015

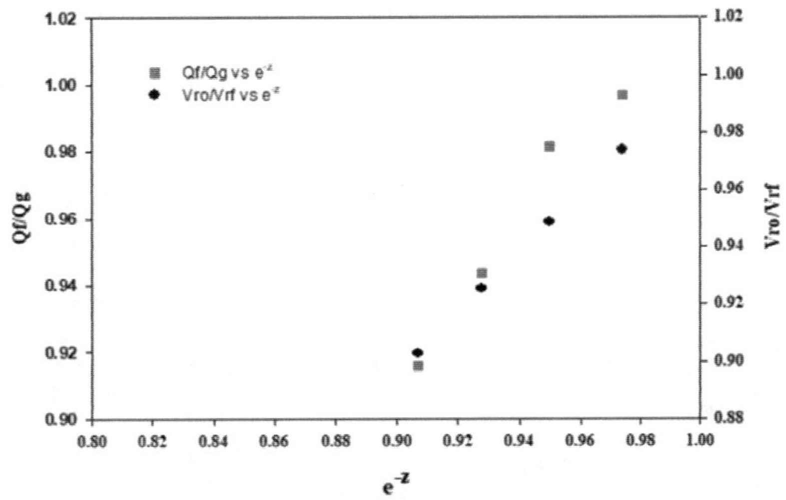

Fig. 3. The Lorentz-Park and Cunneen-Russell plots of azido starch-filled NR films

3.3 Tensile Properties

Fig. 4 shows the effect of incorporation of azido starch on tensile properties of NR latex films. According to Fig. 4, the tensile strength of azido starch-filled NR films increased at 0.5 and 1.0 phr compared to unfilled (control) sample but slightly decreased from 1.5 to 2.0 phr. It can be observed that the optimum tensile strength of the samples was at 1.0 phr. The increase in tensile strength for azido starch filled NR films was attributed from strong filler-rubber interaction and compatibility of NR latex and starch after surface modification of the starch. The azido starch capable of binding active chemical group in latex compounds through the formation of chemical bonding between starch and the NR latex. Thus, improve the interfacial adhesion and compatibility between the starch filler and NR latex. This leads to improve mechanical properties and better stress distribution between filler and latex particles during tensile stress application. Elongation at break (EB) in Fig. 5 slightly reduced with increasing the azido starch loading. Increased in azido starch loading has increased the stiffness of NR latex films hence, reduce its flexibility. The NR latex films become rigid due to increase in non-deformable characteristic of natural filler and less deformable rubber part in the NR films. The opposite trend was observed for tensile modulus (M100 and M300) which shown an increasing trend with increasing azido starch loading as shown in Fig. 6. Moduli of rubber matrix was increased by the enhancement in the surface activity of the fillers, the good dispersion of the filler in the rubber matrix and also better rubber-filler interaction [23]. The chemical surface modification of starch has improved the rubber-filler interaction. The increased of azido starch loading has reduced the chain mobility and increased the rigidity of the NR latex films. Modulus directly indicates the crosslinking density of the NR latex films. These changes related to changes in the main chain scission and crosslink modifications of the original crosslink structure. The crosslinks and bonding between NR latex and azido starch increased with the increment in the azido starch concentration region in the rubber matrix [24].

IOP Conf. Series: Materials Science and Engineering **548** (2019) 012015 doi:10.1088/1757-899X/548/1/012015

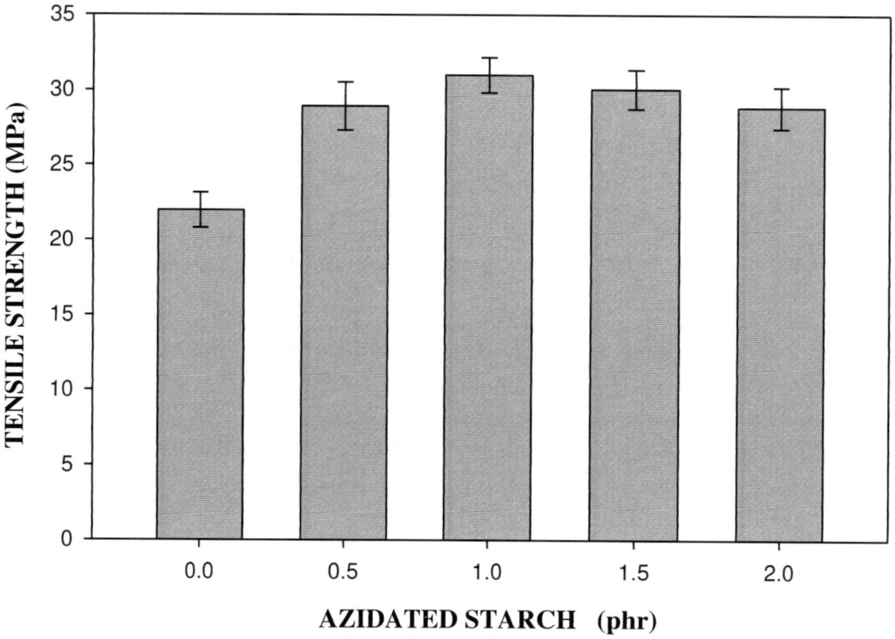

Fig. 4. The tensile strength of azido- starch- filled NR films

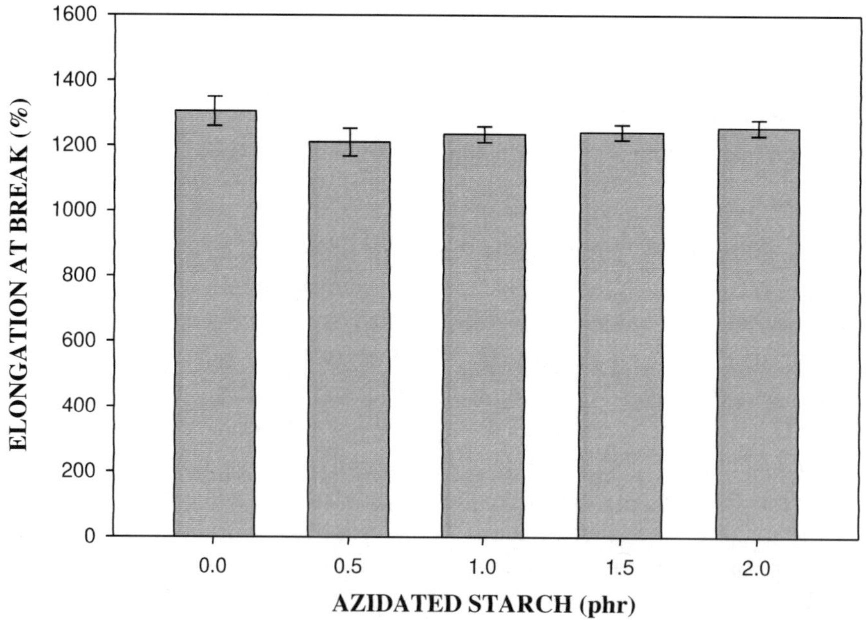

Fig. 5. The elongation at break of azido-starch-filled NR films

IOP Conf. Series: Materials Science and Engineering **548** (2019) 012015 doi:10.1088/1757-899X/548/1/012015

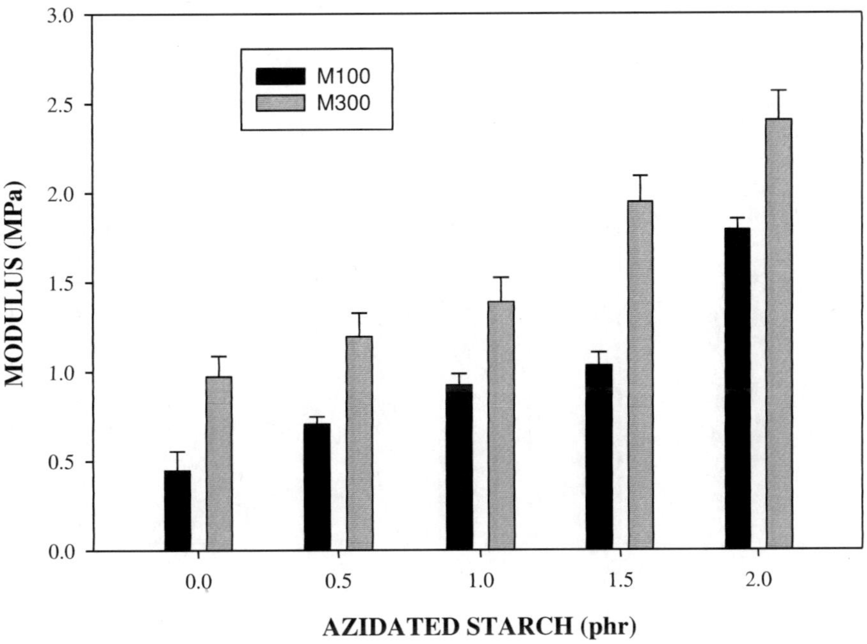

Fig. 6. The modulus at 100% elongation (M100) and at 300% elongation (M300) of azido-starch- filled NR films

3.4 Tear Strength

Fig. 7 shows that the addition of azido starch loading has increased the tear strength of the NR latex films. The tear strength of the NR latex films increased with increasing azido starch loading but abruptly decreased at 1.5 phr azido starch loading. The tear strength in azidated starch-filled NR latex films was attributed by the ability of azido starch to form strong resistance towards crack and propagation where the rubber macromolecules to deviate the crack propagation during the tear test. The azidation surface modification on the starch particle acts as a barrier to the propagation of microcracks, thus provide higher tear strength. The presence of azido starch also contributes to a deviation of the tear path from a straight line, so that the tear becomes rougher which contribute to a higher tear strength. The tear strength also was largely dependent on the stiffness of the specimens, on other words, the higher the crosslinks density of the films, the higher the tear strength [12].

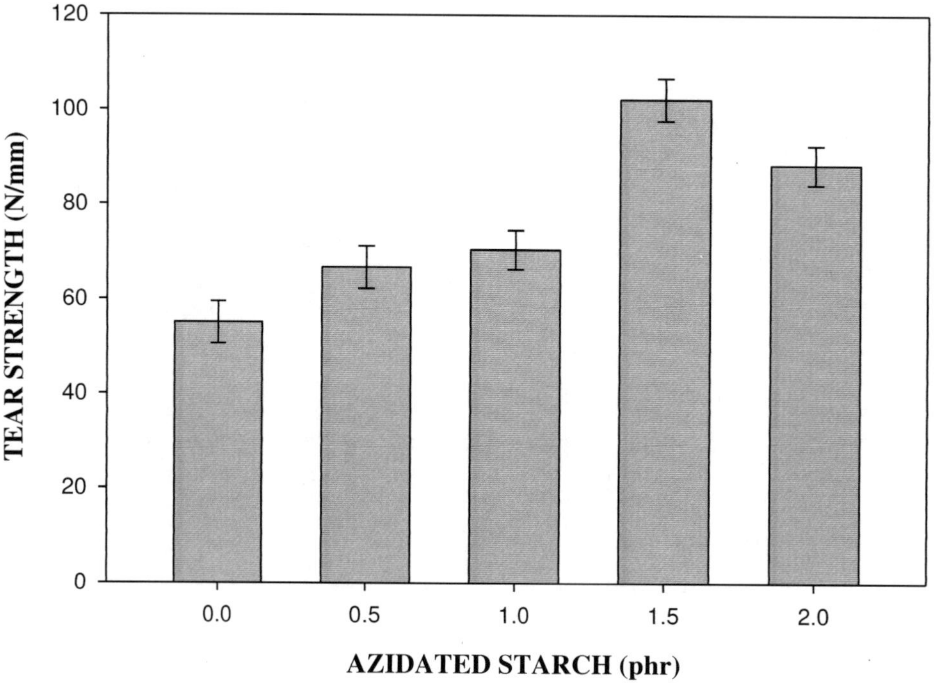

Fig. 7. The tear strength of azido- starch- filled NR films

4. Conclusion

The azidation of sago starch via tosylation and azidation pathway of azidated sago starch overall has shown an increase in the mechanical properties of NR latex films. The tensile strength of azo-starch/NR films was optimum at 1.0 phr azido-starch loading and start to decrease at 1.5 phr azido-starch loading. These were due to increase in modulus which signified the increase in stiffness of the samples as an increase in azido-starch loading. The increased in the stiffness of azo- starch/NR films indicate the presence of filler-rubber crosslinks in addition to chemical crosslinks between NR chains. The greater extent of crosslinks in the rubber chains has reduced the ability of the azo-starch/NR films to resist breakage upon strain, thus, reduced in elongation at break of the samples. The evidence of increased crosslinks and the extent of rubber-filler interaction proved theoretically.

5. Acknowledgement

The authors would like to acknowledge the School of Materials and Mineral Resources Engineering of the Universiti Sains Malaysia for support and research facilities. This work is supported by the Bridging Research Grant (Grant No. 304 / PBAHAN / 6316180).

IOP Conf. Series: Materials Science and Engineering **548** (2019) 012015 doi:10.1088/1757-899X/548/1/012015

References

[1.] Izmar M H et al 2012 Compos. Part. B. Eng. **43**, 2746–2750

[2.] Misman M A and Azura A R 2018 The Mechanical Properties Enhancement of Biodegradable Aminofunctional Starch/Carboxylate Nitrile Butadiene Rubber Latex Films Preprint gr-qc/201804

[3.] Misman M A et al 2016 Procedia Chem. **19**, 770

[4.] Whistler R and Daniel J 2009 Molecular Structure of Starch (Starch: Chemistry and Technology vol 3) ed R Whistler (Oxford: Elsevier) chapter 6 pp.153-160

[5.] Alcazar-Alay S C and Meireles M A A 2015 Food Sci. Technol. **35** 215

[6.] Song D et al 2009 J. Biobased Mater.Bioenergy **3** 321

[7.] Rachtanapun P et al 2012 Int. Food Res. J. **19** 923

[8.] Shey J et al 2006 Carbohydrate. Polymers. **65** 529

[9.] Song D et al 2009 Appl.Polym.Sci. **113** 3019

[10.] Reddy N and Yang Y 2010 Food Chem. **118** 702.

[11.] Elchinger P et al 2014 Ind.Eng.Chem.Res. **53** 7604

[12.] Rice A F A 2001 Raw Mater. Appl **5** 242

[13.] John M J et al 2009 Lignocellulosic Fiber Reinforced Rubber (Natural Fibre Reinforced Polymer Composites : From Macro to Nanoscale) ed T Sabu (USA:Old City Publishing)chapter 10 pp 252–269

[14.] Masina N 2017 Carbohydr. Polym. **157** 1226

[15.] Sevenou O et al J R 2002 Int. J. Biol. Macromol **31** 79

[16.] Bastioli C et al 2013 Starch (Bio-Based Plastics:Materials and Applications) ed S Kabasci and C Stevens (West Sussex:Wiley & Sons) chapter 2 pp. 9-28

[17.] Apopei D F et al 2012 Digest Journal of Nanomaterials and Biostructures **7**707

[18.] Elchinger P et al 2011 Polymers **3** 1607

[19.] Heinze U 2001 New Highly Functionalized Starch Derivatives (Biorelated Polymers: Sustainable Polymer Science and Technology) ed E Chiellini (USA:Springer)chapter 17 pp. 205-210

[20.] Panrat K et al 2012 Procedia. Chem **4** 322

[21.] Barkoula N M et al 2008 Polym. Compos **16** 101

[22.] Taylor G L and Boonstra B B 1965 Rubber Chem. Technol **38** 943

[23.] Egwaikhide A P and Okiemen F E 2013 Sci. J. Chem. **1**50

[24.] Liu C et al 2008 Polymer (Guildf) **49** 2176

The potential use of immobilized *Bacillus cereus* ATCC 14579 for biodegradation of natural rubber latex film added with *Metroxylon Sagu* pith form

N.SAdzami[1], A.R. Azura[1] and H.A. Tajarudin[2,3]

[1] School of Materials and Mineral Resources Engineering, Universiti Sains Malaysia, 14300 Nibong Tebal, Pulau Pinang, Malaysia

[2] School of Industrial Technology, Universiti Sains Malaysia, 14300 Nibong Tebal, Pulau Pinang, Malaysia

[3] Solid Waste Management Cluster, Science and Engineering Research Centre, Universiti Sains Malaysia, 14300 Nibong Tebal, Pulau Pinang, Malaysia

E-mail: azan@usm.my

Abstract. The previous study reported that Bacillus cereus ATCC 14579 able to degrade natural rubber latex films added with Metroxylon Sagu pith waste form (MSPW/NR) and natural rubber (NR) latex films as main source of carbon and energy. Some attempts have been made to degrade them in the newly batch systems rather than disposing of them in landfill. Immobilization technique was performed to adsorb Bacillus cereus onto modified activated carbon. This work aims to study the potential of immobilized Bacillus cereus for biodegradation of MSPW/NR. The results revealed that 30% nitric acid for 2 hours impregnation time showed significant influence on Bacillus cereus adsorption. The highest adsorption capacity of Bacillus cereus was found to be 72.8% under its optimum conditions. The weight loss of MSPW/NR by 12.843 % were obtained after 14th days biodegradation period. Thus, immobilization of Bacillus cereus ATCC 14579 has potency to provide a biotechnological solution to the waste rubber and latex disposal problem.

1. Introduction

The increasing of various diseases outbreak across international borders are making people more concern about their healthcare awareness, such as good personal hygiene. This awareness was identified as one of the main drivers that contribute to the overall production of disposable rubber gloves around the world. However, in recent years, the waste disposal problem due to the growing consumption of latex gloves has created a demand for scarce landfill space and environmental healthcare [1,2]. Many studies have been made to encounter this problem such as the production of biodegradable gloves [3,4] and identification of rubber-degrading bacteria to accelerate biodegradation of gloves [5-11]. However, the space of landfill getting smaller by time to time.

As far as the authors are aware, there is no published analysis of NR products degradation using identified rubber-degrading bacteria. Previously, we managed to identify Bacillus cereus ATCC 14579 as the most effective rubber-degrading bacteria for biodegradation of natural rubber latex films added with *Metroxylon Sagu* pith waste form (MSPW/NR) [12]. It seemed of interest to create a batch system by immobilizing *B.cereus* onto activated carbon (AC) for the degradation of disposable rubber gloves. The application of this process makes bacteria able to reuse for the next biodegradation batch cycle, reduce the odor as well as site-specific targeted method.

Thus, the main objective of this study was to investigate the potential of immobilized Bacillus cereus ATCC 14579 for MSPW/NR latex films. In this study, the commercial AC were performed the chemical modification to enhance functional groups which increase the immobilization interaction of *B.cereus* onto the modified activated carbon (MAC). The immobilized *B.cereus* has tested its functionality towards biodegradation of MSPW/NR latex films as their sole carbon and energy sources.

2. Materials and Methods

Bacillus cereus ATCC 14579 previously isolated from buried MSPW/NR films and was used throughout this study. The culture was maintained on nutrient agar (Merck, Germany) slants and stored at 4°C. During the biodegradation process, mineral salts medium (MSM) was used in this study. The composition of MSM containing (g L^{-1}) 2.0 $NaNO_3$, 0.5 $MgSO_4$, 0.5 KCl, 0.01 $Fe2(SO_4)_3.H_2O$, 0.14 KH_2PO_4, 1.2 K_2HPO_4 and supplemented with 0.02 yeast extract [13]. MSPW/NR latex films were used as sole carbon and energy for culture.

Meanwhile, the modification of the coconut shell based commercial granular AC purchased from Bendosen Laboratory Chemicals, Malaysia was washed with deionized water to remove fine particles and dried at 105°C for 24 h. About 5 g of AC was impregnated with 50 mL of nitric acid (HNO_3) at 30 wt. % concentration in 250 ml conical flask with a reflux condenser, respectively. The samples were placed on a hot plate for the reflux process. The impregnation time was tested at 2 h. After cooling, MAC was separated and washed several times with deionized water until the pH of the filtrate was constant. The washed MAC were dried at 105°C in a hot air oven for 24 h. The MAC was then sieved through a 1-5cm sieve. About 3cm particle size MAC was used as adsorbent and isolated *B.cereus* act as adsorbate in this study.

The MAC was used *B.cereus* was cultivated in nutrient broth on 150 rpm, 30°C for 6 h. The culture concentration OD 1.5 was then centrifuged at 4000 rpm for 5min to recover the cells. The pellet was washed three times with sterile distilled water and was suspended in 25 ml of 1% Sodium chloride (NaCl) (Merck, Germany), pH 7.0 in 100 ml flask. About 2g of MAC was added into the flask. The flask was agitated at 45 rpm, 30°C for 24 h [14]. The MAC was separated and the final concentration of culture was monitored by turbidometry in terms of optical density (OD) using DR 2800 spectrometer at 600 nm wavelength. Scanning Electron Microscope (SEM) analysis was done to observe the attachment of *B.cereus* onto MAC. The MAC which present immobilized *B.cereus* was adhered on a sample stage and then coated with gold before subjected to U-8000, Hitachi Co. Ltd. Japan. The adsorption capacity was used to determine the adsorption of *B.cereus* onto MAC was calculated by following Equation 1,

$$\text{Adsorption capacity (\%)} = [(B_i - B_f)/B_f] \times 100\% \qquad (1)$$

where B_i is the initial concentration of *B.cereus* and B_f is the final concentration of *B.cereus* after 24 h immobilization process.

The biodegradation studies of MSPW/NR and NR latex films were conducted using immobilized *B.cereus* onto MAC. At first, 1 g MSPW/NR and NR latex films were first dissolved in 100 ml of acetone for pre-treatment process. The process left overnight under room temperature. About 300 mg of treated films were mixed with 2 g of immobilized *B.cereus* in 500 mL flask. The flask was filled with 150 ml MSM and cultivated on 45 rpm for 14 days at 30°C. Both studies were performed with three replicates and control. After 14th days of cultivation, the films were rinsed and dried at 40°C until constant weight was recorded. The weight loss of films was calculated based on equation 2,

$$\text{Weight loss (\%)} = [(W_i - W_f)/W_i] \times 100\% \qquad (2)$$

where W_i is the initial dry weight of the sample and W_f is the dry weight of sample after biodegradation.

For storage stability and reusability of immobilized *B.cereus*, about 1 g of immobilized *B.cereus* was added onto 50 ml PBS solution in a 100ml flask. For storage stability study, the samples were stored at 4°C for 30 days. The biodegradation process was repeated for 5 additional cycles to observe the reusability of immobilized *B.cereus*. The bacteria viability of all studies was observed by DR 2800 spectrophotometer.

3. Results and Discussion

The chemical modification of activated carbon procedure is a crucial step of the immobilization of *B.cereus*. It was initially attempted to introduce acidic functional groups onto the surface of the activated carbon [15]. In this study, nitric acid was chosen oxidant for wet oxidation process to create oxygen functional groups on the surface of commercial AC. Previously, 30% w/v HNO3 solution was self-reflux for 2 h impregnation time with the optimum conditions were set at pH 7.0, 45 rpm, 30°C for 24 h. Thus, the highest adsorption capacity was found to be 72.8%.

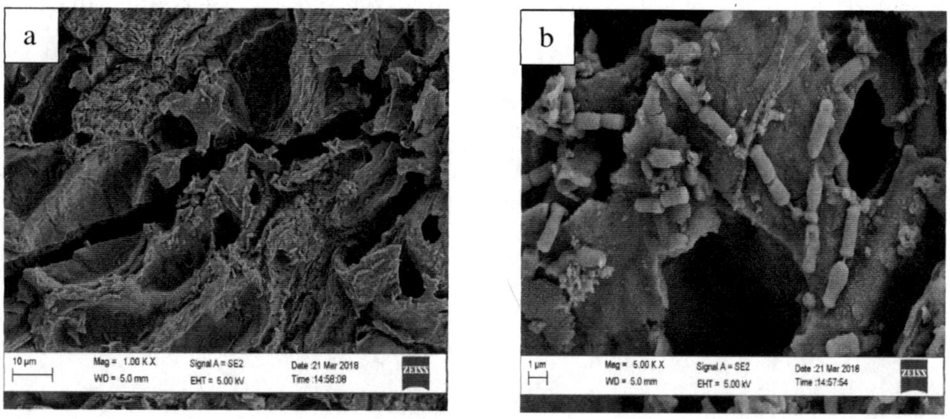

Fig. 1. SEM micrograph, (a) 1500 x MAC and (b) 5000 x MAC after dominant *Bacillus cereus* ATCC 14579 immobilization.

The MAC and immobilized *B.cereus* onto MAC were examined under SEM to observe the surface characteristic of the samples. As shown in the Fig. 1(a) shows the MAC showed the presence of numerous pores and rough surfaces where there is a possibility for the adsorption process. Fig. 1(b) shows *B.cereus* was dominant retained to the surface of MAC. Since *B.cereus* on the surface was direct contact to MSPW/NR latex films as substrates, thus it is able to degrade films as its substrate to grow and proliferate.

During the study, immobilized *B.cereus* was added into MSM media with MSPW/NR latex films in the batch system. The previous study showed *B.cereus* can utilize both starch and rubber particles as sole carbon and energy sources for their growth [12]. The conventional weight loss method is applied to prove the biodegradation of films using immobilized *B.cereus*.

The weight loss of MSPW/NR and NR latex films during incubation with immobilized *B.cereus* are shown in Fig. 2. After 14 days incubation, the weight loss in percentage of films is recorded at 12.843%. The control sample without inoculation, the weight loss of films was 1.418%. Meanwhile, the weight loss of NR latex films is 7.273% and its control is calculated at 1.673%.

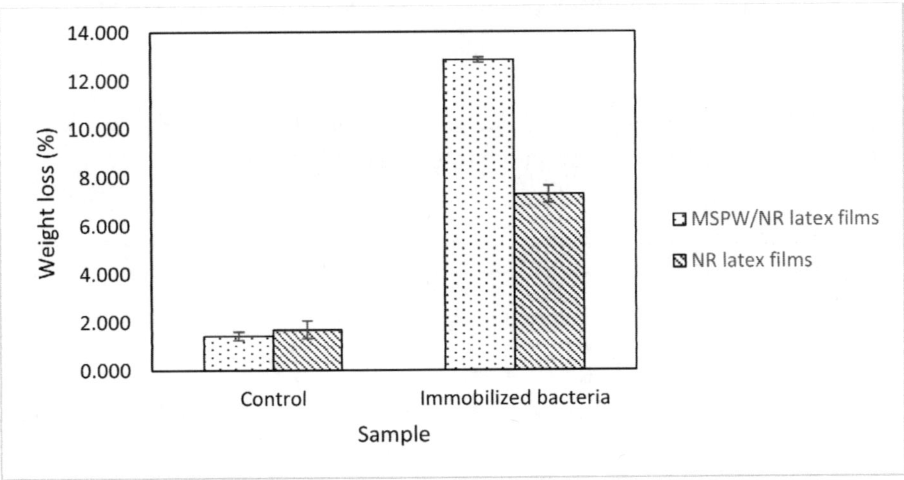

Fig. 2. The weight loss in percentage of MSPW/NR latex films with immobilized *Bacillus cereus* ATCC 14579 after 14th days of biodegradation periods.

The results showed the immobilized *B.cereus* able to improve the biodegradation process of MSPW/NR latex films with compare to NR latex films. Few researchers agreed that the introduction of bio-filler onto latex formulation products can accelerate the biodegradation properties of the products [16-18].

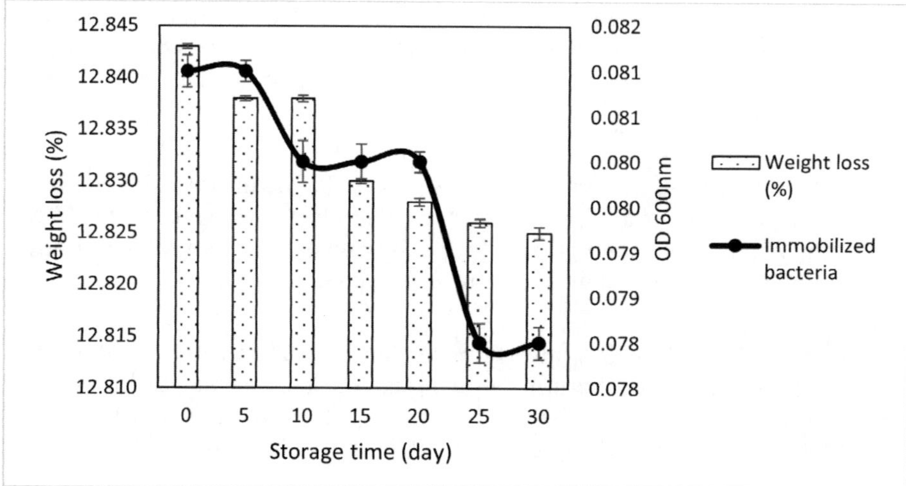

Fig. 3. The weight loss and cell viability of immobilized *Bacillus cereus* ATCC 14579 at 4°C for 30 days.

The viability of bacteria during long-term storage and its reusability were significant for practical application and commercialization of bacterial immobilization.

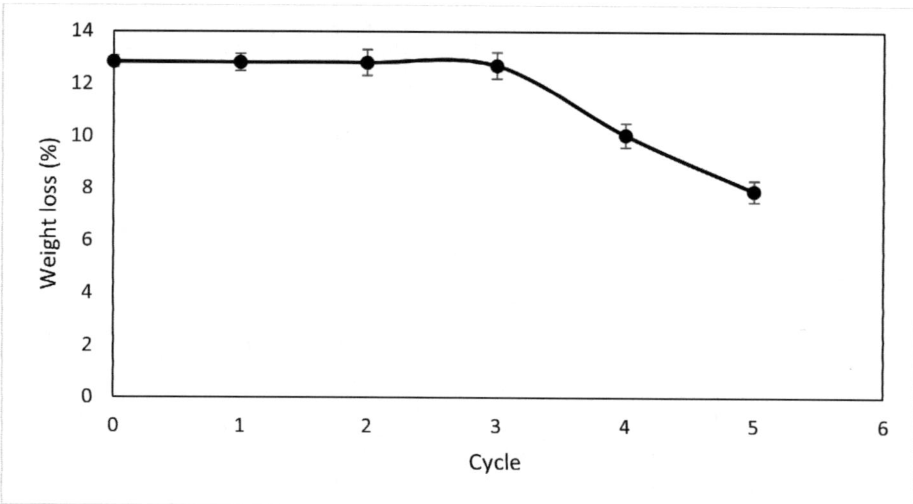

Fig. 4. The reusability of immobilized *Bacillus cereus* ATCC 14579.

The viability of *B.cereus* was examined for the storage period of 30 days. As shown in Fig. 3, the weight loss percentage of films is typically decreased with decreasing cell viability of immobilized *B.cereus*. Initially, the weight loss percentage of films recorded at 12.843% with the cell viability concentration is at OD 0.081. After immobilized bacteria being stored for 30 days at

4°C, the bacteria concentration was decreased slightly at OD 0.078 with the weight loss of films recorded at 10.196%.

For the reusability pattern of the immobilized *B.cereus*, 5 cycles of reuse were tested. The results in Fig. 4 showed the weight loss of films was decreased for higher cycles. After the three cycles of regeneration, weight loss percentage of films was recorded at 10.193%. For fourth regeneration cycles and above, the weight loss percentage reduce to 10.045% and 9.611%. The reduction in the weight loss of films might be due to the detachment of the immobilized bacteria in the washing step. The results of immobilized bacteria reusability suggesting activated immobilized *B.cereus* can maintain their rubber-degrading activity throughout each cycle.

4. Conclusions

The results of the investigation demonstrate that applying the chemical modification of commercial activated carbon under specific conditions able to adsorb Bacillus cereus ATCC 14579 to the maximum state. The SEM micrograph results clearly showed the presence of *B.cereus* on the surface of modified activated carbon. It has been shown that these immobilized *B.cereus* able to utilize MSPW/NR latex films as the sole carbon and energy sources for their growth. The immobilized *B.cereus* was more stable and viable in the developed study. For commercialization, the viability of immobilized *B.cereus* of storage and thermal is stable and desirable. Thus, these results suggested that immobilized on modified activated carbon had a potential for practical application in the treatment of degrading rubber and latex films wastes rather than fills up the landfills.

Acknowledgements

The authors would like to acknowledge the School of Materials and Mineral Resources Engineering of the Universiti Sains Malaysia for support and research facilities. This work is supported by the Research University Grant (Grant No. 1001 / PBAHAN / 814279).

References

[1] Yikmis M and Steinbüchel A 2012 Historical and recent achievements in the field of microbial degradation of natural and synthetic rubber *Appl. Environ. Microbiol.* **78(13)** 4543–4551

[2] Johari A, Alkali H, Hashim H, Ahmed S I and Mat R 2014 Municipal solid waste management and potential revenue from recycling in Malaysia *Modern Appl. Sci.* **8(4)** 37–49

[3] Azura A R, Adzami N S and Tajarudin H A 2017 Utilization of Metroxylan sagu Pith Waste as Biodegradable Filler for Natural Rubber (NR) Latex Films *Solid State Phenom.* **264** 198-201

[4] Afiq M M and Azura A R 2013 Utilization of Starch to Accelerate the Growth of Degrading Microorganisms on the Surface of Natural Rubber Latex Films *J. Chem. Chem. Eng.* **7** 137–144

[5] Chia K H, Nanthini J, Thottathil G P, Najimudin N, Haris M R H M and Sudesh, K 2014 Identification of new rubber-degrading bacterial strains from aged latex *Polym. Degrad. Stab.* **109** 354–361

[6] Hesham A E-L, Mohamed N H, Ismail M A and Shoreit A A M 2015 Degradation of natural rubber latex by new Streptomyces labedae strain ASU-03 isolated from Egyptian soil *Microbiology* **84(3)** 351–358

[7] Ibrahim E M A, Arenskötter M, Luftmann H and Steinbüchel A 2006 Identification of poly(cis-1,4-isoprene) degradation intermediates during growth of moderately thermophilic actinomycetes on rubber and cloning of a functional lcp homologue from Nocardia farcinica strain E1 *Appl. Environ. Microbiol.* **72(5)** 3375–3382

[8] Linos A, Reichelt R, Keller U and Steinbüchel A 2000 A Gram-negative bacterium, identified as Pseudomonas aeruginosa AL98, is a potent degrader of natural rubber and synthetic cis-1,4-polyisoprene *FEMS Microbiol. Lett.* **182(1)** 155–161

[9] Rifaat H M and Yosery M A 2004 Identification and characterization of rubber degrading actinobacteria *Appl.Ecol. Environ Res.* **2(1)** 63–70

IOP Conf. Series: Materials Science and Engineering **548** (2019) 012016 doi:10.1088/1757-899X/548/1/012016

[10] Rose K, Tenberge K B and Steinbüchel A 2005 Identification and Characterization of Genes from Streptomyces sp. Strain K30 Responsible for Clear Zone Formation on Natural Rubber Latex and Poly(cis -1,4-isoprene) Rubber Degradation *Biomacromolecules* **6(1)** 180–188
[11] Subramaniyam R and Vimala R 2012 Solid State and Submerged Fermentation for the Production of Bioactive Substances: A Comparative Study *Intenational Journal of Science and Nature* **3(3)** 480–486
[12] Adzami N S, Azura A R and Tajarudin H A 2018 Biodegradation of natural rubber latex film added with Metroxylan Sagu pith form by Bacillus cereus ATCC 14579 *Malays. J. Microbiol.* **14(2)** 102-107
[13] Heisey R M and Papadatos S 1995 Isolation of microorganisms able to metabolize purified natural rubber *Appl. Environ. Microbiol.* **61** 3092 – 3097
[14] Rivera-utrilla J, Bautisa-toledo I, Ferro-garcia M A and Moreno-castilla C 2003 Bioadsorption of Pb (II), Cd (II), and Cr (VI) on activated carbon from aqueous solutions *Carbon* **41** 323–330
[15] Yin C Y, Aroua M K and Daud W M A W 2007 Review of modifications of activated carbon for enhancing contaminant uptakes from aqueous solutions *Sep. Purif. Technol.* **52(3)** 403–415
[16] Awg-adeni D S, Aziz S A, Bujang K and Hassan M A 2012 Recovery of Glucose from Residual Starch of Sago Hampas for Bioethanol Production *BioMed Res. Int.* 1 - 8
[17] Chew T Y and Shim Y L 1993 Sago Processing Waste in Kuala Lumpur: Waste management in Malaysia- Current Status and Prospects for Bioremediation *Ministry of Science, Technology and the Environment* 159 - 167
[18] Uthumporn U, Wahidah N and Karim A A 2014 Physicochemical Properties of Starch from Sago (*Metroxylon Sagu*) Palm Grown in Mineral Soil at Different Growth Stages *Mater. Sci. Eng.* **62(1)** 012026

IOP Conf. Series: Materials Science and Engineering **548** (2019) 012017 doi:10.1088/1757-899X/548/1/012017

Effect of Latex Compound Dwell Time for the Production of Prototyped Biodegradable Natural Rubber Latex Gloves

M F Ab Rahman[1], A Rusli[1], M T Kuwn[1] and A R Azura[1]

[1] School of Materials and Mineral Resources Engineering, Universiti Sains Malaysia, 14300 Nibong Tebal, Pulau Pinang, Malaysia

E-mail: srazura@usm.my

Abstract. The production of prototyped biodegradable natural rubber (NR) latex gloves with the addition of sago starch as a fillers in latex compounding using Lab Scale Batch Dipping Machine has been investigated. As biodegradable gloves, the progress of biodegradation of NR latex gloves is also being determined through weight loss measurement test. Results showed the optimum dwell time of latex dipping was achieved at 6 sec with optimum tensile properties and within normal examination glove thickness. The weight loss of biodegradable NR latex gloves after one month soil burial showed additional of sago starch into latex compounding can accelerate the biodegradation process.

1. Introduction

Rubber gloves are one of the highest productions of consumable and disposable commodity in the global market. The Malaysian Rubber Glove Manufacturers Association (MARGMA) has estimated the global demand for rubber gloves in 2018 is expected to grow at 15% [1]. The continuing increased in demands of rubber gloves will directly contribute to rubber waste solid disposal problem.

In order to solve the increasing of disposal gloves, production of biodegradable gloves might one of the options to overcome this global issue. Biodegradable matter is material capable of being decomposed by bacteria or other biological means. It has the ability to be broken down into smaller, harmless products by way of the action of living organisms [2]. Moreover, the biodegradable NR latex gloves are usually less harmful to the environment than the commercial gloves. It might able to provide soil nutrients necessary for the regeneration of live.

The biodegradable NR latex gloves were produced by introducing bio-material into the NR latex system. According to Afiq and Azura [3], addition of the carbohydrates polymer such as sago satrch as the bio-material can acts as filler through void spaces between the rubber matrix particles. It is much feasible for the microbial activity to occur during the disposal of biodegradable NR latex gloves. The microorganisms will produce specific enzyme that able to decompose the rubber molecules. Hence, it accelerates the rate of decompose of the biodegradable NR latex gloves.

In order to commercialize it as mass production, the production of the biodegradable NR latex gloves with the addition of sago starch as a fillers in latex compounding in larger scale is highly needed. As latex compound dwell time is one of the important parameters in gloves production, the aim of this study is to investigate the effect of latex compound dwell time toward physical properties of biodegradable NR latex gloves. Dwell time is the time interval between former dipping and withdrawal of former from the latex compound. This study is important to assess the relationship between dwell time, thickness and physical properties of biodegradable NR latex gloves.

IOP Conf. Series: Materials Science and Engineering **548** (2019) 012017 doi:10.1088/1757-899X/548/1/012017

2. Materials and Methods

2.1. Materials

NR latex was purchased from Zarm Scientific and Supplies Sdn. Bhd. with the initial properties of dry rubber content (DRC) of 60.8%, total solid content (TSC) of 61.7%, mechanical stability time (MST) of 1020 sec, and volatile fatty acid number (VFA no.) not exceeding 0.2. Meanwhile, sago starch was obtained from Sago Link Sdn. Bhd. with amylose and amylopectin content ranging from 22 - 31.7% and 68.3 - 78%; respectively and other compounding ingredients such as zinc diethyldithiocarbamate, zinc oxide, potassium hydroxide, antioxidant [2,2'-Methylene-bis(4-methyl-6-tert-butylphenol] and sulphur were obtained from Farben Technique (M) Sdn. Bhd. Whereas calcium nitrate [$Ca(NO_3)_2$], acetic acid and sodium hydroxide were purchased from Merck Sdn. Bhd.

2.2. Preparation of NR latex compound and pre-vulcanization process

The preparation of NR latex compound was carried out in accordance with a previous study [3]. All the ingredients were mixed for 2 hours at 270 rpm. The compound was then pre-vulcanized at temperature of 70^{0}C in a water bath and constantly stirred at speed of 270 rpm until chloroform number 2 (degree of vulcanization of pre-vulcanized NR latex) has achieved. The pre-vulcanized latex was left for maturation process at room temperature for 18 hours.

2.3. Production of gloves

The formers (examination gloves type) were cleaned in the acetic acid and sodium hydroxide with the aid of brushes to remove any dirt attached to the formers. Subsequently, the formers were rinsed with the tap water to remove the chemicals. The temperature of acetic acid solution, tap water and sodium hydroxide solution were maintained at 60°C.

The cleaned formers were first dipped into the coagulant tank for 10 sec and dried in the oven at 70^{0}C for 5 minutes. The formers were then dipped into the pre-vulcanized latex tank with different dwell time (3, 6 and 9 sec) followed by drying in the oven at 100^{0}C. After the first 3 minutes in the oven, the formers were taken out for beading followed by drying in oven again until the gloves were cured. The formers were taken out from the oven and the gloves were stripped with the aid of calcium carbonate powder after the formers were cooled.

2.4. Physical properties test

Tensile and tear tests were carried out according to ASTM D 412 and ASTM D 624 respectively by using Instron Machine (Model 3366) with the crosshead speed of 500 mm/min. Five specimen of each sample were used to obtain average values for tensile strength, elongation at break, modulus of elasticity (M100, M300 and M500) and tear strength.

2.5. Crosslink density

The swelling test was carried out according to ASTM D471 where a test piece weighing about 0.2 g was cut from NR latex films. The films were immersed in pure toluene in universal bottle before transferred into a water bath and heated at 40°C for 48 hours. The test piece was taken out, the loose liquid was rapidly removed by blotting with filter paper, and the swollen weight was immediately measured. The volume fraction of latex (V_r) in the swollen network can be calculated by using equation (1).

$$v_r = \frac{W_B/Pr}{(W_B/P_r)+(W_A-W_B/Ps)} \tag{1}$$

where W_B and W_A are weights of NR latex samples before and after toluene immersion. ρ_r and ρ_s are the density of the NR latex (0.9203 g/cm^3) and toluene (0.862 g/cm^3); respectively. The crosslink density [X] was determined by the Flory-Rehner theory in equation (2) [4].

$$[X] = \frac{-[\ln(1-V_r)+V_r+\chi V_r^{\,2}]}{V_s\,(V_r^{\,1/3} - \frac{V_r}{2})} \tag{2}$$

where V_s is the molecular volume of the toluene (105.91 cm^3/mol) and χ is Flory-Huggins NR-toluene interaction parameter [4].

2.6. Biodegradation procedure and weight loss analysis

The gloves were buried in the soil for one month for biodegradation process. Upon collection, the gloves were rinsed and dried at room temperature to a constant weight and the weight was recorded. Weight loss of the samples with time was used to determine the degradation rate of the samples by following Equation 3:

$$\text{Weight loss (\%)} = [(W_i - W_d)/W_i] \times 100 \tag{3}$$

where W_i is the initial dry weight of the sample and W_d is the dry weight of the sample after biodegradation process.

2.7. Physical properties test

Tensile and tear tests were carried out according to ASTM D 412 and ASTM D 624 respectively by using Instron Machine (Model 3366) with the crosshead speed of 500 mm/min. Five specimen of each sample were used to obtain average values for tensile strength, elongation at break, modulus of elasticity (M100, M300 and M500) and tear strength.

3. Results and Discussion

3.1. Mechanical properties and crosslink density analysis

According to Fig. 1(a), it shows that the tensile strength of the NR latex gloves increased with dwelling time. This is attributed to the increased of films thickness due to increase of latex deposited on the former. This observation were augmented by the results of crosslink density shown in Fig. 1(b). The figure shows that 3 sec dwelling time produced the lowest crosslink

IOP Conf. Series: Materials Science and Engineering **548** (2019) 012017 doi:10.1088/1757-899X/548/1/012017

density and the highest was achieved at 9 sec. As the time for film formation is longer, more latex particles be able to settle to form continuous films and this phenomenon were manifested with the increased of crosslinking formed in cured NR latex films. It is in agreement with the work done by Zhao et al. which study the influence of crosslink density on mechanical properties of NR latex [5].

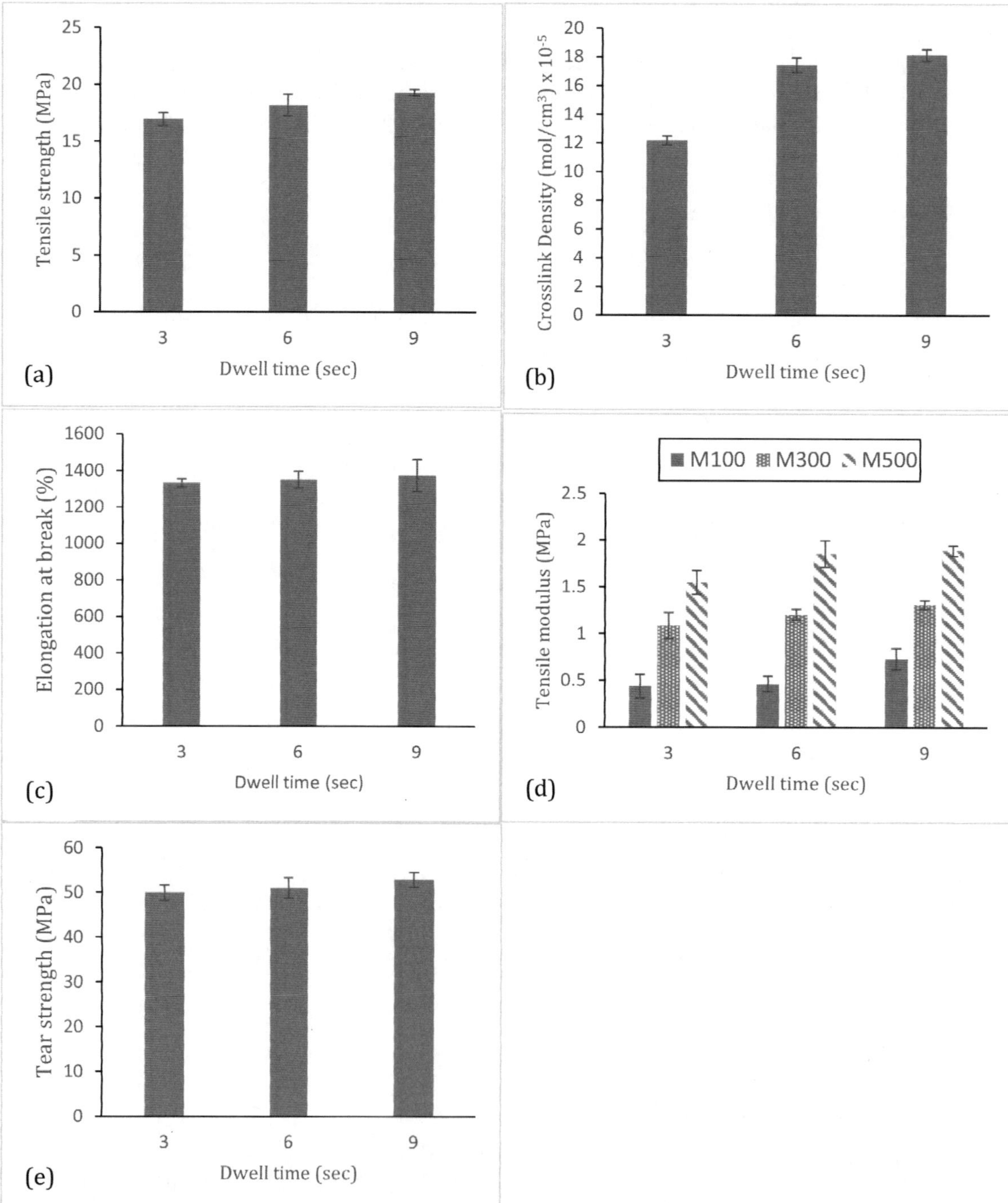

Fig. 1. (a) Tensile strength; (b) crosslink density; (c) elongation at break; (d) tensile modulus; and (e) tear strength of biodegradable NR latex gloves with different latex compound dwell time.

Fig. 1(c) shows the elongation at break of the biodegradable NR latex gloves with respond to different dwell time. It can be seen that the elongation at break slightly increased with increasing of dwell time. As discussed previously, the crosslink density of biodegradable NR latex gloves increased with increasing dwell time. The crosslink joint will provide restriction on the mobility of inter-crosslink chains. In other words, the higher crosslink density leads have to higher strain concentration as more forces needed to pull the specimen [7].

Fig. 1(d) shows the tensile modulus corresponding to different dwell time. It can be seen that the tensile modulus increase with increasing dwell time at 100, 300 and 500% of elongation. It shows similar trend to the tensile strength properties which indicates that the tensile modulus is affected by the crosslink density and the thickness of biodegradable NR latex gloves as discussed earlier. As the crosslink density increase and higher force is needed to deform the biodegradable NR latex gloves resulted in increase of tensile modulus.

Fig. 1(e) shows the effect of dwell time on the tear strength of the biodegradable NR latex gloves. It can be observed that the tear strength increased with increasing of dwell time. A thicker biodegradable NR latex gloves gives a higher strength value than a thin film with the same dosage of vulcanizing ingredients. It is reasonable to expect that a thin biodegradable NR latex gloves is easily ruptured due to incoherent formation of firm due to low amounts of material. It shows similar trend with the finding of Lim et al. who studies in mechanical properties of carboxylated nitrile latex film with varying thickness. They found that tear strength increasing with increased in thickness for carboxylated nitrile latex film [8].

Moreover, the crosslink density also affects the tear strength of the biodegradable NR latex gloves. As longer the dwell time, the higher the crosslink density. Hence higher force is needed to initial the crack and propagation in the specimen resulted in higher tear strength [9].

3.2. Gloves thickness analysis

Based on Table 1, the biodegradable NR latex gloves thickness increased with increasing longer dwell time. The thickness at 9 sec dwell time was the thickest, 0.191 mm, followed by 6 sec dwell time, 0.177 mm and 3 sec dwell time, 0.169 mm. It might be attributed to as the dwell time longer, there was longer time for the NR latex to colloidally de-stabilize on the surface of the hot former. When more NR latex de-stabilized and deposits on the surface of the former, it leads to greater thickness.

For this study, we have used the examination glove typed formers to produce the gloves; hence the right glove thickness need to consider. Normally, the examination gloves has made with thinner materials (0.1–0.18 mm) compared to surgical gloves (0.17–0.3 mm) [10]. Based on Table 1, it can be observed that 9 sec dwell time already exceed the maximum range for common examination glove's thickness. Besides, the aforementioned results between 6 and 9 sec dwell time also showed comparable physical strength and crosslink density. Thus, the dwell time at 6 sec was preferred to be the best dwell time for production of biodegradable NR latex gloves.

Table 1. Average gloves thickness at different dwell time.

Dwell time (sec)	Average gloves thickness (mm)
3	0.169
6	0.177
9	0.191

3.3. Mass loss analysis

Further study has been conducted to monitor the biodegradability of thin NR latex gloves. Fig. 2 shows the effect of biodegradation on biodegradable NR latex gloves on the mass loss assessment. For one month, the mass loss of biodegradable (sago starch filled) NR latex gloves was 11.33% whereas sago starch unfilled NR latex gloves was 7.19%. The mass loss might attribute to the consumption of sago starch by the microorganism in the soil. Due to microbial consumption of carbon in poly-isoprene chain in biodegradable NR latex gloves and the enzymatic cleavage of glucosidic links in sago starch, size and number of empty voids formed increased. This contributes to the increase of mass loss of biodegradable NR latex gloves as biodegradation process occurred [11].

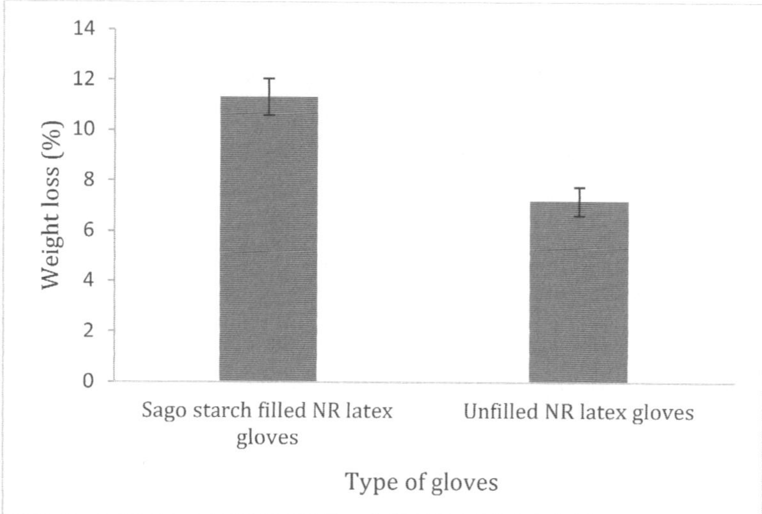

Fig. 2. Biodegradation rate of the gloves after one month of biodegradation periods.

4. Conclusions

In conclusion, the dwell time at 6 sec was preferred as it showed high mechanical properties with a thinner NR latex gloves film. The incorporation of sago starch into latex compound has proven can accelerate the biodegradation of thin NR latex gloves.

Acknowledgments

This work was financially supported by the School of Materials and Mineral Resources Engineering, Universiti Sains Malaysia from research grant (grant no: 203/PBAHAN/6740016).

IOP Conf. Series: Materials Science and Engineering **548** (2019) 012017 doi:10.1088/1757-899X/548/1/012017

References

[1] Rubber Asia 2019 World rubber glove industry; steady growth on rising demand [online] Available at https://rubberasia.com/2018/09/17/world-rubber-glove-industry-steady-growth-rising-demand/ [accessed on March 18th, 2019]

[2] Joutey N T, Bahafid W, Sayel H and Ghachtouli N E 2013 Biodegradation: Involved microorganisms and genetically engineered microorganisms. IntechOpen

[3] Afiq M M and Azura A R 2013 *J. Chem. Chem. Eng.* **7(2)** 137-144

[4] Ho C C and Khew M C, 1999 *Langmuir* **15** 6208-6219

[5] Zhao F, Bi W and Zhao S 2011 *J.Macromol. Sci., Part B: Phys.* **50(7)** 1460-1469

[6] Roslim R and Amir Hashim M Y 2010 *J. Rubb. Res.* **13(2)** 125-138

[7] Zhao J, Yu P and Dong S 2016 *Mater.* **9(4)** 234-243

[8] Lim H M, Vivayganathan K and Amir-Hashim M Y 2012 *J. Rubb. Res.* **15(3)** 167-178

[9] Nuraya A S S, Baharin A, Azura A R, Hakim M H M R, Mazlan I, Adnan M and Nooraziah A A 2012 *J. Rubb. Res.* **15(2)** 124-140

[10] Mäkelä E A, Vainiotalo S and Peltonen K 2003 *Ann. Occup. Hyg.* **47(4)** 313-323

[11] Afiq M M and Azura A R 2013 *Int. Biodeterior. Biodegrad.* **85** 139-149

International Rubber Conference IOP Publishing
IOP Conf. Series: Materials Science and Engineering **548** (2019) 012018 doi:10.1088/1757-899X/548/1/012018

Studies on the Influence of Mixed Culture from Buried Soil Sample for Biodegradation of Sago Starch Filled Natural Rubber Latex Gloves

M F Ab Rahman[1]**, A Rusli**[1]**, N S Adzami**[1] **and A R Azura**[1]

[1] School of Materials and Mineral Resources Engineering, Universiti Sains Malaysia, 14300 Nibong Tebal, Pulau Pinang, Malaysia

E-mail: srazura@usm.my

Abstract. The addition of sago starch into natural rubber (NR) latex compounds was aimed to accelerate the degradation process of the end products including gloves as an alternative to reduce solid waste disposal problem. In this research work, the degradation of sago starch filled natural rubber (SS/NR) latex gloves by a mixed culture from buried soil sample was investigated. The Gram staining test of mixed culture showed the Gram-negative bacteria are the dominant species as an effective bacterium to degrade the SS/NR latex gloves. The starch hydrolysis test was performed to confirm the existence of starch-degrading bacteria in mixed culture that consume sago starch from the gloves. Based on the biodegradation rate results obtained, the existence of starch-degrading bacteria together with rubber-degrading bacteria were observed, which accelerate the biodegradation of SS/NR latex gloves by 53.68%. Meanwhile, the biodegradation rate for sago starch unfilled natural rubber (control) latex gloves were only 50.31%. Thus, the addition of sago starch has shown an increased in the biodegradation rate, which can be used as alternative method to overcome the solid waste disposal problem.

1. Introduction

Latex gloves have been designed for single use, which makes their disposal becomes the worldwide solid waste problem. The rubber waste can be eliminated by either burned or used as landfills, but these processes caused serious pollution [1]. One of the alternative ways to overcome this environmental problem is by addition of natural biodegradable fillers such as starch base materials into the gloves formulation [2]. Although natural rubber (NR) latex is able to biodegrade, the incorporation of starch can accelerate the microorganism activities, thus able to decompose the composites effectively. It is reported that the bacteria as well as *fungi* have the ability to degrade the rubber but the process is slow [3].

Sago starch commonly used as filler or matrix to produce biodegradable composites due to the incorporation of the starch not only assists in the disintegration of polymer composites but also inexpensive and naturally abundant [4]. Polysaccharide chain of amylose and amylopectin from the starch is consumed by microbes in the soil to produce enzymes such as amylases and glucosidases that break the glucosidic linkages in starch [5]. Furthermore, it was suggested that microbes favored to consume simpler glucosidic linkages in starch during biodegradation periods compared to polyisoprene molecules of NR latex alone [3].

In biodegradation process, the rubber-degrading bacteria consume NR as the sole carbon and energy sources for their growth. The rubber has to be cleaved extracellularly into low-molecular-weight compounds; so that it can be transported across the cell membrane and used for metabolism [6]. The structure of NR, which has one functional group of carbon-carbon double bond per isoprene unit, is the primary target of NR-cleaving enzymes called rubber oxygenases

IOP Conf. Series: Materials Science and Engineering **548** (2019) 012018 doi:10.1088/1757-899X/548/1/012018

[7]. The secretion of rubber oxygenases by rubber-degrading bacteria is used to initiate the break-down of the polymer and to consume the generated cleavage products as a carbon source.

To date, the previous studies have shown most of microorganism only degrade latex samples without fillers [8-9] while the studies about the effect of the integration of sago starch into latex compounding on microbial degradation are still limited. Thus, this paper aims to investigate the influence of mixed culture from the buried soil sample on the biodegradation of sago starch filled natural rubber (SS/NR) latex to confirm the existence of starch-degrading bacteria that consumed the sago starch from the gloves which accelerate the biodegradation process.

2. Materials and Methods

2.1. Materials

NR latex with the initial properties of dry rubber content (DRC) of 60.08wt%, total solid content (TSC) of 61.00wt%, mechanical stability time (MST) of 1140 s, and volatile fatty acid number (VFA no.) not exceeding 0.2 was purchased from Zarm Scientific and Supplies Sdn. Bhd. (Penang, Malaysia). Meanwhile, sago starch dispersion was produced in-house and other compounding ingredients such as diethyldithiocarbamate (ZDEC), zinc oxide (ZnO), potassium hydroxide (KOH), 2,2'-Methylene-bis (4-methyl-6-tert-butylphenol) and sulfur were obtained from Farben (Malaysia) Sdn. Bhd. In addition, the Gram staining kit such as crystal violet, Gram's iodine, ethyl alcohol and safranin was purchased from Pro-Lab Diagnostics Inc. (United Kingdom).

2.2. Preparation of NR latex compound and gloves

The NR latex compound was prepared with 31.00 ± 0.20wt% of TSC. The preparation of NR latex compound was carried out in accordance with a previous study [3]. The compound was pre-vulcanized at a temperature of 70°C and the process was stopped until the degree of vulcanization has achieved chloroform number 2. Then, the compound was left overnight under room temperature for the latex maturation process. The formers with pre-layered of calcium nitrate were dipped into the maturated compound and cured in the oven for 25 minutes at 110°C. The gloves were stripped with the assistance of talc powder to avoid the gloves from sticking.

2.3. Biodegradation procedure and weight loss analysis

The gloves were buried in the soil for 5 months for biodegradation process. Upon collection, the films were rinsed and dried at room temperature to a constant weight and the weight was recorded. Weight loss of the samples with time was used to determine the degradation rate of the samples by following Equation 1:

$$\text{Weight loss } (\%) = [(W_i - W_d)/W_i] \times 100 \quad (1)$$

Where W_i is the initial dry weight of the sample and W_d is the dry weight of the sample after biodegradation process.

2.4. Starch hydrolysis test

1 g of each soil buried control and SS/NR latex gloves were mixed with 50 ml of MSM in a 250 ml flask. The flasks were agitated at 150 rpm for 3 days at 30°C. Subsequently, 100 µL of each sample was spread on latex overlay agar. The plates were then incubated at 30°C for 3 days. To detect starch-degrading bacteria, the most abundant dissimilar morphology of the mixed culture from the incubated plates was transferred to a fresh starch agar for the iodine test. The iodine test was conducted by flooding the plates with Gram's iodine for 10 minutes [10].

The latex overlay agar used in the starch hydrolysis test was prepared by the overlay technique according to Braaz and co-workers [11]. The composition of mineral salts medium (MSM) used in latex overlay agar was based on the ingredients described by Heisey and Papadatos [12] and supplemented with 0.02 g/L of yeast extract. In the meantime, 15 g/L of bacteriological agar powder was mixed into the culture medium for agar preparation. During the agar preparation, 0.6% of purified NR latex was added to MSM. Meanwhile, 3 g/L meat extract, 5 g/L peptone, 2 g/L soluble starch and 15 g/L bacteriological agar under pH 7.2 was used to prepare the starch agar.

2.5. Gram staining test

A loopful of the culture has been placed on the slide and it was fixed with the heat in order to kill the bacteria in the smear, adhere the smear to the slide and allow the sample to more readily take up the stain. The staining was done by flooding staining reagents by the order: crystal violet, Gram's iodine, ethyl alcohol and safranin. The results were observed under oil immersion (100x) using a light microscope.

3. Results and Discussion

There are various bacteria can hydrolyze the starch, which called starch-degrading bacteria. The existence of this type of bacteria is the key factor to determine the effectiveness of incorporation of sago starch into latex compounding for biodegradation purposes which was done by measuring the biodegradation rate of the buried gloves.

Figure 1 showed the positive reaction towards iodine on starch agar plates which containing bacteria from SS/NR latex gloves. The formation of the clear zone around the bacteria indicates the starch was hydrolyzed which means the bacteria were capable of breaking down the starch into maltose through the activity of the extra-cellular α-amylase enzyme. Based on the results obtained, it is proven that the mixed culture from buried sample contains the starch-degrading bacteria which are very beneficial for the application of SS/NR latex gloves.

As sago starch being introduced into the NR latex system, the starch employed the void within NR latex colloidal particles and remains as cementing substances for the films [13-14]. The cementing mechanism among starch and NR latex particles depends on the interaction of random coiled amylose and lipids in rubber particles [13]. As the hydrolysis process occurs in amylose, the linkages are broken and reduced the effective interaction with rubber particles and hence decreased the physical integrity of SS/NR latex gloves. Thus, it increased the feasibility of the SS/NR latex gloves to biodegrade.

IOP Conf. Series: Materials Science and Engineering **548** (2019) 012018 doi:10.1088/1757-899X/548/1/012018

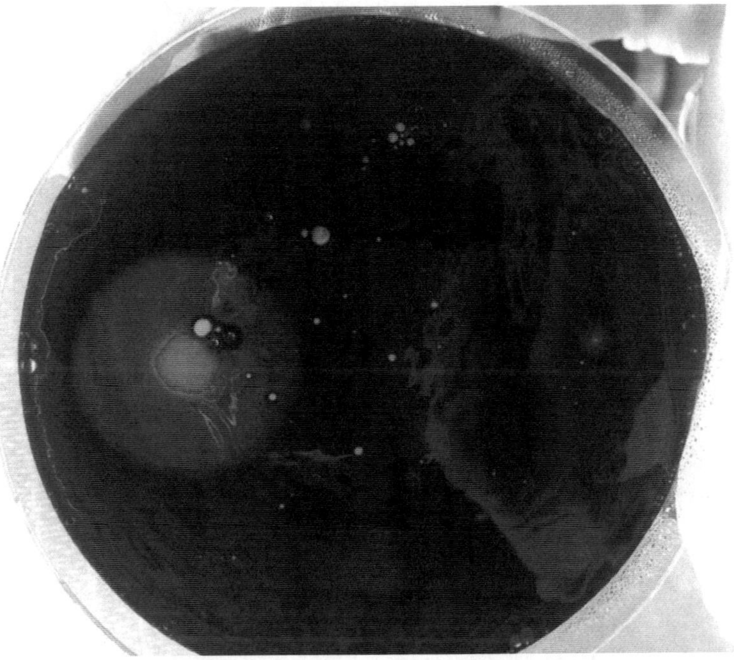

Fig. 1. Clear zone formation on the starch agar plate

Figure 2 showed the biodegradation rate of the control and SS/NR latex gloves throughout the 5 months of cultivation in the soil. Although both types of gloves showed a sign of degradation, the biodegradation of SS/NR latex gloves gave the higher rate. It is proven that the existence of starch-degrading bacteria to consume the starch can accelerate the biodegradation of SS/NR latex gloves by 53.68%. Meanwhile, the absence of starch-degrading bacteria in control latex gloves gave lesser biodegradation rate i.e. 50.31%.

Natural degradation of NR latex is a slow process, as crosslinks between rubber polymer chains, and the many additives and stabilizers within its structure, making it extremely resistant to biodegrade [15]. Therefore, the concept of a combination of starch into latex/rubber compounding gains much interest from other researchers. Kiing et al. [2] studied the potential of incorporating NR latex into chemically modified starch for the making biodegradable blends. They found that the modified starches that blended with 20% NR latex would still be able to biodegrade up to 50% within 12 weeks. Afiq and Azura [3] also found that sago starch is a nutritive substance to accelerate the growth of microbial colonies on the surface of NR latex films during the soil burial test. It was suggested that microorganisms favored consuming simpler glucosidic linkages in starch during biodegradation periods compared to polyisoprene molecules of NR latex films alone.

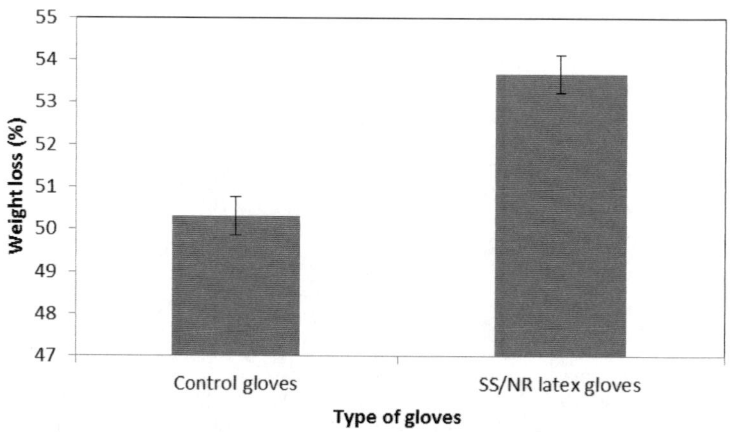

Fig. 2. Biodegradation rate of the gloves after 5 months of biodegradation periods

Further study has been done on the Gram staining to confirm the buried soil has different mixed culture. Based on Figure 3, there are lots of red stains compared to purple stains taken from SS/NR latex gloves. It indicates there have lots of residential starch- and latex-degrading bacteria that are Gram-negative.

The Gram-negative bacteria also have the capability to degrade latex/rubber composites. Linos et al. [16] found that strain AL98; a Gram-negative bacterium, was able to disintegrate NR, either in the raw state as NR latex concentrates or in the vulcanized state as NR latex glove, as well as raw synthetic *cis*-1,4-polyisoprene. Besides, Birke at al. [17] also claimed the *Xanthomonas* sp. 35Y is the only known as Gram-negative for rubber degrader.

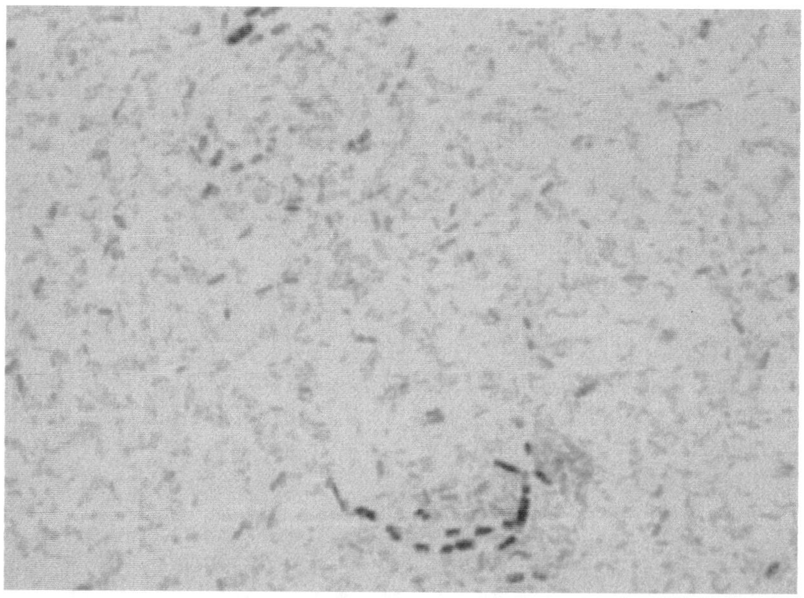

Fig. 3. Cell morphology of mixed culture under a light microscope (Magnification: 100x)

4. Conclusions

In conclusion, it was confirmed that a mixed culture from buried soil showed the potential in degrading NR latex gloves within 5 months. The existences of starch-degrading bacteria, as well as rubber-degrading bacteria, proved that it can accelerate the biodegradation of SS/NR latex gloves.

Acknowledgments

This work is carried out at School of Materials & Mineral Resources Engineering and financially supported by Ministry of Higher Education under Fundamental Research Grant (grant no: 203/PBAHAN/6071353).

References

[1] Nawong C, Sermwittayawong N and Umsakul K 2016 Comparative studies on rubber biodegradation by mixed culture and pure culture isolated from the soil samples *International Journal of Advances in Agricultural and Environmental Engineering* **3(2)** 2349-1523
[2] Kiing S C, Dzulkefly K and Yiu P H 2013 Characterization of biodegradable polymer blends of acetylated and hydroxypropylated sago starch and natural rubber *J. Polym. Environ.* **21(4)** 995-1001
[3] Afiq M M and Azura A R 2013 Utilization of starch to accelerate the growth of degrading microorganisms on the surface of natural rubber latex films *J. Chem. Chem. Eng.* **7** 137-144
[4] Misman M A, Azura A R and Hamid Z A A 2015 Physico-chemical properties of solvent based etherification of sago starch *Ind. Crops Prod.* **65** 397-405
[5] Linos A and Steinbüchel A 2005 Biodegradation of natural and synthetic rubbers. Biopolymers Online. Wiley-VCH Verlag GmbH & Co. KGaA
[6] Birke J, Röther W, Schmitt G and Jendrossek D 2017 RoxB is a novel type of rubber oxygenase that combines properties of rubber oxygenase RoxA and latex clearing protein (Lcp) *Appl. Environ. Microbiol.* **83(14)** 1-13
[7] Jendrossek D and Birke J 2019 Rubber ocygenases *Appl. Environ. Microbiol.* **103(1)** 125-142
[8] Rose K and Steinbüchel A 2005 Biodegradation of natural rubber and related compounds; recent insights into hardly understood catabolic capability of microorganisms *Appl. Environ. Microbiol.* **71(6)** 2803-2812
[9] Cherian E and Jayachandran K 2009 Microbial degradation of natural rubber latex by a novel species of *Bacillus* sp. SBS25 isolated from soil *Int. J. Environ. Res.* **3(4)** 599-604
[10] Abiola C and Oyetayo V O 2016 Isolation and biochemical characterization of microorganisms associated with the fermentation of Kersting's groundnut (*Macrotyloma geocarpum*) *Res. J. Microbiol.* **11** 47-55
[11] Braaz R, Fischer P and Jendrossek D 2004 Novel type of heme-dependent oxygenase catalyzes oxidative cleavage of rubber (poly-*cis*-1,4-isoprene) *Appl. Environ. Microbiol.* **70(12)** 7388-7395
[12] HEeisey R M and Papadatos S 1995 Isolation of microorganisms able to metabolize purified natural rubber *Appl. Environ. Microbiol.* **61** 3092-3097
[13] Blackley D C 1997 Vol. 1: Fundamental and principals, second edition. London, United Kingdom: Chapman & Hall
[14] Joseph L K and Alexander F R 2010 Fundamental of latex film formation. Dordrecht, Netherlands: Springer
[15] Stevenson K, Stallwood B and Hart A G 2008 Tire rubber recycling and bioremediation: a review *Biorem. J.* **12(1)** 1-11
[16] Linos A, Reichelt R, Keller U and Steinbüchel A 2000 A Gram-negative bacterium, identified as *Pseudomonas aeruginosa* AL98, is a potent degrader of natural rubber and synthetic *cis*-1,4-polyisoprene *FEMS Microbiology Letters* **182** 155-161
[17] Birke J, Röther W, Schmitt G and Jendrossek D 2013 Functional identification of rubber oxygenase (RoxA) in soil and marine myxobacteria *Appl. Environ. Microbiol.* **79(20)** 6391-6399

IOP Conf. Series: Materials Science and Engineering **548** (2019) 012019 doi:10.1088/1757-899X/548/1/012019

Production of Poly(carbonate-co-ether) via Catalytic Polymerization of Epoxidized Linseed Oil, Propylene Oxide and Carbon Dioxide

F W Shaarani[1,2], J J Bou[2] and R N Hakim[3]

[1] Section of Chemical Engineering, Universiti Kuala Lumpur, Malaysian Institute of Chemical and Bioengineering Technology, Lot 1988 Kawasan Perindustrian Bandar Vendor, Taboh Naning, 78000 Alor Gajah, Melaka, Malaysia

[2] Department of Chemical Engineering, Escola Tècnica Superior d'Enginyeria Industrial de Barcelona, Universitat Politècnica de Catalunya, Barcelona Tech, Diagonal 647, 08028 Barcelona, Spain

[3] Section of Polymer Engineering Technology, Universiti Kuala Lumpur, Malaysian Institute of Chemical and Bioengineering Technology, Lot 1988 Kawasan Perindustrian Bandar Vendor, Taboh Naning, 78000 Alor Gajah, Melaka, Malaysia

E-mail: farrashaarani@unikl.edu.my

Abstract. Poly-(carbonate-co-ether) was synthesized via catalytic polymerization between epoxidized linseed oil (ELO), propylene oxide (PO) and carbon dioxide (CO_2) using heterogeneous catalyst namely Co-Zn double metal cyanide (Co-Zn DMC). The influence of polymerization time was studied. It was revealed that the properties of the resultant polymers were affected by the polymerization time. Within the range of polymerization time studied, the maximum average molecular weight (Mn) of the resultant polymer obtained was 6.21×10^5 g.mol^{-1} with polydispersity index (PDI) of 1.05 at reaction time of 24 hours.

1. Introduction

The transformation of carbon dioxide (CO_2) into polycarbonates via copolymerization with epoxides is perceived as an interesting strategy to utilize the abundance of greenhouse gases. After the first breakthrough made by Inoue and coworkers [1] in the year of 1969, copolymerization of CO_2 with epoxides has been studied intensively by researchers whereby diverse kinds of catalysts [2, 3], epoxides [4, 5] and operating conditions [6, 7] have been reported. To date, the most common epoxides used in the copolymerization reaction are propylene oxide (PO), cyclohexene oxide and styrene oxide which all are derived from petroleum feedstocks. However, due to the repeated oil crisis associated with the high price and fear of the exhaustion of the crude oil, more attention was given to the identification of alternative renewable sources, based on natural raw material.

Vegetable oil has become one of the most interesting renewable resources for polymer synthesis because of its universal availability, inherent biodegradability and low toxicity [8-11]. The major components of vegetable oil are triglycerides which comprise of three fatty acid chains combined with glycerol through ester linkage. The length of fatty acid ranges from 8 to 22 carbon, and some of them have chemical functionalities such as hydroxyl groups, epoxide groups, and unsaturation (carbon–carbon double bond) [12]. The naturally occurring functional groups present in triglycerides can be chemically modified prior to polymerization and in some cases use directly to produce vegetable oil-based polymer [13].

Amongst miscellaneous vegetable oils, linseed oil is an excellent option for the synthesis of polymeric materials. This is owing to its chemical structure, which has a large number of unsaturations with an average of 6.6 double bonds per molecule [14], which could be chemically modified to introduce new functional groups that readily polymerize. For example, the unsaturation present in linseed oil can be chemically modified to introduce functional group like epoxide via epoxidation reactions and subsequently the epoxidized linseed oil (ELO) can be used as a starting material in the copolymerization reaction with CO_2 to produce polycarbonate (Scheme 1).

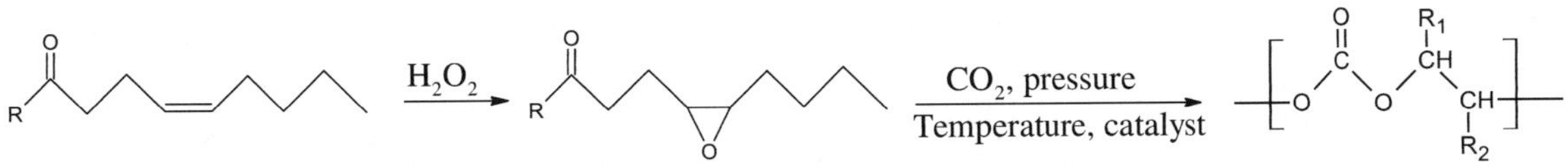

Scheme 1. Reactions to produce natural polycarbonate from natural fat/fatty acid

This research was undertaken to study the viability of utilizing ELO in the terpolymerization reaction with PO and CO_2 to produce polycarbonate. The effect of reaction time in correlation with the yield and molecular weight of the resultant polymers was also studied

2. Materials and Methods

2.1. Materials

Potassium hexacyanocobaltate (III)($K_3Co(CN)_6$), zinc chloride ($ZnCl_2$), and tertiary butyl alcohol (tert-butanol) were used for catalyst preparation without further purification. For polymerization reaction, ELO used was supplied by Traquisa (Barbera del Valles, Barcelona Spain) whereas PO (>99%) was purchased from Sigma Aldrich. CO_2 of 99.99% purity was used as received. Methanol (CH_3OH) and methylene chloride (CH_2Cl_2) of analytical grade were supplied by Scharlau and Panreac respectively and directly utilized in the experiments.

2.2. Methods

In this work, a typical heterogeneous catalyst used in CO_2–epoxide copolymerization that is Co-Zn double metal cyanide (Co-Zn DMC) is used. The catalyst preparation followed the method as being described by Shaarani et al. [15].

Polymerization took place in a 100 mL stainless steel reactor (Autoclave Engineers, Erie, PA USA) equipped with mechanical stirrer and an automatic temperature controller system. 0.2 g Co-Zn DMC was placed into the dried autoclave together with 10 mL of ELO and 10 mL of PO. The reactor was purged twice with CO_2 and then slowly pressurized to 5.0 MPa. Next, the temperature was raised and maintained at 60°C and the stirring speed was kept constant at 500 rpm to initiate the polymerization reaction. Meanwhile the reaction time was varied between two to 24 hours. Once the reaction ends, the autoclave was cooled down to room temperature, slowly depressurized and opened. A small aliquot of the reaction mixture was removed from the reactor for proton nuclear magnetic resonance (^1H NMR) analysis. The remaining products obtained were purified by dissolving in dichloromethane, precipitated by excess methanol and then dried at room temperature to a constant weight.

Spectroscopic analyses of products were performed using Fourier transform infrared spectroscopy (FT-IR)(Model: Perkin Elmer 1000) and a Bruker NMR spectrometer (Model: Bruker

AMX 300) with ^{1}H probe and deuterated chloroform ($CDCl_3$) as the solvent. The number average molecular weight (Mn), the weight average molecular weight (Mw) and polydispersity index (PDI) of the polymer products were estimated using a gel permeation chromatography (GPC) system (Model: Agilent HPLC). Hexafluoroisopropanol was used as an eluent.

3. Results and Discussion

In this study, ELO was used as one of the precursors in the catalytic polymerization reaction with CO_2. The yield of products and productivity of the catalyst from the terpolymerization reaction between ELO, PO and CO_2 prepared at different reaction time are shown in Table 1. As can be seen from Table 1, no product was collected at reaction time of 2 h which indicates that this polymerization reaction had a long induction period. Nevertheless, as the time rose to 4 h, more meaningful activity was observed. The yield of resultant polymer recorded at 4 h was 1.562 g and increased to 3.720 g at 24 h. Similar trend was observed for the catalytic activity which shown an increment as the reaction time prolonged from 4 h to 24 h. The values of Mn, Mw and PDI estimated from the GPC are also tabulated in Table 1. The results obtained showed that both Mn and Mw of the resultant polymers are affected by the reaction temperature and fall between 5.02 x 10^5 to 6.21 x 10^5 g.mol^{-1} and 5.89 x 10^5 to 6.54 x 10^5 g.mol^{-1} respectively. Meanwhile, PDI of the resultant polymers showed a little discrepancy between 1.05 to 1.18. The narrow PDI values indicate that the synthesized polymers have a narrow molecular weight distribution (MWD).

Table 1. Effect of polymerization time on terpolymerization of ELO, PO and CO_2

Sample	Reaction Time (h)	Yield^a (g)	Productivity (g product/ g catalyst)	Mn/Mw^{b,*} x10^5	PDI^{b,*}
ELO2	2	/	/	/	/
ELO4	4	1.562	7.81	5.02/5.89	1.17
ELO6	6	1.906	9.53	5.04/5.93	1.18
ELO16	16	2.411	12.06	5.08/5.93	1.17
ELO24	24	3.720	18.60	6.21/6.54	1.05

Reaction conditions: volume of ELO = 10 mL, volume of PO = 10 mL, catalyst loading = 0.2 g, reaction temperature = 60°C, pressure of CO_2 = 5.0 MPa.
[a] Weight of the polymer obtained
[b] Determined by gel permeation chromatography (GPC)
*Two peaks were observed in the GPC spectrums of the resultant polymers, representing the highest molecular weight of the polymer (quasi-crosslinked sample) also the oligomeric mixtures containing polyether and the unreacted ELO. The molecular weight reported here was from the quasi-crosslinked sample.

The formation of polymeric product was confirmed by both FTIR and ^{1}H NMR analyses. Fig. 1 displays the FTIR spectrum of product obtained at reaction time of 24 h. As can be seen from Fig. 1, the resultant polymer exhibits a peak at 1740 cm^{-1} which corresponds to the characteristic absorption of the carbonate group (C=O), confirming the formation of polycarbonate. Nevertheless, in addition to carbonate absorption, a peak at 1231 cm^{-1} confirms the presence of the ether linkage ν(C–O–C), suggesting that the homopolymer part of the epoxide also exists in the synthesized polymer. Besides, the peak which correspond to epoxide group also still visible indicating not all epoxide was consumed in the polymerization process. In general, the FTIR spectrum of product prepared at polymerization time of 4 h, 6 h and 16 h (Fig. not shown) also demonstrated similar characteristic absorption.

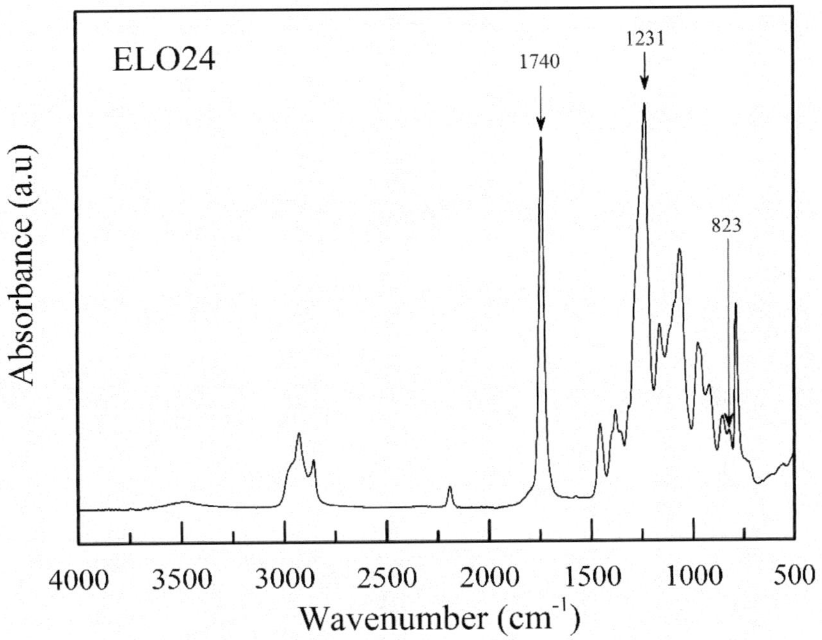

Fig. 1. FTIR spectra of product prepared at reaction time of 24 h (ELO24)

These results were further confirmed by ^1H NMR analysis. Fig. 2 illustrates the comparison of ^1H NMR spectra for ELO monomer and crude ELO24. The appearance of peaks associated to polycarbonate can be seen in the ^1H NMR spectrum alongside with a broad peak at δ 3.5 ppm correspond to ether unit. Also prominent in the ^1H NMR spectrum of the resultant polymer are peaks around δ 2.8–3.2 ppm region associated to epoxy proton signifying that epoxides are not fully reacted in the polymerization reaction most likely due to the steric hindrance owing to the long chain structure of triglyceride. Also notable are the methine proton $-CH_2-\mathbf{CH}-CH_2-$ of the glycerol backbone at δ 5.1–5.3 ppm and methylene protons $-\mathbf{CH_2}-CH-\mathbf{CH_2}-$ of the glycerol backbone at δ 4.1–4.3 ppm which revealed triglyceride structure of ESO is not disturbed [16]. Based on the ^1H NMR spectrum, it can be concluded that the polymer obtain from the terpolymerization of ELO, PO and CO_2 is poly(carbonate co-ether) bonded to the triglyceride chain.

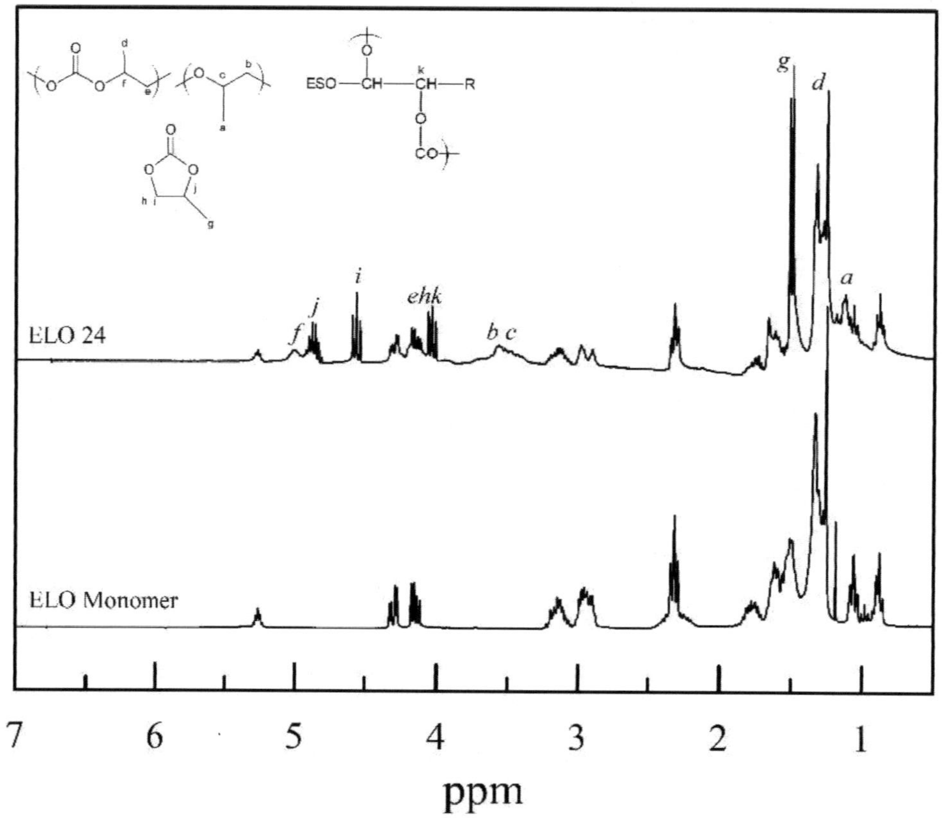

Fig. 2. Comparison of ^{1}H NMR spectra of ELO monomer and crude ELO24

4. Conclusions

In summary, a bio-based poly(carbonate-co-ether) bonded to the triglyceride chain of ELO has been successfully synthesized via catalytic polymerization of ELO a product from natural resources with PO and CO_2. The presence of each component in the resultant products was verified by means of FTIR and ^{1}H NMR analyses. The reaction time plays a vital role on the catalyst activity and the properties of the resultant polymers such as yield, Mn, Mw and PDI.

Acknowledgments

The authors are grateful for the financial support by Spanish Ministry of Economy and Competitiveness (Program RETOS, Grant No. MAT2016-80045-R) and Universiti Kuala Lumpur, Malaysia for funding a fellowship to make this study possible.

References

[1] Inoue S, Koinuma H and Tsuruta T 1969 Copolymerization of carbon dioxide and epoxide *J. Polym. Sci., Part B: Polym. Lett.* **7** 287-92

[2] Ang R-R, Tin Sin L, Bee S-T, Tee T-T, Kadhum A A H, Rahmat A R and Wasmi B A 2015 A review of copolymerization of green house gas carbon dioxide and oxiranes to produce polycarbonate *J. Cleaner Prod.* **102** 1-17

[3] Klaus S, Lehenmeier M W, Anderson C E and Rieger B 2011 Recent advances in CO2/epoxide copolymerization—New strategies and cooperative mechanisms *Coord. Chem. Rev.* **255** 1460-79

[4] Zhang X-H, Wei R-J, Zhang Y Y, Du B-Y and Fan Z-Q 2015 Carbon Dioxide/Epoxide Copolymerization via a Nanosized Zinc–Cobalt(III) Double Metal Cyanide Complex: Substituent Effects of Epoxides on Polycarbonate Selectivity, Regioselectivity and Glass Transition Temperatures *Macromolecules* **48** 536-44

[5] Masoumeh T and P. P P 2014 Green polycarbonates prepared by the copolymerization of CO2 with epoxides *J. Appl. Polym. Sci.* **131**

[6] Oh H J and Ko Y S 2013 Effect of polymerization conditions on the polymer properties of CO2-cyclohexene oxide copolymer prepared by double metal cyanide catalyst *Ind. Eng. Chem.* **19** 1939-43

[7] Zhang X-H, Wei R-J, Sun X-K, Zhang J-F, Du B-Y, Fan Z-Q and Qi G-R 2011 Selective copolymerization of carbon dioxide with propylene oxide catalyzed by a nanolamellar double metal cyanide complex catalyst at low polymerization temperatures *Polymer* **52** 5494-502

[8] Zhang C, Garrison T F, Madbouly S A and Kessler M R 2017 Recent advances in vegetable oil-based polymers and their composites *Prog. Polym. Sci.* **71** 91-143

[9] Stemmelen M, Lapinte V, Habas J-P and Robin J-J 2015 Plant oil-based epoxy resins from fatty diamines and epoxidized vegetable oil *Eur. Polym. J.* **68** 536-45

[10] Alam M, Akram D, Sharmin E, Zafar F and Ahmad S 2014 Vegetable oil based eco-friendly coating materials: A review article *Arabian J. Chem.* **7** 469-79

[11] Miao S, Wang P, Su Z and Zhang S 2014 Vegetable-oil-based polymers as future polymeric biomaterials *Acta Biomaterialia* **10** 1692-704

[12] Liu K, Madbouly S A and Kessler M R 2015 Biorenewable thermosetting copolymer based on soybean oil and eugenol *Eur. Polym. J.* **69** 16-28

[13] Montero de Espinosa L and Meier M A R 2011 Plant oils: The perfect renewable resource for polymer science?! *Eur. Polym. J.* **47** 837-52

[14] Xia Y and Larock R C 2010 Vegetable oil-based polymeric materials: synthesis, properties, and applications *Green Chem.* **12** 1893-909

[15] Shaarani F W and Bou J J 2017 Synthesis of vegetable-oil based polymer by terpolymerization of epoxidized soybean oil, propylene oxide and carbon dioxide *Sci. Total Environ.t* **598** 931-6

[16] Liu Z, Shah S N, Evangelista R L and Isbell T A 2013 Polymerization of euphorbia oil with Lewis acid in carbon dioxide media *Ind. Crops Prod.* **41** 10-6

IOP Conf. Series: Materials Science and Engineering **548** (2019) 012020 doi:10.1088/1757-899X/548/1/012020

Effect of Pyrolysis Temperature and Time on Properties of Palm Kernel Shell-Based Biochar

MH Mohd Hasan[1] RT Bachmann[1] SK Loh[2] S Manroshan[3] and SK Ong[1]

[1] Universiti Kuala Lumpur, Malaysian Institute of Chemical and Bioengineering Technology, Lot 1988, Kawasan Perindustrian Bandar Vendor, Taboh Naning, 78000 Alor Gajah, Melaka, Malaysia

[2] Malaysian Palm Oil Board, Bandar Baru Bangi, 43000 Kajang, Selangor, Malaysia

[3] Malaysian Rubber Board, RRIM Research Station, 47000 Sungai Buloh, Selangor, Malaysia

E-mail: skong@unikl.edu.my

Abstract. Pyrolysis is a thermal decomposition of biomass occurring in the absence of oxygen. Biochar (solid), bio-oil (liquid) and biogas (gas) are the typical products from pyrolysis of biomass. This endothermic process produced biochar with high carbon content. During pyrolysis, the material is heated up from ambient to a peak temperature and remains for a defined residence time. Therefore, the pyrolysis peak temperature and the residence time are the key parameters for pyrolysis. Studies on the effect of these parameters on the biochar characteristics and the pyrolysis products composition are numerous. However, there are limited findings of these parameters with palm kernel shell (PKS) as the biomass. This study focusses on the effect of the pyrolysis temperature and residence time of PKS on the yield of biochar produced and the biochar physio-chemical properties. The results showed that biochar yield decreased as the peak temperature and residence time increased. This finding is consistent with the findings by other researchers. However, those factors do not have distinct influence on biochar's carbon content as found in other study which the peak temperature has the bigger impact instead of residence time. The effect of peak temperature or residence time on grindability; i.e. particle size of biochar after wet ball milled is insignificant. Smaller size of biochar may improve its function as reinforcing filler. As a conclusion, the optimum setup of pyrolysis is needed for a balance production in yield and biochar's properties. It is recommended to produce biochar at higher peak temperature and shorter residence time to increase the total production. Additional analysis e.g. physical testing on the final polymer product can be used to investigate the effect of pyrolysis peak temperature and residence time.

1. Introduction

Outstanding physical and mechanical properties made plenty different type of composites based on polymer matrix available on the market [1]. The interest on carbon nanotubes and graphene as fillers in polymers has increased [2]–[7]. However, reduction in the cost of production and the cost of filler are important towards commercialization. Attention has been given to composites derived from environmental-friendly materials. Areas such as Malaysia [8] has abundance source of biomass i.e., palm kernel shells (PKS) produced from palm oil mill. It has potential to be utilize for new applications [9]–[12].

Pyrolysis is a major thermal decomposition process to convert biomass into three main products; biochar (solid), bio-oil (liquid) and biogas (gas) [13]–[15]. The biomass is heated up from ambient to a set peak temperature and remained for a defined residence time during the pyrolysis process in the absence of oxygen [16]. Biochar can be used as alternative filler since it is carbon rich.

The pyrolysis temperature and the residence time play a role in the yield and governing the physio-chemical properties of the biochar. Increasing pyrolysis temperature is expected to give

IOP Conf. Series: Materials Science and Engineering **548** (2019) 012020 doi:10.1088/1757-899X/548/1/012020

lower biochar yield [17] while the residence time has no significant effect [18]. For proximate analysis especially on ash and fixed carbon content, pyrolysis temperature has remarkable impact compared to time [19]. Hydrogen/carbon (H/C) and oxygen/carbon (O/C) ratio based on determined C, H, nitrogen (N), sulphur (S) and O from ultimate analysis showed decreasing trend as the pyrolysis temperature increased [20]. Changes on pH value of biochar substantially influenced by pyrolysis temperature with time has statistically non-significant impact [21].

Studies showed that the biochar particle size affects it interaction with the rubber as well as the rheological properties of the biochar dispersion [22-23]. Thus, the grindability of the biochar is crucial and the grindability depends on the pyrolysis parameters [24-25]. Apart from the pyrolysis parameters, the biochar particle size is limited by temperature, loading, rotational speed and others on wet ball milling technique [26]–[28]. This study focuses on the effect of pyrolysis temperature and the pyrolysis residence time on the physico-chemical properties of PKS biochar and its grindability; i.e. particle size after wet ball milling as well as its dispersion rheological properties.

2. Experimental

2.1. Material

PKS as the raw biomass was used for pyrolysis process. It is the waste shell after removal of the nut in palm oil mill. PKS was collected from Palm Oil Mill Technology Centre (POMTEC), Malaysia. For each pyrolysis process, 20 kg of PKS was sieved to remove sand, dirt and stones before dried at 105°C for 24 hours to reduce the moisture content.

2.2. Pyrolysis process

A multi-mode pyrolysis machine, Biochar Experimenter's Kit (BEK) from All Power Labs (APL) was used for pyrolysis process to produce biochar as the main product. After drying, 17 kg of PKS biomass was fed into the BEK. A combination of propane gas and compressed air were used to ignite the flame in the burner during start-up. The heat energy does indirect heating to the PKS biomass from ambient; 25°C to the desired temperature; 400 or 500°C and maintained at a set residence time; 30, 45, 60 or 75 minutes. After the temperature was above 100°C, the pyrolysis sustained by burning of produced biogas without propane gas support. The pyrolysis products; biochar and bio-oil were weighed to obtain the yield using the equation below where $W_{BC/BO}$ is weight of biochar or bio-oil (kg) and W_{RP} is weight of raw PKS (kg).

$$\mathrm{Yield}_{BC/BO}(\mathrm{wt.\%}) = \frac{\mathrm{W}_{BC/BO}}{\mathrm{W}_{RP}} \times 100\%$$

2.3. Physico-chemical characterization

Prior to the physico-chemical characterisation, the PKS biochar was ground using Golden Bull SY-10 grinder and sieved to the size of 105 µm. Ground biochar undergo pH analysis used modified dilution of 1:20 biochar: deionized water (w:v) with 90 minutes in a shaker for equilibrium [29] using a calibrated BP3001 Trans Instruments pH meter. Proximate analysis was done using thermogravimetric analyser Leco TGA 701 according to ASTM D5142 and ultimate analysis using CHNS analyser Leco 628 according to ASTM D5373. Oxygen content calculated using equation below. The raw PKS biomass was characterized using the same methods.

$$\mathrm{O\ (wt.\%)} = 100 - \mathrm{C\ (wt.\%)} - \mathrm{H\ (wt.\%)} - \mathrm{N\ (wt.\%)} - \mathrm{S\ (wt.\%)} - \mathrm{Ash\ (wt.\%)}$$

IOP Conf. Series: Materials Science and Engineering **548** (2019) 012020 doi:10.1088/1757-899X/548/1/012020

2.4. *Wet ball milling*

Only biochar from pyrolysis temperature of 400 and 500°C with pyrolysis time of 30 and 60 minutes were chosen to proceed with wet ball milling process. Deionized water was mixed with the biochar to produce 50 wt.% of biochar dispersion. Mill jar of 2.5 liters capacity was used with alumina grinding ball media and biochar dispersion occupied 60% of the total volume. The mill jar positioned horizontally on a pair of rods on the mill machine. The ball milling process was run at 50 rpm for 24 hours.

2.5. *Particle size analysis*

After ball milling, the biochar's particle size was analyzed using particle size analyzer Malvern Mastersizer 2000. The stirrer speed was set at 1500 rpm. The system was clean using distilled water until laser level reach minimum 76%. The biochar dispersion was put into the dispersion cuvette until laser obscuration level reached 5%. Before measurement was done, sonication was made for 30 seconds.

2.6. *Scanning electron microscopy morphological analysis*

Morphology analysis on the particle size of PKS biochar before and after ball milling were done using scanning electron microscopy (SEM) Hitachi SU-3500 Microscope. The biochar mounted on a 12 mm aluminium sample holder (stub) using carbon double-sided tape for adhesion.

2.7. *Rheological properties analysis*

Steady state experiment was done on the 50 wt.% of biochar dispersion using Anton Paar MCR 302 rheometer. The shear rate studied was 0 to 1000 s-1 with frequency fixed at 1 Hz. Cone and plate setup were used with temperature set at 25°C. The biochar dispersion was pre-sheared for 2 minutes followed by rest for 10 minutes and continued with shear for another 2 minutes.

3. Results and Discussion

3.1. *Analysis of PKS*

Table 1 showed volatile content of raw PKS is the highest from the proximate analysis. The pyrolysis process is intended to remove this volatile matter to increase its grindability in the later stage. The carbon content is the highest from the ultimate analysis. This carbon is the important part of reinforcement with rubber particles.

Table 1. Proximate and ultimate analysis of PKS

Proximate analysis	Value (wt.%)	Ultimate analysis	Value (wt.%)
Moisture content	4.86	Carbon	48.7
Volatile content	70.74	Hydrogen	6.21
Fixed carbon	22.23	Nitrogen	0.057
Ash content	2.17	Sulphur	[a]BDL
		[b]Oxygen	45.06

[a]Below detectable limit [b]By difference

3.2. Yield of PKS biochar

Fig. 1 showed the yield of PKS biochar after pyrolysis process. 25% increase in pyrolysis temperature from 400 to 500°C had decreased the biochar yield from 43.13% to 30.69% with 12.44% in difference. For pyrolysis temperature of 400°C, the pyrolysis time need to be increased for 150% from 30 to 75 minutes to have biochar yield decreased with 13.3% difference. For temperature of 500°C, increasing the time for 150% will only decrease the biochar yield with 5.19% difference. As reported by Titiladunayo and co-workers [30], decreased in biochar yield with the increasing of pyrolysis temperature is due to more volatile matter were released. Increasing pyrolysis time across the same pyrolysis temperature also has the same effect but only at lower pyrolysis temperature because higher temperature predominantly alters the biochar's internal structure and surface [31].

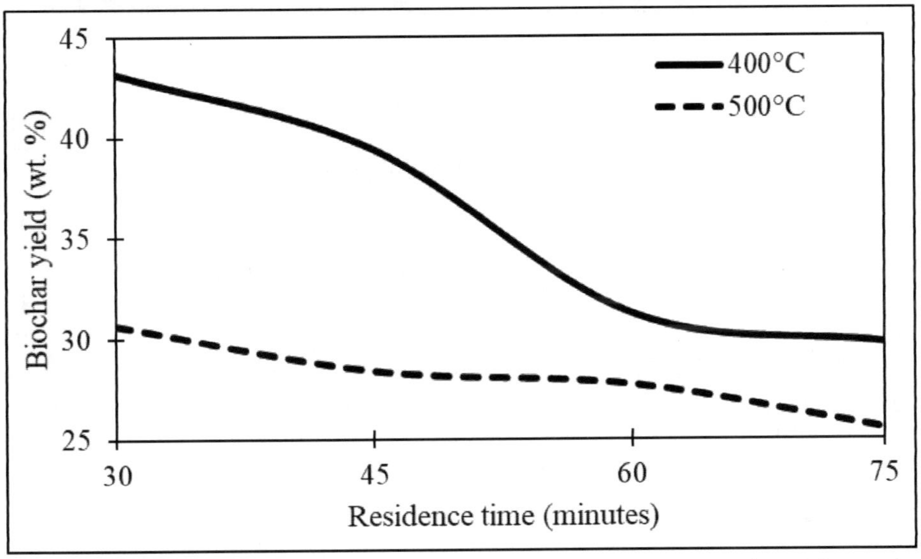

Fig. 1. Yield of PKS biochar at different pyrolysis temperature and residence time.

3.3. Fixed carbon, ash, moisture and volatile content

There is no significant difference on the moisture and ash content as the pyrolysis temperature or time increased. At fixed pyrolysis temperature of 500°C, both volatile matter and fixed carbon content has no significant difference across pyrolysis time of 30 to 75 minutes. However, for pyrolysis temperature of 400°C, it took 100% increase in pyrolysis time or 25% increase in pyrolysis temperature to have volatile matter and fixed carbon content of below 20% and above 70% respectively. This results are in agreement with Ronsse and co-workers' findings as the pyrolysis temperature has greater impact on the volatile matter and fixed carbon content compared to time because of devolatilization reaction is developing more carbon [19].

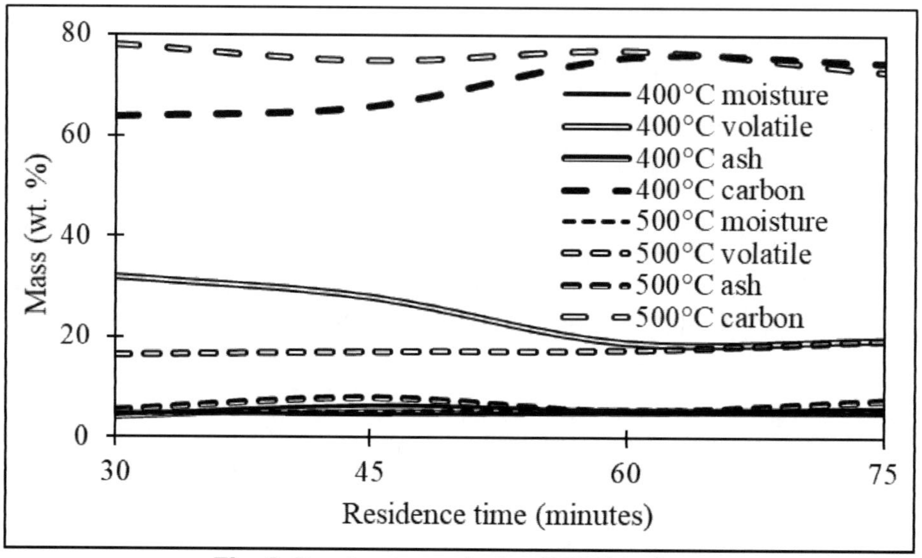

Fig. 2. Proximate analysis of PKS biochar.

3.4. *C, H, N, S and O content*

The C, H, N, S and O content from ultimate analysis were used to obtain H/C and O/C ratio as shown in Van Krevelen of Fig. 3. PKS biochar produced from pyrolysis parameters of 400°C at 30 minutes and 400°C at 45 minutes lies outside of maximum H/C ratio of 0.7. Those biochars did not meet the standard by Initial Biochar Initiative (IBI). All PKS biochars produced using temperature of 500°C positioned within the IBI standard in the Van Krevelen diagram. The pyrolysis temperature has significant impact to reduce H/C and O/C ratio compared to pyrolysis time due to better deoxygenation and dehydration as pyrolysis temperature increases [32].

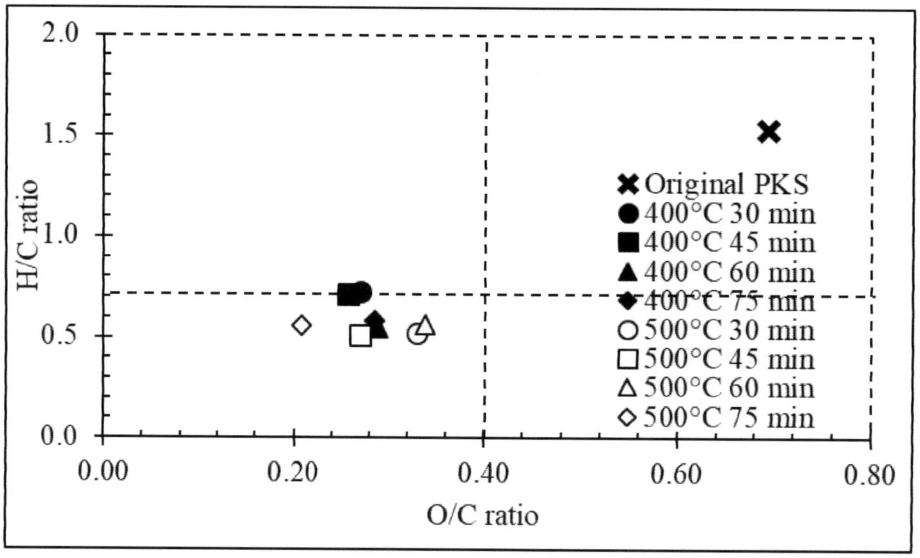

Fig. 3. Van Krevelen of PKS biochar at different pyrolysis temperature and residence time.

3.5. pH of biochar

Fig. 4 shows the effect pf pyrolysis temperature and pyrolysis residence time on the biochar's pH. The pH of biochar from pyrolysis temperature of 400°C is consistently lower than those of 500°C within the pyrolysis residence time studied.

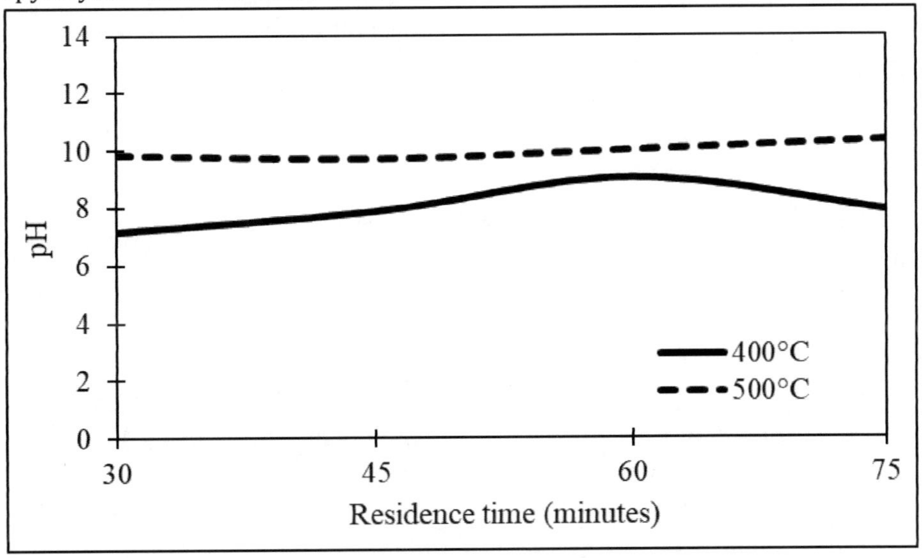

Fig. 4. pH of PKS biochar obtained from different pyrolysis temperature and residence time.

To produce biochar with pH at around 10, a 25% increase in pyrolysis temperature from 400 to 500°C is needed. However, 150% increase in pyrolysis time from 30 to 75 minutes still unable to produce biochar with pH at around 10. This showed pyrolysis temperature has greater impact on pH of produced biochar compared to pyrolysis time. Increasing pyrolysis temperature will increase the pH of produced biochar as it will lead to higher ash content. Higher ash content in biomass produced biochar with higher pH value. This also depends on the characteristics of the raw biomass [33].

3.6. Particle size of wet ball milled biochar

From Fig. 5, PKS biochar produced from pyrolysis temperature of 400°C requires 100% increase in pyrolysis time to have particle size comparable to biochar from temperature of 500°C prior to 12 hours of wet ball milling. This showed pyrolysis temperature has more influence on grindability of biochar compared to pyrolysis time for ball milling lesser than 12 hours. After 12 hours of ball milling until 24 hours, no significant difference was observed on particle size of all biochars. There is significant improvement on grindability between raw biomass and biochar but there is minimal improvement with biochars produced at different pyrolysis temperature [23].

IOP Conf. Series: Materials Science and Engineering **548** (2019) 012020 doi:10.1088/1757-899X/548/1/012020

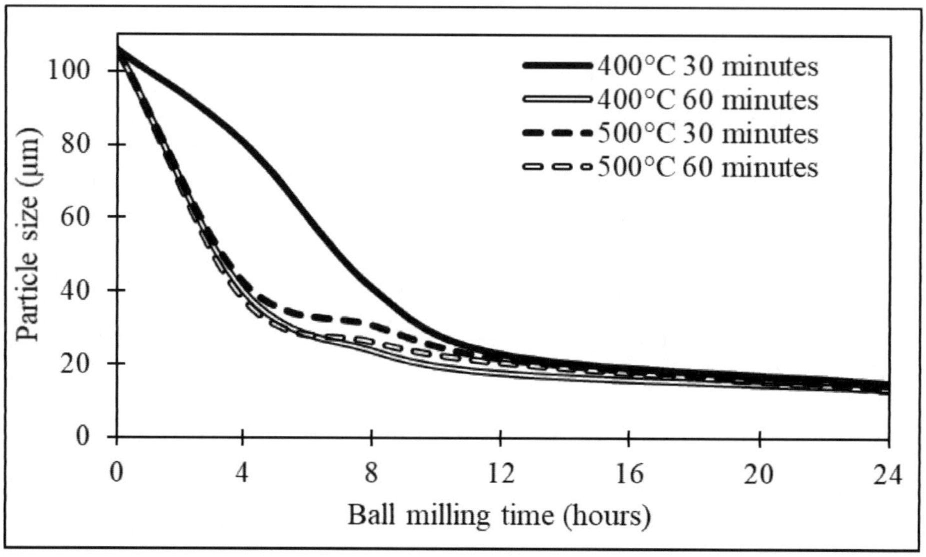

Fig. 5. Particle size of PKS biochar during wet ball milling measured using Mastersizer 2000.

Fig. 6 shows PKS biochar before wet ball milling process while Fig. 7 shows PKS biochar after 24 hours of wet ball milling. The morphological images showed in both Fig. 6 and Fig. 7 is in agreement with the biochar size measured using Mastersizer 2000. Before ball milled, the particle size was at 105 μm and between 13-15 μm after ball milled for 24 hours.

Fig. 6. PKS biochar before ball milling at <105 μm. (a) 400°C; 30 minutes (b) 400°C; 60 minutes (c) 500°C; 30 minutes (d) 500°C; 60 minutes.

Fig. 7. PKS biochar after ball milling at 13 - 15 µm. (a) 400°C; 30 minutes (b) 400°C; 60 minutes (c) 500°C; 30 minutes (d) 500°C; 60 minutes.

3.7. Rheological properties of biochar dispersion

Shear stress against shear rate results is shows in Fig. 8 while viscosity against shear rate results is shows in Fig. 9. From the results it showed that there is no clear influence of the pyrolysis temperature or residence time on the steady state rheological properties. However, it is noticeable that all biochar dispersion exhibit shear-thinning behaviour. Addition of biochar dispersion into other suspension such as natural rubber latex would lead to poor mixing thus, this rheology study is important to investigate the flow behaviour of PKS biochar dispersion due to its characteristic came from its flocculated nature [34], [35].

IOP Conf. Series: Materials Science and Engineering **548** (2019) 012020 doi:10.1088/1757-899X/548/1/012020

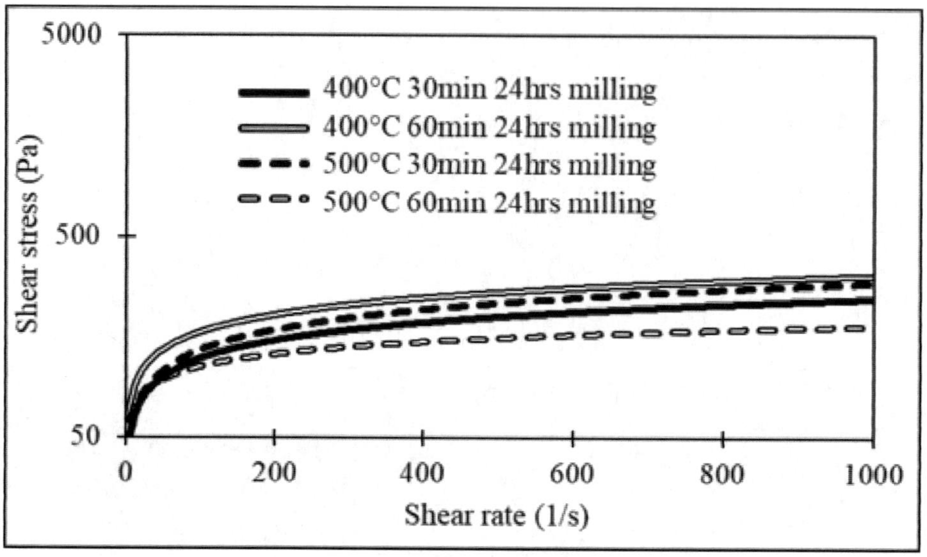

Fig. 8. Shear stress against shear rate of PKS biochar dispersion.

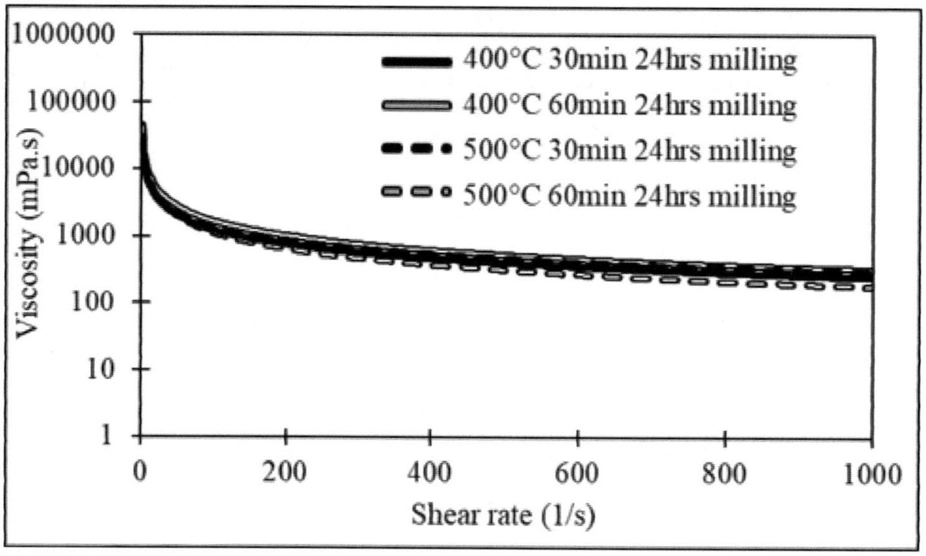

Fig. 9. Viscosity against shear rate of PKS biochar dispersion.

4. Conclusion

Pyrolysis parameters such as temperature and time affects the biochar yield. Higher temperature and longer time give lower yield of biochar. Other properties that are decreasing as the pyrolysis temperature and time increased are volatile matter content, moisture content, H/C and O/C ratio. In contrast, pH value, fixed carbon content and carbon content increased as the pyrolysis temperature and time increased. Increasing pyrolysis temperature and time also improved grindability of biochar prior to 12 hours of wet ball milling. However, no clear effect on the rheological properties of biochar dispersion as the pyrolysis temperature and time increased. Based on the results, it can be concluded that the pyrolysis temperature has greater impact on the biochar physico-chemical characteristics compared to pyrolysis time and subsequently affect the grindability.

5. Acknowledgement

The authors acknowledge the assistance provided by Universiti Kuala Lumpur (UniKL), Malaysian Palm Oil Board (MPOB) and Malaysian Rubber Board (MRB) in funding and facilities that resulted in this article. The authors would like to thank the Plastics and Rubber Institute of Malaysia (PRIM) on providing the platform to share these findings.

References

[1] Yashas Gowda T G, Sanjay M R, Subrahmanya Bhat K, Madhu P, Senthamaraikannan P, and Yogesha B 2018 Polymer matrix-natural fiber composites: An overview *Cogent Eng* **5** pp. 1–13
[2] Dul S, Id A P, and Fambri L 2018 Effects of the Nanofillers on Physical Properties of Acrylonitrile-Butadiene-Styrene Nanocomposites: Comparison of Graphene Nanoplatelets and Multiwall Carbon Nanotubes *Nanomaterials* **8** p. 674
[3] Silva M, Vale D, Rocha J, Rocha N, and Santos R M 2018 Synergetic effects of carbon nanotube- graphene nanoplatelet hybrids in carbon fibre reinforced polymer composites, *5th Int. Conf. Eng. Against Fail. (ICEAF-V 2018)* **188** pp. 6–13
[4] Abbasi H, Antunes M, and Velasco J I 2018 Effects of carbon nanotubes/graphene nanoplatelets hybrid systems on the structure and properties of polyetherimide-based foams *Polymers (Basel)* **10**
[5] Lee Y J *et al.* 2018 Highly Dispersible Buckled Nanospring Carbon Nanotubes for Polymer Nano Composites, *Sci. Rep.* **8** pp. 1–10
[6] Hardy A, Dix J, Williams C D, Siperstein F R, Carbone P, and Bock H 2018 Design Rules for Graphene and Carbon Nanotube Solvents and Dispersants *ACS Nano* **12** pp. 1043–1049
[7] Song S H 2018 The Effect of Carbon Black / Multiwall Carbon Nanotube Hybrid Fillers on the Properties of Natural Rubber Nanocomposites *Int. J. Polym. Sci.* **2018** p. 8
[8] Chuah T G, Wan Azlina A G K, Robiah Y, and Omar R 2006 Biomass as the renewable energy sources in Malaysia: An overview *Int. J. Green Energy* **3** pp. 323–346
[9] Nanda S, Dalai A K, Berruti F, and Kozinski J A 2016 Biochar as an Exceptional Bioresource for Energy, Agronomy, Carbon Sequestration, Activated Carbon and Specialty Materials *Waste and Biomass Valorization* **7** pp. 201–235
[10] Tan X *et al.* 2015 Application of biochar for the removal of pollutants from aqueous solutions *Chemosphere* **125** pp. 70–85
[11] Raclavsky K, Raclavska H, Kovalova L, Skrobankova H, and Frydrych J 2015 Utilization of carbon produced by torrefaction of grass for energy purposes and related risks *2015 IEEE 15th Int. Conf. Environ. Electr. Eng. EEEIC 2015 - Conf. Proc.* pp. 1972–1976
[12] El-Hendawy A N A 2005 Surface and adsorptive properties of carbons prepared from biomass *Appl. Surf. Sci.* **252** pp. 287–295
[13] Dawoud B, Amer E, and Gross D 2007 Experimental investigation of an adsorptive thermal energy storage *Int. J. energy Res.* **31** pp. 135–147
[14] Zabaniotou A, Ioannidou O, Antonakou E, and Lappas A 2008 Experimental study of pyrolysis for potential energy, hydrogen and carbon material production from lignocellulosic biomass *Int. J. Hydrogen Energy* **33** pp. 2433–2444
[15] Phan A N, Ryu C, Sharifi V N, and Swithenbank J 2008 Characterisation of slow pyrolysis products from segregated wastes for energy production *J. Anal. Appl. Pyrolysis* **81** pp. 65–71
[16] de Figueredo N A, da Costa L M, Melo L C A, Siebeneichlerd E A, and Tronto J 2017 Characterization of biochars from different sources and evaluation of release of nutrients and contaminants *Rev. Cienc. Agron.* **48** pp. 395–403
[17] Crombie K, Mašek O, Sohi S P, Brownsort P, and Cross A 2013 The effect of pyrolysis conditions on biochar stability as determined by three methods *GCB Bioenergy* **5** pp. 122–131
[18] Mohamed A R, Hamzah Z, Daud M Z M, and Zakaria Z 2013 The effects of holding time and the sweeping nitrogen gas flowrates on the pyrolysis of EFB using a fixed bed reactor *Procedia Eng.* **53** pp. 185–191
[19] Ronsse F, van Hecke S, Dickinson D, and Prins W 2013 Production and characterization of slow pyrolysis biochar: Influence of feedstock type and pyrolysis conditions *GCB Bioenergy* **5** pp. 104–115
[20] Fryda L and Visser R 2015 Biochar for Soil Improvement: Evaluation of Biochar from Gasification and Slow Pyrolysis *Agriculture* **5** pp. 1076–1115
[21] Rehrah D *et al.* 2014 Production and characterization of biochars from agricultural by-products for use in soil quality enhancement *J. Anal. Appl. Pyrolysis* **108** pp. 301–309
[22] Liu P, Zhu M, Leong Y K, Zhang Y, Zhang Z, and Zhang D 2017 An Experimental Study of the Rheological Properties and Stability Characteristics of Biochar-Algae-Water Slurry Fuels *Energy Procedia* **105** pp. 125–130
[23] Fu S Y, Feng X Q, Lauke B, and Mai Y W 2008 Effects of particle size, particle/matrix interface adhesion and particle loading on mechanical properties of particulate-polymer composites *Compos. Part B Eng.* **39** pp. 933–961
[24] Wang L *et al.* 2017 Impact of Torrefaction on Woody Biomass Properties *Energy Procedia* **105** pp. 1149–1154
[25] Abdullah H and Wu H 2009 Biochar as a fuel: 1. Properties and grindability of biochars produced from the pyrolysis of mallee wood under slow-heating conditions *Energy and Fuels* **23** pp. 4174–4181
[26] Hlabangana N, Danha G, and Muzenda E 2018 Effect of ball and feed particle size distribution on the milling efficiency of a ball mill: An attainable region approach *South African J. Chem. Eng.* **25** pp. 79–84

[27] Zhao Q, Yamada S, and Jimbo G 1989 The Mechanism and Grinding Limit of Planetary Ball Milling **302** pp. 297–302

[28] Bartos C, Szabó-Révész P, Bartos C, Katona G, Jójárt-Laczkovich O, and Ambrus R 2016 The Effect of an Optimized Wet Milling Technology on the Crystallinity, Morphology and Dissolution Properties of Micro- and Nanonized Meloxicam *Molecules* **21**

[29] Rajkovich S, Enders A, Hanley K, Hyland C, Zimmerman A R, and Lehmann J 2012 Corn growth and nitrogen nutrition after additions of biochars with varying properties to a temperate soil *Biol. Fertil. Soils* **48** pp. 271–284

[30] Titiladunayo I F, McDonald A G, and Fapetu O P 2012 Effect of temperature on biochar product yield from selected lignocellulosic biomass in a pyrolysis process *Waste and Biomass Valorization* **3** pp. 311–318

[31] Sun J, He F, Pan Y, and Zhang Z 2017 Effects of pyrolysis temperature and residence time on physicochemical properties of different biochar types *Acta Agric. Scand. Sect. B Soil Plant Sci.* **67** pp. 12–22

[32] Zhao B *et al.* 2018 Effect of pyrolysis temperature, heating rate, and residence time on rapeseed stem derived biochar *J. Clean. Prod.* **174** pp. 977–987

[33] Tag A T, Duman G, Ucar S, and Yanik J 2016 Effects of feedstock type and pyrolysis temperature on potential applications of biochar *J. Anal. Appl. Pyrolysis* **120** pp. 200–206

[34] Tadros T F 1992 Control of stability/flocculation and rheology of concentrated suspensions *Pure Appl. Chem.* **64** pp. 1715–1720

[35] Goodwin J W and Reynolds P A 1998 The rheology of flocculated suspensions *Curr. Opin. Colloid Interface Sci.* **3** pp. 401–407

IOP Conf. Series: Materials Science and Engineering **548** (2019) 012021 doi:10.1088/1757-899X/548/1/012021

Purification of Monomers Leads to High-Quality Lignin Macromonomers

S.A.R. Syed Azhar[1,2] and Dong June Ahn[1,3*]

1. Department of Chemical and Biological Engineering, College of Engineering, Korea University, Seoul 02841, Republic of Korea.
2. Universiti Kuala Lumpur, Malaysian Institute of Chemical & Biological Technology, Lot 1988, Vendor City Industrial Area, 78000 Alor Gajah, Malacca, Malaysia
3. KU-KIST Graduated School of Converging Science & Technology, Korea University, Seoul 02841, Republic of Korea.

E-mail: ahn@korea.ac.kr

Abstract. Purification is critical in any chemical process. The removal of impurities will produce the product in better quality and high standard. In this study, a new type of monomer was prepared by condensation polymerization of alkali lignin (AL) and methacryloyl chloride (MAC). The effect of AL/MAC ratio and the purification of MAC were investigated. The physical and chemical properties of the product obtained which is lignin methacrylate (LMA) were characterized by Fourier Transformed Infrared spectroscopy (FT-IR) and Nuclear Magnetic Resonance (^{1}H-NMR). The result reveals that the purification was successfully removed the hydroxyl groups from the stabilizer in MAC and it is found that all the hydroxyl groups are coming from AL. The single interaction between AL and MAC has successfully enhanced the product obtain which does show no hydroxyl groups. The ^{1}H-NMR data also showed that the purification of MAC influences the outcome. The noise and impurities were eliminated after the purification of MAC, and more pure products were obtained. This new monomer (LMA) synthesis that can be further utilized for various applications.

1. Introduction

Paper and pulping industries produce a vast amount of by-products during the stage of its production. Lignin, being one of the by-products, is often considered as waste [1-4] and not having proper utilization. An estimated 50 million tons of lignin were generated every year [5-6] with less than 10% being satisfied exploited [3, 7]. Inappropriate disposal of lignin into the water source gave rise to an escalation in carbon oxygen demand (COD) [8-11] as well as a contribution towards water pollution on the environment. The exploration of lignin is still the main topics of research even though it is reported lacking in the proper applications [12-14]. The majority of lignin are burned to produce fuel. Meanwhile, some industries utilize lignin to generate the steam, electricity, and heat [15-17]. Even though the fuel made provides more energy than cellulose, this activity will exacerbate global warming since carbon dioxide is released into the environment and will facilitate the pollution of air [13]. Recycling or modification of lignin into useful products is the best ways to minimize the waste and foster revenue for the industries [1].

Lignin as the second largest polymer after the cellulose [18-20] is amorphous, polyphenolic and is a highly cross-linked polymer [1, 6, and 14]. Lignin is a three-dimensional polymer, polymer with infinite molecular weights and high surface area [20-22]. It is consist of 16-30% [5-6, 23] of the total biomass, and this percentage depends on the species and extraction methods. Lignin bonded to xylan (hardwood) and galactoglucomann (softwood) through covalent bonds [24-25]. The main advantage of lignin is that it is capable of withstanding extreme weather conditions, vigorous chemical reactions and it also has exceptional resistance to microorganism [25-26].

IOP Conf. Series: Materials Science and Engineering **548** (2019) 012021 doi:10.1088/1757-899X/548/1/012021

Nowadays the modification of polymer into value-added product seems to be more attractive to the scientists. The change of lignin can increase the number of functional groups and consequently, increases the specific chemical moieties [27-28]. The functional groups that initiated by lignin such as carboxyl, amine, hydroxyl, methoxyl and aldehyde introduce variations in the modification [14, 19, 29]. Besides that, lignin also reported having the ability to be as an adsorbent for metal ions removal [30]. Unmodified, modified and hybrid lignin show high capacity to remove metal ions such as lead, mercury, cadmium, copper, zinc, nickel, chromium, and cobalt [2-3, 5-23]. Though the lignin mechanism for the adsorption is still not well understood, the capability of lignin to adsorb the metal ions is even getting prominent considerations [5-6]. Up-to-today the discovery of lignin are shifted to more broadening and value-added products such as antibacterial, drug delivery, carbon fibre nanoparticles, epoxy resin, polyurethane foams and biodegradable film [26,31-34]. Lignin can also be used to replace commercial chelating agent since it is low toxicity, biodegradable, inexpensive and due to its continuous availability [24-25, 35-36].

Lignin-based macro-monomer published in several articles. The macro-monomer can improve the chemical and physical properties of the lignin. Recently, many researchers have reported the synthesis of a macro-monomer with different types of lignin such as biobutanol lignin [37], organosolv lignin [38], and hydroxypropyl lignin [39-40]. The macro-monomer synthesis from lignin with methacryloyl chloride remains an attractive topic among researchers where the protocol to produce this specific type of monomer has been highlighted [41]. However, during their recent studies, it found that the chemical used have not undergone further purification.

In this present study, we investigated the effect of chemicals purification and the ratio of lignin to a methacryloyl chloride and its influence onto the lignin segments. To the best of our knowledge, this goal has never attained yet. The monomer is synthesis similarly to the previous studies with some minor modifications [41]. The monomer produced is then characterized portrayal by using Fourier Transformed Infrared spectroscopy (FT-IR) and Nuclear Magnetic Resonance (^{1}H-NMR). The results show that noise and impurities successfully removed in the monomer produced after the purification of the chemicals.

2. Experimental

2.1. Materials

Alkali lignin (AL), triethylamine (TEA), tetrahydrofuran (THF) and methacryloyl chloride (MAC, 97%) were supplied by Sigma-Aldrich Korea Ltd. All this chemical were utilized without any purification. Prior used for the synthesis of lignin methacrylate, MAC was used with and without purification.

2.2. Preparation of lignin methacrylate (LMA) monomer

First, 0.1 g of lignin was dissolved in 40 mL THF. Then, 100.0 g of TEA was added under continuous stirring. Next, 100.0 g of MAC solution was added dropwise into the solution. Upon completion, the flask was immediately sealed, and the esterification process was performed at room temperature for 24 h. Then, the resultant TEA salt precipitate was separated using the filtration process, and the remaining THF solution was turned into a concentrate via the evaporation process. The resultant residue was dissolved in chloroform and an aqueous amount of sodium hydrogen carbonate. After that, water was used to rinse the remaining salts and methacrylic acid from the synthesized salt.

Finally, the final product was dried over magnesium sulfate and was evaporated under vacuum for 24 h. The resulting product was labeled as LMA monomer. Two different mass ratios of lignin and MAC were studied in this experiment which is 1:100 and 1:200. Besides, the purification effect of MAC on the lignin framework also was investigated.

IOP Conf. Series: Materials Science and Engineering **548** (2019) 012021 doi:10.1088/1757-899X/548/1/012021

2.3. Physico-chemical characterisation

Fourier Transform Infrared (FT-IR) measurements for all samples were recorded in the range between 4000 and 400 cm^{-1} by the spectrometer (Spectrum GX, Perkin Elmer). 1 mg of sample was mixed with 100 mg of KBr pellet during the preparation. The FT-IR spectra were then set to the transmission mode, and the analysis was performed at resolution 2.0 cm^{-1}. The nuclear magnetic resonance (^{1}H-NMR) spectra for LMC were recorded by a nuclear magnetic resonance spectrometer (Varian NMR Systems 500MHz) using DMSO-d$_6$ as the solvent. For the ^{1}H-NMR, the samples of LMA (100 mg) was dissolved in 0.5 ml of DMSO-d$_6$ for analysis of purposes.

3. **Results and discussion**

3.1. Characteristics of purified and unpurified methacryloyl chloride (MAC)

The interaction of MAC and lignin was investigated by the studies on the purification effect of MAC. The reaction scheme of LMA synthesis is shown in Fig.. 1.

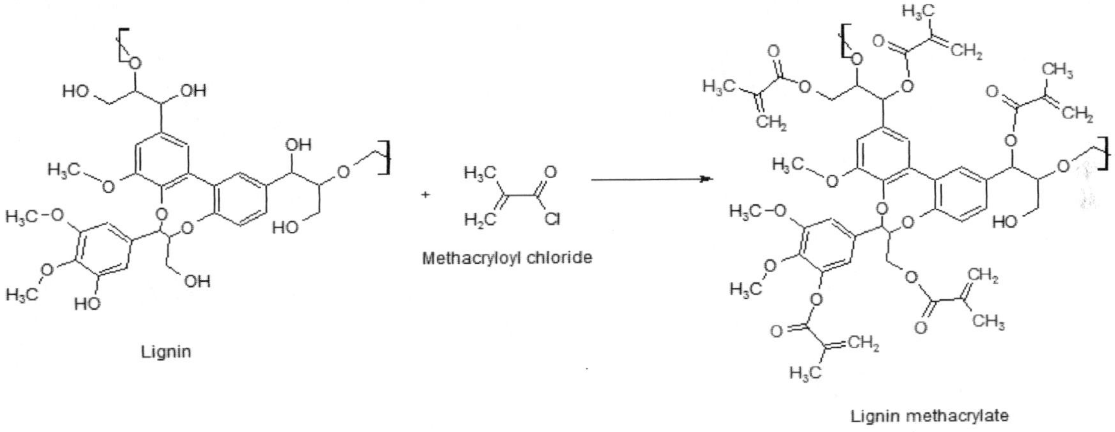

Fig. 1. The reaction scheme for the synthesis of lignin methacrylate

Fig. 2 showed the FT-IR spectra of the standard and purified MAC. At a band of 3200–3500 cm^{-1} regions, the conventional MAC spectra (a) had absorption peaks at 3460 cm^{-1}, which is assigned to OH stretching band. Meanwhile, the purified MAC spectra (b) shows that the stretching band of the hydroxyl group was disappeared. The appearance of new peaks in this region at 3552 cm^{-1}, 3479 cm^{-1} and 3420 cm^{-1} were assigned to the methyl group (-CH_3) of MAC segments. Also, we noted that the peak intensity at 1617 cm^{-1} bending mode of the carbonyl group of MAC segments became stronger and more apparent as compared to that of with FT-IR spectra in (a). This evidence implied that the OH group of monomethyl ether hydroquinone segments successfully removed from MAC by the distillation process. After the purification, it found that all the hydroxyl groups are coming from AL and not from the MAC. The ^{1}H-NMR spectra for purified MAC in DMSO-d$_6$ shown in Fig. 3.

IOP Conf. Series: Materials Science and Engineering **548** (2019) 012021 doi:10.1088/1757-899X/548/1/012021

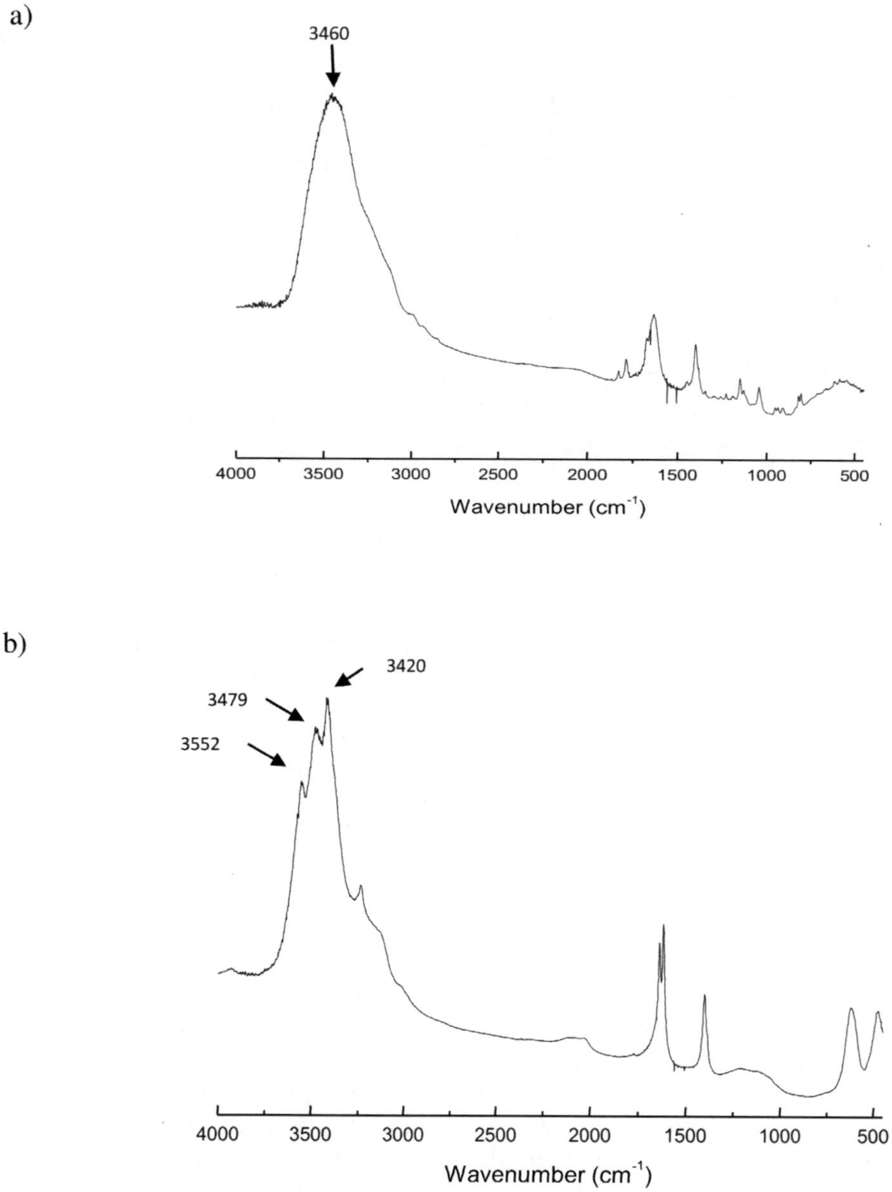

Fig. 2. Comparison of FT-IR spectra of a) conventional and b) MAC with purification.

IOP Conf. Series: Materials Science and Engineering **548** (2019) 012021 doi:10.1088/1757-899X/548/1/012021

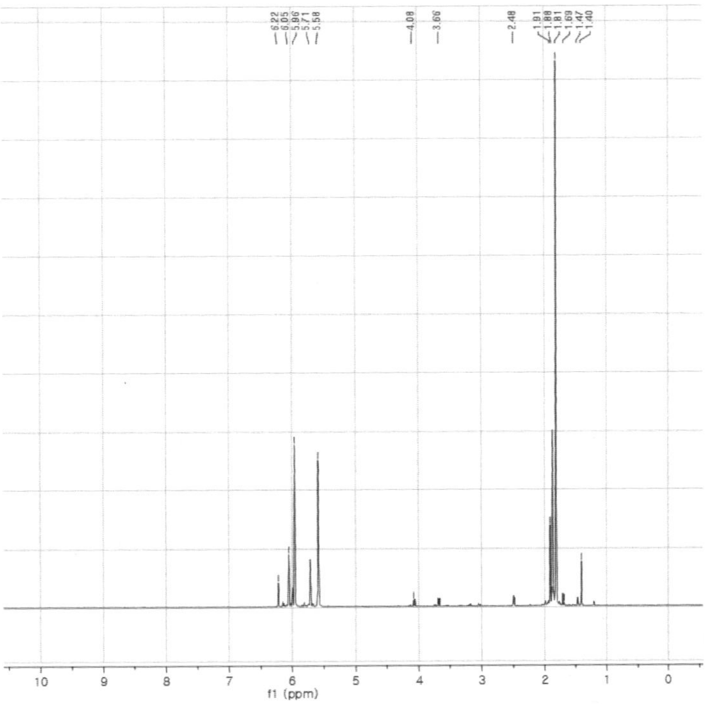

Fig. 3. ^{1}H-NMR spectra of alkali lignin

3.2. Characteristics of lignin methacrylate (LMA) by FT-IR spectroscopy

FT-IR spectra of lignin methacrylate prepared at a mass ratio of 1:100 and 1:200 between lignin and MAC were compared and used to confirm the presence of the MAC segments in the lignin framework.

As per shown in Fig.s 4 and 5, the IR band appeared at 1725 cm^{-1} and 1637 cm^{-1}. These bands assigned to C=O stretch which originates from MAC segments while -C=C- stretch from a vinyl group respectively. In the framework of LMA, it found the group of -C=O from MAC was present. The vinyl stretch together with ester group also observed at 1637 cm^{-1} and 1298 cm^{-1} respectively. The vinyl group intensity at 3235 cm^{-1} increases meanwhile, the band at 3500 cm^{-1} for hydroxyl group was unable to be detected in both cases of LMA, which explains the occurrence of interaction between MAC within the lignin structure. These indicated that the LMA was successfully synthesized using esterification process.

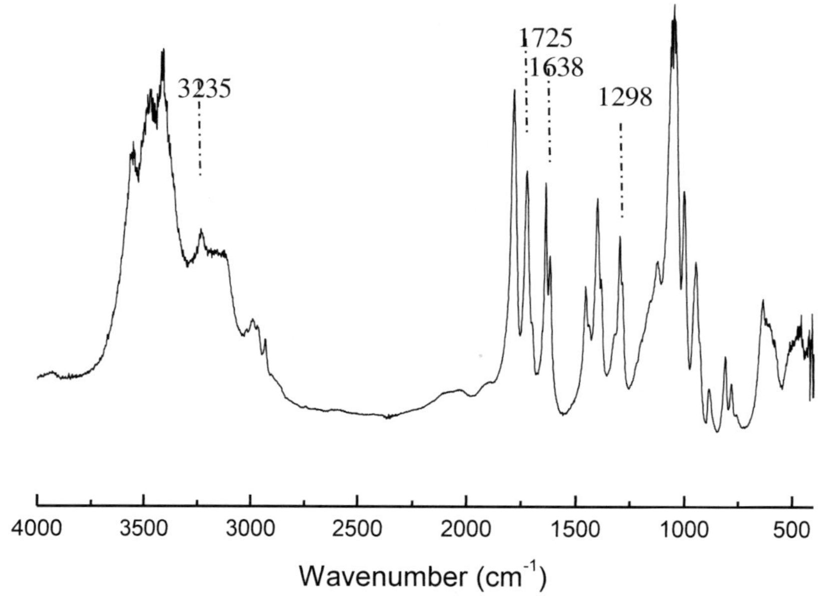

Fig. 4. FT-IR spectra of lignin methacrylate (LMA) with the ratio (1:100 wt%)

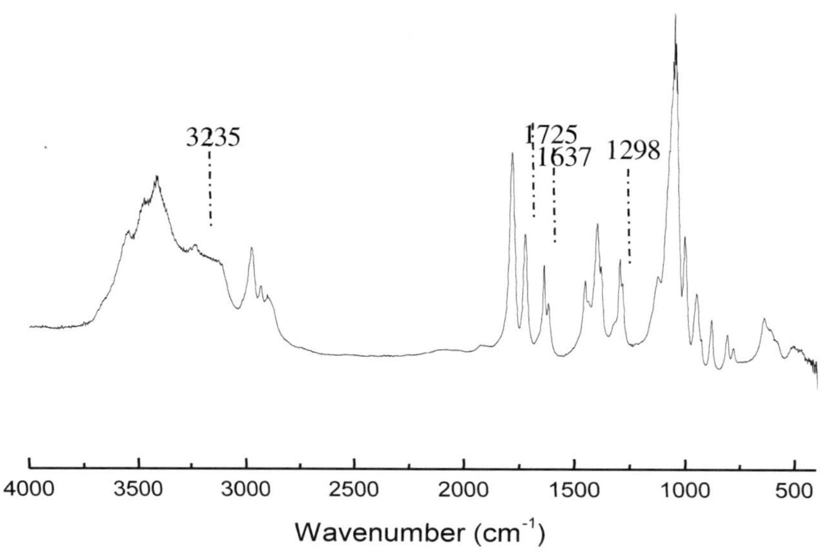

Fig. 5. FT-IR spectra of lignin methacrylate (LMA) with the ratio (1:200 wt%)

The comparison of LMA with a different mass ratio of 1:100 and 1:200 using purified and unpurified of MAC investigated by using FT-IR analysis. Fig.. 6 shows that the peak at the range 1290 cm^{-1} representing ester group as a product of the reaction between MAC and lignin. The height of this product can indicate in the FT-IR spectra after purification process The LMA monomer formation verified by the absence of some characteristic signals of a hydroxyl group and the appearance of the double bond signals in the IR spectra. Besides that, it observed that in the LMA monomer, there are two indications of absorption bands which can show at 953 cm^{-1} (C-H bending of a vinyl group) and another one at bands at 1637 cm^{-1} (C=C stretching vibration). These two bands verify the formation of lignin methacrylate from the primary reaction between lignin and MAC.

a)

b)

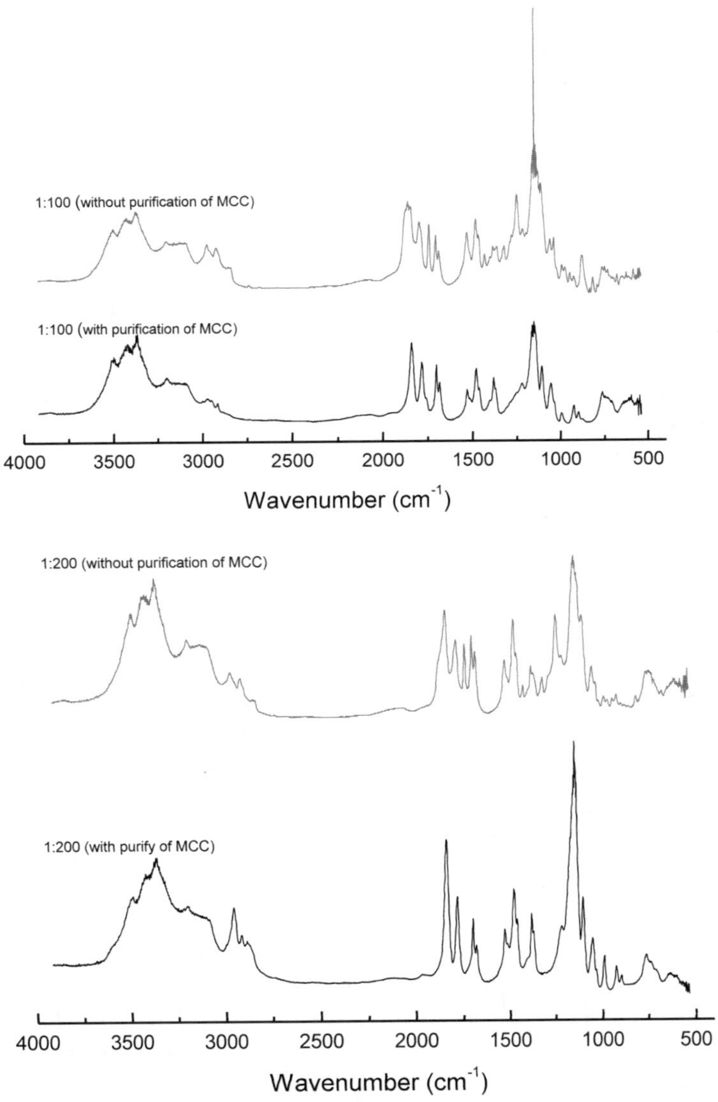

Fig. 6. FT-IR of LMA with mass ratio 1:100 (a) and 1:200 (b) with and without purification

3.3. Characteristics of lignin methacrylate by 1H-NMR spectroscopy

The interaction behavior between MAC and lignin structure analyzed by using NMR spectroscopy. Fig. 7 presents a comparison of the ^{1}H-NMR spectra for lignin methacrylate of the different ratio of MAC with and without purification. Fig. 7 a (ii) and b (ii) show that the signals at 6.24 and 6.02 ppm were indicated the vinyl protons in the lignin methacrylate. However, the intensity of vinyl proton peak in spectra a (i) and b (i) is lower compared to that of lignin methacrylate with the purification of MAC. The peak at 1.93 ppm for both ratios of lignin methacrylate show the presence of protons in the aromatic group from the lignin segments. Meanwhile, as for the lignin methacrylate without purification of MAC, the impurities can be detected at the region of 1.93 ppm as shown in spectra a (i) and b (i). The signal presence at 6.02 ppm (1H), 6.24 ppm (2H), 1.93 ppm (3H, -CH_3) and 2.51 ppm (2H, -CH_2) and 3.59-3.58 ppm (1H, -C-O-CH) prove the formation of lignin methacrylate (LMA).

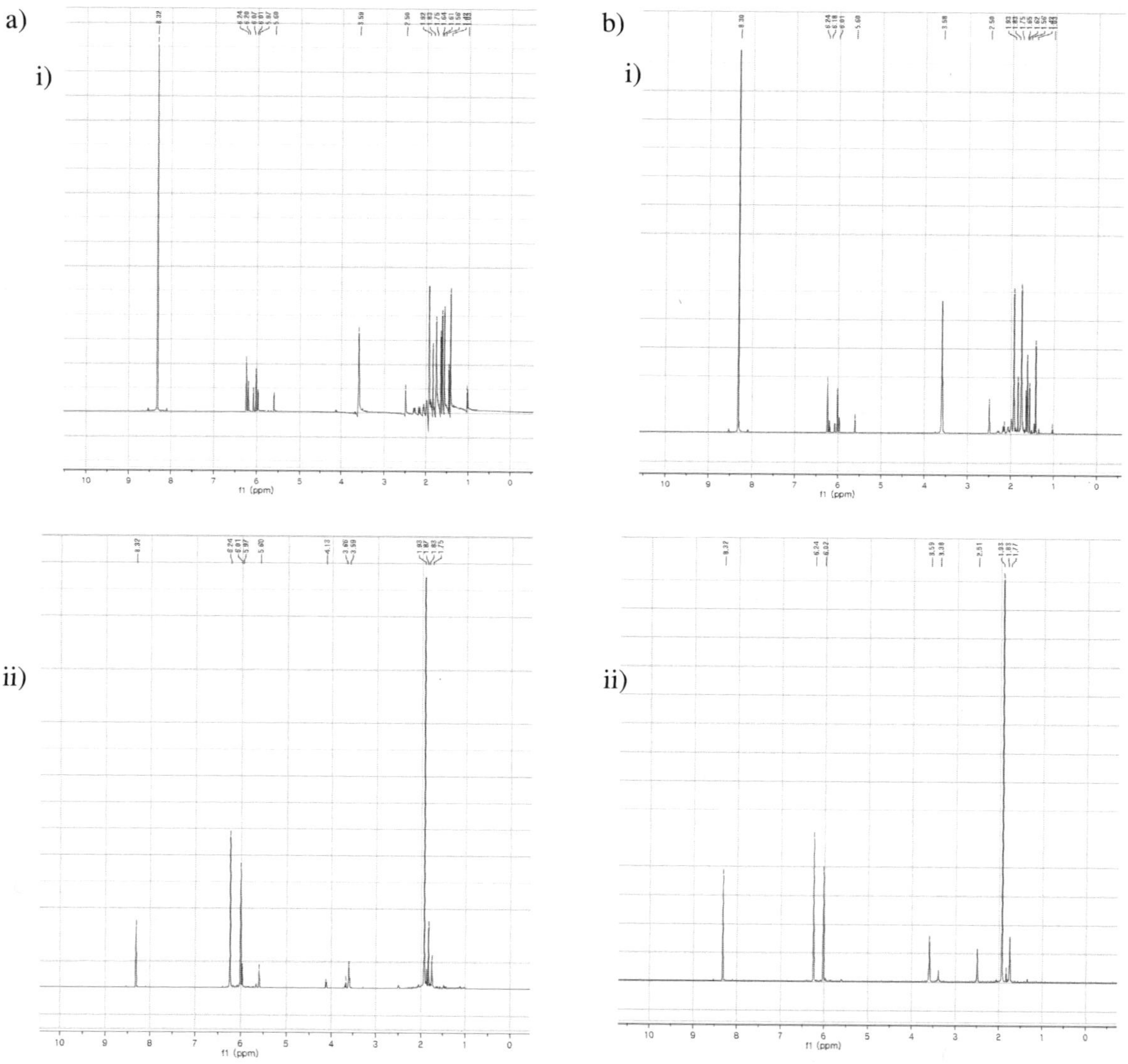

Fig. 7. ^{1}H NMR of LMA at mass ratio 1:100 (a) and 1:200 (b) with (i) and without (ii) purification of MAC

4. Conclusions

A new monomer, lignin methacrylate (LMA) was successfully synthesized by the condensation of methacryloyl chloride (MAC) with alkali lignin (AL) in the presence of triethylamine (TEA) in anhydrous tetrahydrofuran under N_2 gasses at room temperature. FT-IR spectra show that the purification of MAC is successfully removed the hydroxyl groups from the stabilizer (monomethyl ether hydroquinone). After the purification, we found that all the hydroxyl groups are coming from the AL segments and not from MAC solvent. The final product which is lignin methacrylate does not show any hydroxyl groups in the FT-IR spectra shows the reaction successfully enhanced by the individual response between AL and MAC. FT-IR spectra also indicate that vinyl group from MAC segments successfully interact with lignin **s**tructure for both ratio 1:100 and 1:200. However, the peak of the ester group detected with the purification of MAC. The LMA with a ratio of 1:200 then was considered as the best parameter in the production of lignin methacrylate with the purification of MAC solvent. The ^{1}H-NMR result provides evidence that the purification of MAC will reduce the noise and impurities when compared to the one in the case of without purification. Both results from FT-IR and ^{1}H-NMR show that the purification of MAC is crucial and plays a significant role in the lignin segment during the synthesis of better LMA macromonomer.

5. Acknowledgments

This work was supported by the Majlis Amanah Rakyat (MARA) and National Research Foundation of Korea (NRF, 2017R1A2B3006770, 2017M3D1A1039421), KU-KIST Graduate School of Converging Science and Technology and a Korea University Grant.

References

[1] Garcia-Valls R Hatton T A 2003 Metal ion complexation with lignin derivatives *Chem. Eng. J.* **94** 99-105
[2] Wysokowski M Klapiszewski L Moszynski D Bartczak P Szatkowski T Majchrzak I Siwinska- Stefanska K Bazhenov V V Jesionowski T 2014 Modification of chitin with kraft lignin and development of new biosorbents for removal of cadmium (II) and nickel (II) ions, *Mar. Drugs* **12** 2245-2268
[3] Sciban M B Klasnja M T Antov M G 2011 Study of the biosorption of different heavy metal ions onto Kraft lignin *Ecological Engineering* **37** 2092-2095
[4] Stewart D 2008 Lignin as a base material for materials applications: chemistry, application and economics *Ind. Crops Prod.* **27** 202-207.
[5] Guo X Zhang S Shan X-Q 2008 Adsorption of metal ions on lignin *J. Hazard. Mater.* **151** 134-142
[6] Mohan D Pittman Jr C U Steele P H 2006 Single, binary and multi-component adsorption of copper and cadmium from aqueous solutions on Kraft lignin-a biosorbent *J. Colloid Interface Sci.* **297** 489-504
[7] Liu X Zhu H Qin C Zhou J Zhao J R Wang S 2013 Adsorption of heavy metal ion from aqueous single metal solution by aminated epoxy-lignin *BioResource* **8**(2) 2257-2269.
[8] Wu Y Zhang S Guo X Huang H 2008 Adsorption of chromium (III) on lignin *Bioresource Technol.* **99** 7709-7715
[9] Maldhure A V Ekhe J D 2011 Preparation and characterizations of microwave assisted activated carbons from industrial waste lignin for Cu(II) sorption *Chem. Eng. J.* **168** 1103-1111
[10] Gao Y Yue Q Gao B Sun Y Wang W Li Q Wang Y 2013 Preparation of high surface area-activated carbon from lignin of papermaking back liquor by KOH activation for Ni(II) adsorption *Chem. Eng. J.* l **217**345-353
[11] Lv J Luo L Zhang J Christie P Zhang S 2012 Adsorption of mercury on lignin: combined surface complexation modelling and X-ray absorption spectroscopy studies *Environ. Pollut.* **162** 255-261
[12] Merdy P Guillon E Aplincourt M Dumonceau J Vezin H 2002 Copper sorption on a straw lignin: experiments and EPR characterization *J. Colloid Interface Sci.* **245** 24-31
[13] Lu Q-F Huang Z-K Liu B Cheng X 2012 Preparation and heavy metal ions biosorption of graft copolymers from enzymatic hydrolysis lignin and amino acid, *Bioresource. Tech.* **104** 111-118
[14] Mohammad Ibrahim M N Wan Ngah W S Norliyana M S Wan Daud W R 2009 Copper (II) biosorption on soda lignin from oil palm empty fruit bunches (EFB) *Clean* **37**(1):80-85
[15] Dizhbite T Zakis G Kizima A Lazareva E Rossinskaya G Jurkjane V Telysheva G Viesturs U 1999 Lignin-a useful bioresource for the production of sorption-active materials, *Bioresource. Tech.* **67**(3) 221-228
[16] Kriaa A Hamdi N Srasra E 2010 Removal of Cu(II) from water pollutant with Tunisian activated lignin prepared by phosphoric acid activation *Desalination* **250** 179-187
[17] Harmita H Karthikeyan K G Pan X J 2009 Copper and cadmium sorption onto Kraft and organosolv lignins, *Bioresource. Tech.* **100** 6183-6191
[18] Wysokowski M Klapiszewski L Moszynski D Bartczak P Szatkowski T Majchrzak I Siwinska-Stefanska K Bazhenov V V Jesionowski T 2014 Modification of chitin with Kraft lignin and development of new biosorbents for removal of cadmium (II) and nickel (II) ions *Mar. Drugs.* **12** 2245-2268

[19] Lalvani S B, Hubner A, Wiltowski T S 2000 Chromium adsorption by lignin, *Energy. Sour.* 22:45-56.
[20] Sciban M B, Klasnja M T, Antov M G 2011 Study of the biosorption of different heavy metal ions onto Kraft lignin, *Ecol. Eng.* **37** 2092-2095
[21] Demirbas A, 2007 Adsorption of Co(II) and Hg(II) from water and wastewater onto modified lignin, *Energy Sour. Part A* **29** 117-123
[22] Acemioglu B Samil A Alma M H Gundogan R 2003 Copper (II) removal from aqueous solution by organosolv lignin and its recovery *J Appl. Polym. Sci.* **89** 1537-1541
[23] Thakur V K Thakur M K 2015 Recent advances in green hydrogels from lignin: a review, International *J. Biol. Macromol.* **72** 834-847
[24] Celik A Demirbas A 2005 Removal of heavy metal ions from aqueous solutions via adsorption onto modified lignin from pulping wastes, *Energy. Sour.* **27** 1167-1177
[25] Demirbas A 2004 Adsorption of lead and cadmium ions in aqueous solutions onto modified lignin from alkali glycerol delignification, *J. Hazard. Mater. B* **109** 221-226.
[26] Wang J Manley R St J Feldman D 1992 Synthetic polymer-lignin copolymers and blends *Prog. Polym. Sci.* **17** 611-646
[27] Gregorova A Sedlarik V 2016 Characterization of structural and physical properties of dichloromethane-and methanol-fractionated Kraft lignin and its adsorption capacity of Cu(II) and NI(II) ions, *Desalin. Water Treat.* 10655-10663
[28] Ge Y Qin L Li Z 2016 Lignin microspheres: an effective and recyclable natural polymer-based adsorbent for lead ion removal, *Mater. Design.* **95** 141-147
[29] Lalvani S B,Wiltowski T S Murphy D Lalvani L S 1997 Metal removal from process water by lignin, *Environ. Technol.* **18** 1163-1168
[30] Albadarin A B Al-Muhtaseb A H Al-Laqtah N A Walker G M Allen S A Ahmad M N M 2011 Biosorption of toxic chromium from aqueous phase by lignin: mechanism, effect of other metal ions and salts, *Chem. Eng. J.* **169** 20-30
[31] Tan E Vermerris W 2015 Recent developments in polymers derived from industrial lignin, *J. Appl. Polym. Sci.* 42069-42069
[32] Laurichesse S Averous L 2014 Chemical modification of lignins: towards biobased polymers, *Prog. Polym. Sci.* **39**1266-1290
[33] Lora J H Glasser W G 2002 Recent industrial applications of lignin: a sustainable alternative to renewable materials *J. Polym. Env.* **10** 39-48
[34] Feldman D Lacasse M Beznaczuk L M 1996 Lignin-polymer systems and some applications *Prog. Polym. Sci* **12** 271-299
[35] Basso M C Cerrella E G Cukierman A L 2014 Cadmium uptake by lignocellulosic materials: effect of lignin content, *Sep. Sci. Technol.* **39** 1163-1175
[36] Crist D R Crist R H Martin J R 2003 A new process for toxic metal uptake by a Kraft lignin *J. Chem. Technol. Biotechnol.* **78** 199-202
[37] Liu X Wang J Yu J Zhang M Wang C Xu Y Chu F 2013 Preparation and characterization of lignin based macromonomer and its copolymers with butyl methacrylate *J. Bio. Macromol.* **60** 309-315
[38] Rials T G Glasser W F 1989 Multiphase materials with lignin: IV. Blends of hydroxypropyl cellulose with lignin, *J. Appl. Polym. Sci.* **37** 2399-2415
[39] Ciemniecki S L Glasser W G 1988 Multiphase materials with lignin: 2. Blends of hydroxypropyl lignin with poly(vinyl alcohol), *Polymer* **29**(6) 1030-1036
[40] Ciemniecki S L Glasser W G 1988 Multiphase materials with lignin: 1. Blends of hydroxypropyl lignin with poly(methyl methacrylate) *Polymer* **29**(6)1021-1029
[41] Cunha C D Deffieux A Fontanille M 1993 Synthesis and polymerization of lignin-based macromonomers. III. Radical copolymerization of lignin based macromonomers with methyl methacrylate *J. Appl. Polym. Sci.* **48** 819-831

IOP Conf. Series: Materials Science and Engineering **548** (2019) 012022 doi:10.1088/1757-899X/548/1/012022

Migration of N-nitrosamines from rubber gloves for handling food - Effect of extraction media

O Pinprayoon[1] **and W Mae**[1]

[1] Rubber and Rubber Products Group, Department of Science Service, Bangkok, Thailand

E-mail: porawan@dss.go.th

Abstract. Rubber gloves are used daily in variety applications including for food handling purposes. In order to assess possible exposure of the wearers to N-nitrosamines as well as possible release of N-nitrosamines to foods being contacted to these gloves, effects of extraction media are investigated. Artificial sweats (both alkaline and acid solutions), artificial saliva, and some food simulants are used for the extractions and twelve N-nitrosamines including NDMA, NDEA, NDPA, NDiBA, NDBA, NPIP, NPYR, NMOR, NEPhA, NMPhA, NDiNA, and NDBzA are analysed using GC-TEA. By using different extraction media, the levels of the released N-nitrosamines and N-nitrosatable substances are different leading to the differences in the judgement for the EU Directive compliance. By comparing with data of daily intake of N-nitrosamines from foods, the intake of N-nitrosamines migrated from gloves used for food handling under the most unfavourable circumstances could lead to an exceeding of dietary intake. Therefore possible risks associated with the use of rubber gloves for food handling are considered high.

1. Introduction

Considerable interest has been shown in literatures concerning the release of nitrosamines and nitrosatable substances from rubber products [1-16, 19-20]. Elastomer teats and soothers were amongst the first rubber products of interest since they involved child use and care articles. Extensive studies as well as risk assessments had been conducted and limits of nitrosamines and nitrosatable substances released from teats and soothers were later imposed by authorities in order to minimize risk being associated with these products [17-18]. Balloon and condom were also the rubber products of which risk for human health relating N-nitrosamine has been concerned. However, there are number of studies supporting the conclusion that risk for an induction of tumors from nitrosamines in balloons and condoms was very low [8-10, 21-23]. For rubber gloves, there are several works investigating on the release of N-nitrosamines and N-nitrosatable substances form these products using artificial sweat as well as artificial saliva as the extraction media [11-14]. However, a literature focusing on the effects of extraction media is lacking. When rubber gloves are used, particularly in food handling applications, two possible exposures to N-nitrosamine have to be concerned. For glove users, the exposure may occur through the wearer skin whilst for food consumers, the possible exposure may occur through the release of N-nitrosamines to foods being contacted to the gloves.

In this study, an effect of extraction media used to extract nitrosamines from rubber gloves was investigated. Artificial saliva, artificial sweats, and some food simulants were used as the extraction media. This work is intended to gather information needed for an establishment of test method suitable for the determination of N-nitrosamines released from rubber gloves used for food handling purposes as well as for the assessment of possible risk associated with the use of these products.

IOP Conf. Series: Materials Science and Engineering **548** (2019) 012022 doi:10.1088/1757-899X/548/1/012022

2. Materials and methods

2.1 Nitrosamine standards

N-nitrosamine standards used in this study are N-nitrosodimethylamine (NDMA, CAS No. 62-75-9), N-Nitrosodiethylamine (NDEA, CAS No. 55-18-5), N-Nitrosodipropylamine (NDPA, CAS No. 621-64-7), N-Nitrosodiisobutylamine (NDiBA, CAS No. 997-95-5), N-Nitrosodibutylamine (NDBA, CAS No. 924-16-3), N-Nitrosopiperidine (NPIP, CAS No. 100-75-4), N-Nitrosopyrrolidine (NPYR, CAS No. 930-55-2), N-Nitrosomorpholine (NMOR, CAS No. 59-89-2), N-nitroso N-ethyl N-phenylamine (NEPhA, CAS No. 612-64-6), N-nitroso N-methyl N-phenylamine (NMPhA, CAS No. 614-00-6), N-nitoso-N,N-di(3,5,5-trimethylhexyl)amine (NDiNA, CAS No. 1207995-62-7), N-Nitrosodibenzylamine (NDBzA, CAS No. 533-53-8) and N-Nitrosodiisopropylamine (NDiPA, CAS No. 601-77-4).

2.2 Extraction media

The artificial saliva was prepared in accordance with EN 12868:2017 [24] and used within five days. The artificial sweats were prepared as described in ISO 105-E04:2013 [25] and used within the day of preparation. 10% ethanol and 4% acetic acid solutions were used as aqueous food simulants.

2.3 Column for solid phase extraction

A pre-filled, single-use cartridge was used for solid phase extraction (SPE). The cartridges were HyperSep SLE pH 9 and HyperSep SLE pH 7 purchased from Thermo-Scientific.

2.4 Glove samples

Four glove samples comprised of two powder-free gloves made of natural latex and two powder-free gloves made of nitrile latex. Sample codes and average weights of the samples are given in Table 1.

Table 1. Sample codes and average weights glove samples

No.	Sample code	Material	Weight (g/glove)
1	NR1	Natural rubber	5.1
2	NR2	Natural rubber	36.4
3	NT1	Nitrile rubber	4.8
4	NT2	Nitrile rubber	2.8

2.5 Sample preparation and migration experiment

Rubber gloves were cut into pieces having sizes approximately 2 cm x 2 cm. Cut pieces of about five grams were extracted using different media (2.2). The migration experiment was carried out as prescribed in EN 12868 except for the boiling process and the migration time was shortened from 24 to 4 hours in order to mimic the real use of rubber glove in food industry. After migration process, the extract was purified through SPE cartridge, followed by concentrated to a volume of about 5 ml in water bath, rinsed with 1 ml of absolute ethanol and further concentrated to a volume of 1.0 ml by gently passing a flow of nitrogen over its surface. The concentrated extract was transferred to a GC vial for analysis.

2.6 Calibration solutions for GC-TEA

Calibration solutions were prepared from stock solutions of certified mixtures of twelve N-nitrosamines. At least five different calibration levels with concentration covering ranges from 2 ng/ml to 500 ng/ml for each N-nitrosamine standard were prepared by dilution with ethanol and addition of 100 ng/ml internal standard solution. A calibration blank solution as well as a sample blank solution also contains approximately 100 ng/ml of the internal standard

2.7 Analysis of twelve target N-nitrosamines

The determination of N-nitrosamines and N-nitrosatable substances released from samples were done in duplicate and the average values of each nitrosamine per sample weight (μg/kg) were calculated. The releases of the twelve target N-nitrosamines were determined using GC-TEA system and the results confirmed by GC-MS/MS. The GC-TEA analysis was performed on Thermo-Scientific Trace 1300 Gas Chromatography connected with Ellutia 800 Series Thermal Energy Analyser (TEA). The GC column used was TR-FFAP (0.32 μm internal diameter, 30 m length, 0.25 μm film thickness). The GC-MS/MS analysis was performed on Thermo-Scientific Trace 1300 GC System connected with Thermo-Scientific TSQ 8000 Evo Triple Quadrupole Mass Spectrometer. The GC column used was TG-WAXMS (0.32 μm internal diameter, 30 m length, 0.5 μm film thickness).

3. Results and discussion

3.1 Released of N-nitrosamines and their precursors

N-nitrosamine migration from four samples after extraction at 40 °C for 4 hours using different extraction media were analysed by GC-TEA. The amount of N-nitrosamines and N-nitrosatable substances, determined as N-nitrosamines, expressed in μg/kg of rubber are shown in Table 2. In general, NDBA and its precursor were detected in all samples with different levels depending on the extraction media used. Other target N-nitrosamines were also detected, i.e. NDMA and NDEA in Sample NT1 and Sample NR1 extracted with artificial alkaline-sweat, and NMOR in Sample NT2 extracted with artificial saliva. For the rest of target N-nitrosamines i.e. NDPA, NDiBA, NPIP, NPYR, NEPhA, NMPhA, NDiNA and NDBzA, the compounds were not detected.

Considering the migration levels with no respect to the extraction media, the total N-nitrosamines varied from 2.3 to 325 μg/kg whilst the total N-nitrosatable amines varied from 0 to 3,697 μg/kg. Of these released compounds, NDMA were found in range from 0 to 9.3 μg/kg, NDEA from 5.1 to 314 μg/kg and NDBA from 2.3 to 41.0 μg/kg. In addition, the N-the nitrosatable NDMA and the N-nitrosatable NDEA were not detected while the nitrosatable-NDBA were found in range from 0 to 3697 μg/kg.

Comparing to work of Feng *et al* [11-12], the authors reported that the detection frequency of NDMA, NDEA and NDBA in all latex glove samples were 23%, 62% and 62%, respectively. The samples being investigated were 13 latex gloves including seven household gloves, three sterile gloves and three examination gloves using similar extraction media and conditions, i.e. extraction with artificial sweat (pH 5.5) at 37 °C for 4 hour. The migration level of NDMA ranged from 23.02 to 151.99 μg/kg, NDEA ranged from 10 to 301.95 μg/kg and NDBA ranged from 57.00 to 1,909.31 μg/kg. The total N-nitrosamines varied from 15.98 to 1,919.23 μg/kg of which the highest level had exceeded approximately 192 times of the recommended limit of the EU Directive 93/11/EEC.

Table 2. N-nitrosamines and the N-nitrosatable substances released from glove samples using different extraction media

Sample	Compound Amount released, μg/kg	Extraction medium				
		Saliva[1]	Alk- sweat[2]	Ac-sweat[3]	10% Ethanol	3% Acetic acid
NT1	Total N-nitrosamines	***2.3***	***11.2***	***325***	***2.8***	***3.8***
	- NDBA	2.3	11.2	7.4	2.8	3.8
	- NDMA	*nd*	*nd*	4.0	*nd*	*nd*
	- NDEA	*nd*	*nd*	314	*nd*	*nd*
	Total N-nitrosatable substances	***13.6***	***1.0***	***12.6***	***3.4***	***0.4***
	- NDBA	13.6	1.0	12.6	3.4	0.4
NT2	Total N-nitrosamines	***12.3***	***16.8***	***12.7***	***9.2***	***6.1***
	- NDBA	5.1	16.8	12.7	9.2	6.1
	- NMOR	7.2	*nd*	*nd*	*nd*	*nd*
	Total N-nitrosatable substances	***59.0***	***1.6***	***77.4***	***2.4***	***0.3***
	- NDBA	59.0	1.6	77.4	2.4	0.3
NR1	Total N-nitrosamines	***20.7***	***41.0***	***266***	***40.9***	***24.3***
	- NDBA	20.7	41.0	30.1	40.9	24.3
	- NDMA	*nd*	*nd*	9.3	*nd*	*nd*
	- NDEA	*nd*	*nd*	227	*nd*	*nd*
	Total N-nitrosatable substances	***3697***	***52.2***	***164***	***55.9***	***0***
	- NDBA	3697	52.2	164	55.9	*nd*
NR2	Total N-nitrosamines	***11.2***	***21.6***	***25.3***	***40.8***	***11.3***
	- NDBA	11.2	21.6	25.3	40.8	11.3
	Total N-nitrosatable substances	***25.0***	***0***	***5.7***	***0***	***2.6***
	- NDBA	25.0	*nd*	5.7	*nd*	2.6

[1] Artificial saliva, [2] artificial alkaline-sweat, [3] artificial acid-sweat, "*nd*" not detected.

Generally, nitrosamines are formed as a result of a reaction between nitrosating agents and secondary amines in rubber. Sources of secondary amines are from accelerators used in cure systems. The amines being breakdown products produced as a result of the chemical reactions tacking place during vulcanization. From this study which NDBA, NDEA, NDMA and NMOR were detected, it can be implied that some certain accelerators being employed could be one or more compounds given in Table 3.

Table 3. Nitrosamines derived from some commonly used accelerators

Accelerator	Secondary amine	N-nitrosamine
TMTD, TMTM and ZDMC	Dimethylamine	NDMA
ZDEC and TETD	Diethylamine	NDEA
ZDBC	Dibutylamine	NDBA
MBS	Morpholine	NMOR

3.2 A closer look on the effect of extraction media

The effect of extraction media on the migration of N-nitrosamine and N-nitrosatable substance was closer looked by focusing on the migration levels of NDBA and its precursor. At the controlled migration condition i.e. temperature at 40 °C for 4 hours, the power of extraction media are illustrated in Figure 1.

It can be seen from Figure 1(A) that, for nitrile rubber gloves (Sample NT1and NT2) the extractable NDBA were migrated to the extraction medium from the highest to the lowest levels by extraction with alkaline-sweat, acid-sweat, 10 % ethanol, 3 % acetic acid and artificial saliva, respectively.

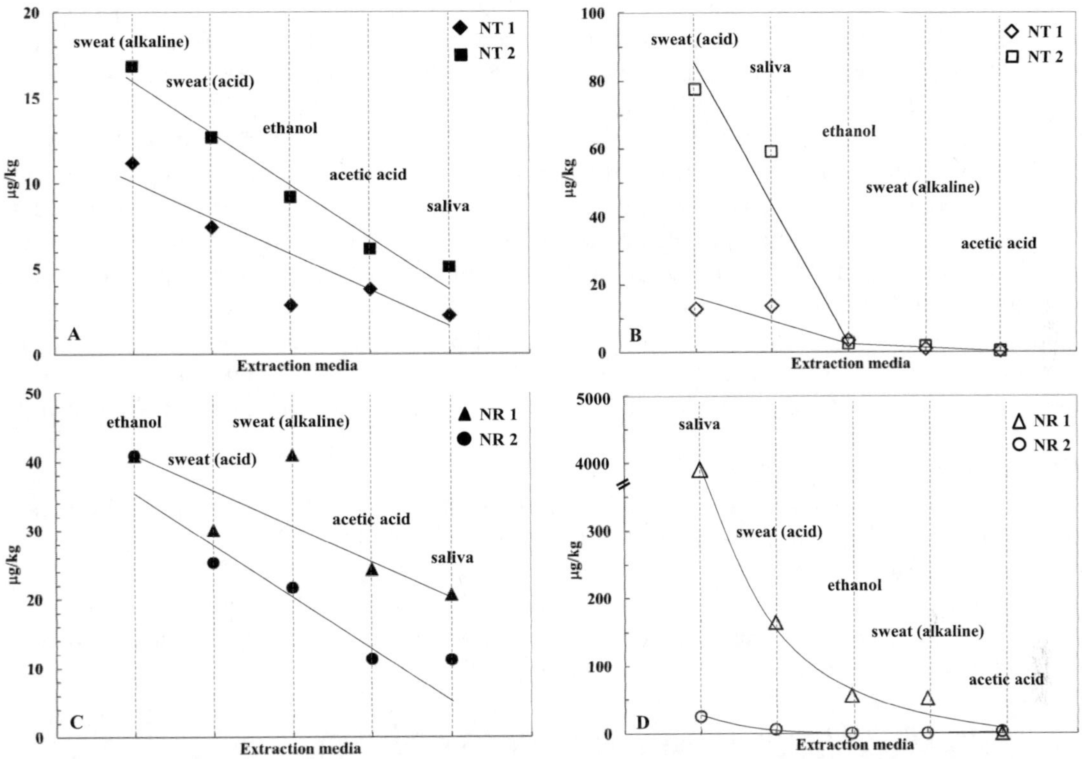

Fig. 1. NDBA and its N-nitrosatable substance released from glove samples using different extraction media after extraction at 40 °C for 4 hours (A) NDBA from nitrile gloves (B) nitrosatable-NDBA from nitrile gloves (C) NDBA from natural rubber gloves (D) nitrosatable-NDBA from natural rubber gloves. Solid lines are drawn for eye-guide only.

Interestingly, the artificial saliva which was regarded as the EU reference test solution for determination of N-nitrosamines and N-nitrosatable substances in teats and soothers are the least powerful to leach out NDBA. For the N-nitrosatable substance, Figure 1(B) shows that the NDBA-precursor was able to migrate from the samples in higher levels using acid-sweat than those of artificial saliva, 10 % ethanol and artificial alkaline-sweat, respectively.

Considering gloves made of natural rubber (Sample NR1and NR2), Figure 1(C), the released NDBA were detected with the highest to the lowest levels by extraction with 10 % ethanol, alkaline-sweat / acid-sweat, 3 % acetic acid and artificial saliva, respectively. Again, the artificial saliva was found to be the least powerful extraction medium for the NDBA. For the N-nitrosatable

IOP Conf. Series: Materials Science and Engineering **548** (2019) 012022 doi:10.1088/1757-899X/548/1/012022

substance, Figure 1(D) indicates almost the same trend as found for nitrile glove samples except the swap order between the acid-sweat and the artificial saliva.

Special attention had been paid on the effect of using acetic acid as the extraction medium. It was discovered that the amount of the NDBA detected in a test portion for the N-nitrosatable substance and N-nitrosamines (refer to EN 12868, Solution A and B, respectively) were very close and almost in the same level. For example, the NDBA detected in Solution A (Sample NT1) was found to be 4.2 μg/g whilst in Solution B the NDBA detected was found to be 3.8 μg/g. According to EN 12868, the nitrosatable-NDBA released from Sample NT1 was calculated to be 0.4 μg/g (see Table 2). In some particular occasions i.e. Sample NR1, the NDBA detected in Solution A was less than that of Solution B in a very close gap, leading to the result of a negative value of the nitrosatable-NDBA (in Table 2, shown as "not detected"). To elaborate this effect, it is essential to briefly explain the extraction procedure prescribed in EN 12868. In EN 12868, while Solution B is basic solution (pH 9), Solution A was made acidic by adding hydrochloric acid in order to enable nitrosation process. Consequently, the amount of nitrosamines detected in Solution A is expected to have higher value than that of Solution B (nitrosatable substances = nitrosamines detected in Solution A - nitrosamines detected in Solution B). In this work, the pH of the 3% acetic acid solution was approximately 2-3, therefore both solutions were already in acidic pH from the earlier state of the migration step. Consequently, the N-nitrosatable substances migrated from the sample during the migration period were probably began nitrosation process immediately after leaching out leading to the result in a similar amount of NDBA detected in Solution A and Solution B. This finding suggested that when the acetic acid is used as the extraction medium instead of artificial saliva, N-nitrosamines and N-nitrosatable substances may not be reported separately.

3.3 Compliance to the EU directive

Since there is no distinct standard test method and specific requirement for N-nitrosamines and N-nitrosatable substances in rubber products except for teats and soothers, it is often that the limits set in EU Directive 93/11/EEC is required and the EN 12868 is therefore being referred. If this also applied to rubber latex gloves and assuming that some modifications on the migration conditions are allowed, the total N-nitrosamines and N-nitrosatable amines released from the products shall not exceed the limits as the following values:

0.01 mg (10 μg) in total of N-nitrosamines /kg of rubber glove
0.1 mg (100 μg) in total of N-nitrosatable substances/kg of rubber glove

Using data in Table 2, judgement for compliance to EU Directive are listed in Table 4.

It can be seen from Table 4 that by using different extraction media, the levels of the released N-nitrosamines and N-nitrosatable substances would be different and the judgment for the EU Directive compliance would consequently be affected. For example, when using artificial saliva as the extraction medium, only Sample NT1 complied with the EU Directive. However when alkaline and acid sweats were used, none of samples complied with the Directive.

Table 4. Levels of N-nitrosamines released from samples and judgements for the EU Directive compliance

Sample	Compound Amount released, μg/g	Extraction medium Saliva	Alk- sweat	Ac-sweat	10% Ethanol	3% Acetic acid
NT1	Total N-nitrosamines	*2.3*	*11.2*	*325*	*2.8*	*3.8*
	Compliance to EU Directive	**Yes**	**No**	**No**	**Yes**	**Yes**
	Total N-nitrosatable substances	*13.6*	*1.0*	*12.6*	*3.4*	*0.4*
	Compliance to EU Directive	**Yes**	**Yes**	**Yes**	**Yes**	**Yes**
NT2	Total N-nitrosamines	*12.3*	*16.8*	*12.7*	*9.2*	*6.1*
	Compliance to EU Directive	**No**	**No**	**No**	**Yes**	**Yes**
	Total N-nitrosatable substances	*59.0*	*1.6*	*77.4*	*2.4*	*0.3*
	Compliance to EU Directive	**No**	**Yes**	**No**	**Yes**	**Yes**
NR1	Total N-nitrosamines	*20.7*	*41.0*	*266*	*40.9*	*24.3*
	Compliance to EU Directive	**No**	**No**	**No**	**No**	**No**
	Total N-nitrosatable substances	*3697*	*52.2*	*164*	*55.9*	*0*
	Compliance to EU Directive	**No**	**No**	**No**	**No**	**Yes**
NR2	Total N-nitrosamines	*11.2*	*21.6*	*25.3*	*40.8*	*11.3*
	Compliance to EU Directive	**No**	**No**	**No**	**No**	**No**
	Total N-nitrosatable substances	*25.0*	*0*	*5.7*	*0*	*2.6*
	Compliance to EU Directive	**No**	**Yes**	**Yes**	**Yes**	**Yes**

3.4 Possible risks associated with the use of rubber gloves for handling food

An attempt had been made to evaluate risk associated with the use of gloves for food handling purposes. Two possible risks were considered, the first being the risk of glove user to exposure nitrosamines through skin contact and the second being the risk of food consumer to exposure nitrosamines by consuming nitrosamine-contaminated foods.

In order to estimate the level of nitrosamines released per one glove, data in Table 1 were used and the level of total N-nitrosamines and N-nitrosatable substances in Table 2 were re-calculated and the results are shown in Table 5.

Table 5. Levels of N-nitrosamines released from samples expressed in μg/glove

Sample	Compound μg/glove	Extraction medium Saliva	Alk- sweat	Ac-sweat	10% Ethanol	3% Acetic acid
NT1	Total N-nitrosamines	11	53	1548	13	18
	Total N-nitrosatable substances	65	5	60	16	2
NT2	Total N-nitrosamines	448	612	462	335	222
	Total N-nitrosatable substances	2148	58	2818	87	11
NR1	Total N-nitrosamines	105	208	1350	208	123
	Total N-nitrosatable substances	18766	265	832	284	-
NR2	Total N-nitrosamines	408	786	921	1485	411
	Total N-nitrosatable substances	910	-	208	-	95

Using data in Table 5 and further assuming that one is wearing one pair of glove, for example, sample NT1, and work continuously in food production line for four hours. In worst case scenario, the entire N-nitrosamine contents were migrated from gloves in which half of them penetrated into the wearer skin and another half migrated into food they are being contacted. With these assumptions, the first possible exposure of glove user to nitrosamines through skin contact was estimated to be a maximum of 58 μg N-nitrosamines [one pair of glove contains (53+5) x 2 μg, so half of them equal to 58 μg] for worker whose sweat is alkaline, or a maximum of 1608 μg N-nitrosamines (1548+60) for worker whose sweat is acidic. For the possible exposure of food consumer to N-nitrosamines by consuming nitrosamine-contaminated foods, it was estimated to be a maximum of 29 μg N-nitrosamines (16+13) for alcoholic foods, or a maximum of 20 μg N-nitrosamines (18+12) for acidic foods. By using the same assumptions outlined above the possible exposures to N-nitrosamines for food handling purposes could also be estimated for Sample NT2, NR1 and NR2.

Comparing to the data of dietary intake of nitrosamines reported by the German Nutrition Society (the Nutrition Report 1988) that, for adults, the intake of N-nitrosamines from foods was between 0.2 (women) and 0.3 (men) μg N-nitrosamines per day [21]. This can be then implied that the use of rubber gloves, NT1, NT2, NR1 or NR2 for food handling purposes under the most unfavorable circumstances could lead to the exceeding of dietary intake so that risks associated with using gloves for food handling purposes are considered high.

4. Conclusion

The effects of extraction media used for extraction of N-nitrosamines from rubber gloves made of natural rubber and nitrile rubber lattices have been investigated. Artificial saliva, artificial sweats (both alkaline and acid solutions) and aqueous food simulants (10 % ethanol and 3 % acetic acid) were used as the extraction media at the control migration condition i.e. temperature at 40 °C for 4 hours. It was found that the NDBA and its N-nitrosatable amine were detected in all samples with different levels depending on the extraction media used. NDMA, NDEA, and NMOR were occasionally detected in some samples also depending on the extraction media. For the rest of N-nitrosamines being analysed i.e. NDPA, NDiBA, NPIP, NPYR, NEPhA, NMPhA, NDiNA and NDBzA, they were not detected.The effect of extraction media on the migration of NDBA and its N-nitrosatable amine was closer looked. At the control migration, the power of extraction media are different for gloves made of nitrile rubber from gloves made of natural rubbers. For nitrile rubber gloves, the extractable NDBA were migrated to the extraction medium from artificial alkaline-sweat > acid-sweat >10% ethanol > 3% acetic acid and > artificial saliva. For their N-nitrosatable NDBA, the substances were migrated to the extraction medium from artificial acid-sweat > artificial saliva > 10% ethanol and > artificial alkaline-sweat. For natural rubber gloves, the released NDBA were detected with the highest to the lowest levels by extraction with 10% ethanol, alkaline-sweat/acid-sweat, 3% acetic acid and artificial saliva, respectively. For their N-nitrosatable substances, similar trends to those of nitrile glove samples were found except the swap order between the artificial acid-sweat and the artificial saliva. This finding leads to the conclusion that by using different extraction media, the levels of the released N-nitrosamines and N-nitrosatable substances would be different and the judgment for the EU Directive compliance would be consequently affected. An attempt had been made to evaluate risks associated with the use of these rubber gloves for food handling purposes. By comparing with data of daily intake of N-nitrosamines from foods reported by the German Nutrition Society, it can be concluded that the intake of N-nitrosamines resulted from using gloves in this study for food handling purposes under the most unfavorable circumstances could lead to the exceeding of dietary intake. This means that the possible risks associated with the use of rubber gloves for food handling are considered high.

Acknowledgement

The authors gratefully acknowledge The Department of Science Service, Ministry of Science and Technology, Thailand, for supporting this research activity.

References

[1] Mutsuka M *et al* 2013 Analysis of N-nitrosamine migration from rubber teats and soothers *Am. J. Anal. Chem.* **4** 277

[2] Vieira E R *et al* 2006 Determination of N-nitrosamines and N-nitrosatable substances in rubber teats and soothers by GC-TEA *Braz. Arch. Biol. Technol.* **49** 73

[3] Voloshenko A *et at* 2011 GC determination of N-nitrosamines by supersonic molecular beam MS equipped with triple quadrupole analyzer, GC/SMB/QQQ/MS *Anal. Chim. Acta.* **685** 162

[4] Havery D C and Fabio T 1982 Estimation of volatile N-nitrosamines in rubber nipples for babies' bottles. *Food Chem. Toxicol.* **20**(6) 939

[5] Sen N P *et al* 1984 Volatile N-nitrosamines in baby bottle rubber nipples and pacifiers: Analysis, occurrence and migration *IARC Sci. Publ.* **57** 51

[6] Sen N P *et al* 1985 Nitrosamines in baby bottle nipples and pacifiers: Occurrence, migration, and effect of infant formulas and fruit juices on in'Vitro formation of nitrosamines under simulated gastric conditions *J. Agric. Food Chem.* **33** 420

[7] Park S J *et al* 2018 Release of N-nitrosamines and N-nitrosatable substances from baby bottle teats and rubber kitchen tools in Korea. *Food Sci Biotechnol.* **9**(52) 1

[8] Altkofer W *et al* 2005 Migration of nitrosamines from rubber products – Are balloons and condoms harmful to the human health*? Mol. Nutr. Food Res.* **49** 235

[9] Proksch E 2001 Review Toxicological evaluation of nitrosamines in condoms *Int. J. Hyg. Environ. Health.* **204** 103

[10] Biaudet H *et al* 1997 Migration of nitrosamines from condoms to physiological secretions *Bull. Environ. Contam. Toxicol.* **59** 847

[11] Feng D *et al* 2009 Detection and toxicity assessment of nitrosamines migration from latex gloves in the Chinese market *Int. J. Hyg. Environ. Health.* **212** 533

[12] Feng D *et al* 2012 Evaluation of simulant migration of volatile nitrosamines from latex gloves and balloons by HS-SPME–GC–MS *J. Chromatogr. Sci.* **50** 733

[13] Eng A H 2001 Effects of chlorination on nitrosamines and nitrosatable substances in natural rubber gloves *J. Rubb. Res.* **4**(1) 11

[14] Fiddler W *et al* 1985 Investigation of volatile nitrosamines in disposable protective gloves *Am. Ind. Hyg. Assoc. J.* **46**(8) 463

[15] Fiddler W *et al* 1998 Nitrosamine formation in processed hams as related to reformulated elastic rubber netting *J. Food Sci.* **63**(2) 276

[16] Fiddler W *et al* 1992 Nitrosamines in rubber bands used for orthodontic purposes *Food Chem. Toxicol.* **30**(4) 325

[17] Commission Directive 93/11/EEC 1993 Concerning the release of the N-nitrosamines and N- nitrosatable substances from elastomer or rubber teats and soothers

[18] U.S. Food and Drug Administration 1984 CPG Sec. 500.450 Volatile N-nitrosamines in rubber baby bottle nipples

[19] Bouma K and Schothorst R C 2003 Identification of extractable substances from rubber netting used to package meat products *Food Addit. Contam.* **20**(3) 300

[20] Pensabene J W and Fiddler W 1994 Gas chromatographic/thermal energy analyzer method for N-nitrosodibenzylamine in hams processed in elastic rubber netting *J. AOAC Int.* **77**(4) 981

[21] BgVV 2002 Risk assessment of N-nitrosamines in balloons http://www.bfr.bund.de

[22] VWA/KvW Report nr. ND04o063/02 Migration of N-nitrosamines and N-nitrosatable substances from latex balloons , January 2005 http://www.nwa.nl

[23] SCCP Opinion on the presence and release of nitrosamine and nitrosatable compounds from rubber balloons, December 2007 https://ec.europa.eu/health/home_en

[24] EN 12868:2017, Child use and care articles-Methods for determining the release of N-nitrosamines and N-nitrosatable substances from elastomer or rubber teats and soothers, European Committee for Standardization

[25] ISO 105-E04:2013, Textiles - Tests for colour fastness - Part E04: Colour fastness to perspiration, International Organization for Standardization, Geneva, Switzerland, www.iso.org

[26] ISO 3696:1987, Water for analytical laboratory use - Specification and test methods, International Organization for Standardization, Geneva, Switzerland, www.iso.org

Modeling and Forecasting of Volatility using ARMA-GARCH: Case Study on Malaysia Natural Rubber Prices

I M Md Ghani[1] and H A Rahim[1]

[1] School of Informatics and Applied Mathematics, Universiti Malaysia Terengganu, 21030 Kuala Nerus, Terengganu Darul Iman

E-mail: intanmartina23@gmail.com

Abstract. Malaysia is one of the largest producers of natural rubber in the world. Among the various types of natural rubber which contribute to the country's agricultural sector is the Standard Malaysian Rubber Grade 20 (S.M.R 20). Since 2008, the rubber price has received attention of investors and Malaysia Rubber Board due to price fluctuation. The price of rubber is characterized by the existence of heavy tails and volatility clustering. These properties play a significant impact on parameter estimation and forecasting performance resulting from S.M.R 20 rubber price data. The approach used in modeling S.M.R 20 rubber price data, is Generalized Autoregressive Conditional Heteroscedasticity (GARCH) model. The aims of this paper are to find the best ARMA-GARCH model by using different specifications structures and to forecast the daily price for 20 days ahead. There are 20 models produced from different specifications in ARMA(R,M) dan GARCH(p,q) models. In this study, 1953 daily price data of S.M.R 20 are taken into consideration. The validity comparison of diagnostic checking and forecasting performance are based on AIC, AICC, SBC, HQC, MSE, RMSE and MAPE. The results reveals that ARMA(1,0)-GARCH(1,2) model is the best volatility modeling in S.M.R 20 rubber price. Based on the implications of the results, the scope of the future research directions has been widen.

1. Introduction

Financial time series usually include of floating and volatility models. The floating component always modelled using mean models such as ARIMA, moving average and many other models. The volatility of financial time series will be model when the series have stylized facts. It can be modelled using ARCH/GARCH. Several studies investigating volatility have been carried out on rubber price [1], oil price [2], electricity price [3] and petroleum [4]. Furthermore, most volatility in business and economic data influenced by stylized facts.

The stylized facts that give significant impact on modeling financial time series are fat tails, volatility clustering and nonlinear dependence[5]. With the existing of volatility clustering in economic data, it can affect the forecasting performance of the mean models. In order to achieve accurate forecast, most researchers applied Generalized Autoregressive Conditional Heteroscedasticity (GARCH) to model volatility.

More recent studies have selected GARCH(1,1) model to analyze time series data. Some references consensus that GARCH(1,1) model is popular among others specifications because it is the simplest and most robust among volatility models [6], fits many data series well [7] and sufficient to capture the volatility clustering in the data [8]. Furthermore, Olson and Wu [9] mentioned that analysis can adequate with only one lag for each variable.

The research to date has tended to focus an exchange rate rather than production and prices from agricultural data. Goh et al. [1] and Isa and Jamil [10] investigated on the Standard Malaysia Rubber 20 (S.M.R 20) price. Both studies was selected GARCH(1,1) model in modeling and forecasting. [10]found that GARCH(1,1) model is appropriate in modeling S.M.R 20 and Ribbed

IOP Conf. Series: Materials Science and Engineering **548** (2019) 012023 doi:10.1088/1757-899X/548/1/012023

Smoked Sheet Grade I (RSS 1) price during the period of 1980 to 2002. In another study, [1]examined the comparison between symmetry and asymmetry GARCH model during pre and post-global financial crisis in 15 years. According to both studies, whereby preferred GARCH(1,1) model in analysis S.M.R. 20 rubber price, one unresolved question is whether GARCH(1,1) model accurate in current S.M.R 20 rubber price. Therefore, in this study S.M.R 20 rubber price using GARCH model with combination ARMA model based on different specifications. Adding to that, the study indicated daily forecasted for S.M.R 20 for 20 days ahead.

The GARCH model [1] is one of the furthermost statistical technique applied in volatility. A large and growing body of literature has investigated using GARCH(1,1) model [1-2, 12-17]. However not all of these literature reported GARCH(1,1) is more appropriate in analyzing. Only [12] shown that GARCH(1,1) has predictive power in modeling daily exchange rate in the nation of Tanzania. Another study by [14] found that ARMA(1,1) with GARCH(1,1) and GARCH(2,1) is applicable in Dhaka Stock Exchange.

The paper is organized as follows. In the next section, few models are briefly described which are the ARMA model, ARCH model and GARCH model. Third session will be the result and discussion of ARMA-GARCH model based on S.M.R 20 from the duration 4th January 2010 to 29th December 2017. Last but not least are the summarized conclusion of the research.

2. Materials and Methods

2.1. Data

In this section, 1953 daily price data of Standard Malaysia Rubber 20 (S.M.R 20) are used. These data consist of period from 4th January 2010 to 29th December 2017. All observations were taken from official portal of Malaysia Rubber Board. Based on the observations, the data are split into two parts which is the in-sample part (4th January 2010 to 4th August 2015) and the out-of-sample part (5th August 2015 to 29th December 2017).

2.2. Methods

There are three types of time series models such as Autoregressive Moving Average (ARMA) model, Autoregressive Conditional Heteroscedasticity (ARCH) model and Generalized Autoregressive Conditional Heteroscedasticity (GARCH) model.

In 1976, Box and Jenkins [18], proposed ARIMA(m,D,n) models where m is the number of autocorrelation terms, D is the number of differencing elements and n is the number of moving average terms. The letter "I" in ARIMA used to differentiate when the series are not stationary. However, when the time series is stationary, we can model it using three classes of time series process: autoregressive (AR), moving-average (MA) and mixed autoregressive and moving-average (ARMA).

An autoregressive model of order m, denoted as AR(m), can be expressed as

$$\gamma_t = \mu + \phi_1 \gamma_{t-1} + \phi_2 \gamma_{t-2} + \cdots + \phi_m \gamma_{t-m} + u_t \tag{1}$$

The moving average of order n which denoted as MA(n) can be expressed as follows

$$\gamma_t = \mu + u_t + \theta_1 u_{t-1} + \theta_2 u_{t-2} + \cdots + \theta_n u_{t-n} \tag{2}$$

where $u_t (t = 1,2,3, \dots)$ is a white noise disturbance term with $E(u_t) = 0$ and $\text{var}(u_t) = \sigma^2$.

The combination of AR(m) model and MA(n) model formed of ARMA(m,n) model which expressed as

$$\gamma_t = \mu + \phi_1 \gamma_{t-1} + \phi_2 \gamma_{t-2} + \cdots + \phi_m \gamma_{t-m} + \theta_1 u_{t-1} + \theta_2 u_{t-2} + \cdots + \theta_n u_{t-n} + u_t \quad (3)$$

or in sigma notation

$$y_t = \mathrm{C} + \sum_{i=1}^{m} \phi_i y_{t-i} + \sum_{j=1}^{n} \theta_j \varepsilon_{t-j} \quad (4)$$

where y_t is the daily rubber S.M.R 20 prices, C is a constant term, ϕ_i are the parameter of the autoregressive component of order m, θ_j are the parameters of the moving average component of order n, and ε_t is the error term at time t. The order m and n are non-negative integers.

There are two time-varying volatility models that popular among researchers: ARCH model and GARCH model. The aims of ARCH model that developed by Engle [6] is to predict the conditional variance of return series.

$$y_t = \mathrm{C} + \varepsilon_t \quad (5)$$

$$\varepsilon_t = z_t \sigma_t \quad (6)$$

Where y_t is an observed data series, C is a constant value, ε_t is residual, z_t is the standardized residual with independently and identically distributed with mean equal to zero and variance equal to one and σ_t is the square root of the conditional variance with non-negative process. The general form of ARCH(q) model with the first q past squared innovations can be expressed as follows [6]:

$$\sigma_t^2 = \eta + \sum_{j=1}^{q} \alpha_j \varepsilon_{t-j}^2 \quad (7)$$

The constraints of parameters are $\eta > 0$ and $\alpha_j \geq 0 (j = 1, \ldots, q)$, which to ensure the conditional variance, σ_t^2 is non-negative. Although the ARCH model is simple model and widely used among researchers, but it has weakness. When modeling volatility using ARCH, there might be a need for a large value of the lag q, hence a large number of parameters to be estimated. This may result the difficulties to estimate parameters.

After four years an extension from ARCH model was developed by [11] namely GARCH model. The GARCH model is more parsimonious (use fewer parameters) than ARCH model [19]. There are two part that consist in GARCH model which are mean equation, y_t; see Equation (4) and variance equation, σ_t^2; see Equation (8). The general form for GARCH(p,q) model can be written as follows:

$$\sigma_t^2 = \eta + \sum_{i=1}^{p} \beta_i \sigma_{t-1}^2 + \sum_{j=1}^{q} \alpha_j \varepsilon_{t-j}^2 \quad (8)$$

where η is the long-run volatility with condition $\eta > 0$, $\beta_i \geq 0; i = 1, \dots, p$ and $\alpha_j \geq 0; j = 1, \dots, q$. If $\beta_i + \alpha_j < 1$, then GARCH(p,q) model is covariance stationary. The unconditional variance of the error terms

$$\mathrm{var}(\varepsilon_t) = \frac{\eta}{1 - \beta - \alpha} \tag{9}$$

From the general form of GARCH(p,q) model, the GARCH(1,1) model can defined as

$$\sigma_t^2 = \eta + \beta\sigma_{t-1}^2 + \alpha\varepsilon_{t-1}^2 \tag{10}$$

2.3. Model Selection

When comparing among different specification of ARMA-GARCH models, then we select an appropriate model based on Akaike Information Criteria (AIC) [20], the corrected Akaike Information Criteria (AICC), Schwarz's Bayesian Information Criterion (SBC) [21] and the Hannan-Quinn Information Criterion (HQC). The AIC, AICC, SBC and HQC can be computed as

$$\mathrm{AIC} = -2\ln(\mathrm{L}) + 2\mathrm{k} \qquad \mathrm{AICC} = \mathrm{AIC} + 2\frac{\mathrm{k}(\mathrm{k}+1)}{\mathrm{N}-\mathrm{k}-1}$$

$$\mathrm{SBC} = -2\ln(\mathrm{L}) + \ln(\mathrm{N})\mathrm{k} \qquad \mathrm{HQC} = -2\ln(\mathrm{L}) + 2\ln\big(\ln(\mathrm{N})\big)\mathrm{k}$$

Where L is the value of the likelihood function evaluated at the parameter estimates, N is the number of observations, and k is the number of estimated parameters. The minimum value of AIC, AICC, SBC and HQC was selected as the better model when comparing among models.

2.4. Model Evaluations

The performance of forecasting models are evaluated using three measures: Mean Square Error (MSE), Root Mean Square Error (RMSE) and Mean Absolute Percentage Error (MAPE).

$$\mathrm{MSE} = \frac{1}{\mathrm{T}}\sum_{t=\mathrm{T}_1}^{\mathrm{T}}(\sigma_t^2 - \hat{\sigma}_t^2)^2 \qquad \mathrm{RMSE} = \sqrt{\frac{1}{\mathrm{T}}\sum_{t=\mathrm{T}_1}^{\mathrm{T}}(\sigma_t^2 - \hat{\sigma}_t^2)^2} \qquad \mathrm{MAPE} = \frac{1}{\mathrm{T}}\sum_{t=\mathrm{T}_1}^{\mathrm{T}}\left|\left(\frac{\sigma_t^2 - \hat{\sigma}_t^2}{\sigma_t^2}\right) \times 100\right|$$

Where T is the number of total observations and T_1 is the first observation in out-of-sample. The σ_t^2 and $\hat{\sigma}_t^2$ is the actual and predicted conditional variance at time t, respectively. When comparing among ARMA-GARCH models, the smallest value of MSE, RMSE and MAPE are chosen as the best accurate forecast model.

3. Results and Discussion

3.1. In-Sample Part

In Fig. 1 (left panel) there is a clear trend of 1367 daily observations prices of the S.M.R 20 in Malaysia from 4th January 2010 to 4th August 2015. When daily prices converted to log returns, the plot in Fig. 1 (right panel) illustrates there are large negative values especially on March and October 2011. Both Fig.s can explained there exist volatility clustering in daily returns on S.M.R 20.

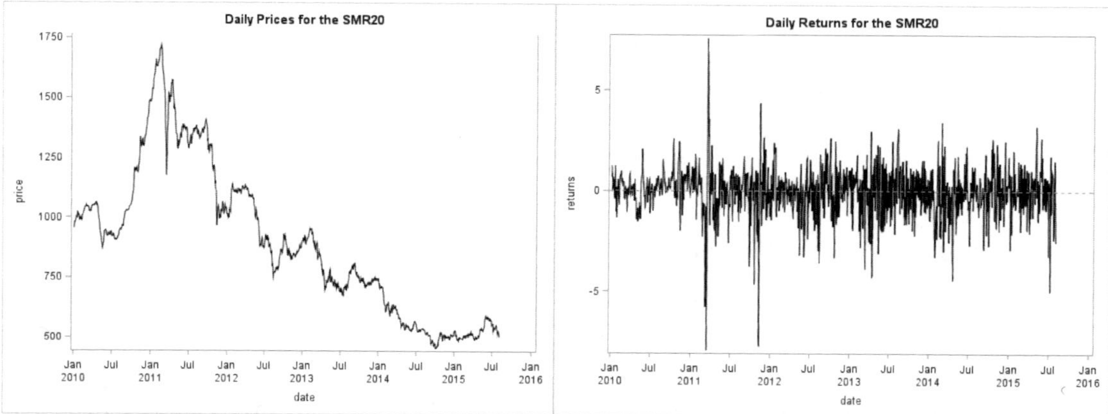

Fig. 1. In-sample results of daily prices (left panel) and daily returns (right panel) of S.M.R 20

The descriptive statistics of the daily returns for S.M.R 20 are shown in Table 1. There is excess kurtosis in daily returns, which is 5.5298 larger than the normal value of 3. This can explain that there exist heavier tails in the data and distribute as leptokurtic.

Table 1. Descriptive Statistics of Daily Returns of S.M.R 20

Mean	Variance	Standard Deviation	Kurtosis	Skewness
-0.0469	1.3963	1.1817	5.5298	-0.6102

Source: Author's calculation using data S.M.R 20

The first step in time series data is to test the stationarity. The SAS output based on Phillips-Perron (PP) test [22], Augmented Dickey-Fuller (ADF) test [23] and Kwiatkowski-Phillips-Schmidt-Shin (KPSS) test [24] are shown in Table 2, 3 and 4, respectively. As Table 2 shows, we reject the null hypothesis of a unit root for the three cases (Zero mean, Single mean and Trend) at the 5 percent level of significance. This is because the "Pr<Tau" values are <.0001. These indicates that the series is stationary where there is no unit root.

Table 2. SAS output based on the PP test

Type	Lags	Rho	Pr<Rho	Tau	Pr<Tau
Zero mean	7	-817.1034	<.0001	-24.2783	<.0001
Single mean	7	-816.8883	0.0019	-24.2876	<.0001
Trend	7	-817.0899	0.0008	-24.2994	<.0001

Source: Author's calculation using data S.M.R 20

In addition, at the 5 percent level of significance the Tau test in Table 3 also rejects the null hypothesis of a unit root in the three cases. The finding of ADF test is consistent with the PP test. Therefore, both PP test and ADF test results imply that the returns are stationary and the series has no unit root. Meanwhile in Table 4, the Eta test do not reject the null hypothesis of a unit root in the two cases (Single mean and Trend) at 5 percent level of significance. These indicates that the series is stationary and there is no unit root. Because all three tests proved the series is stationary and has no unit root, therefore we proceed next step.

Table 3. SAS output based on the ADF test

Type	Lags	Rho	Pr<Rho	Tau	Pr<Tau	F	Pr>F
Zero mean	0	-830.9948	<.0001	-24.3937	<.0001		
Single mean	0	-832.3303	<.0001	-24.4155	<.0001	298.0640	<.0010
Trend	0	-833.6864	<.0001	-24.4367	<.0001	298.5768	<.0010

Source: Author's calculation using data S.M.R 20

Table 4. SAS output based on the KPSS test

Type	Lags	Eta	Pr<Eta
Single mean	23	0.1963	0.2754
Trend	23	0.0861	0.2298

Source: Author's calculation using data S.M.R 20

When the series are stationary, the next testing is done on the statistical independence. In this step two models was tested: AR(1) model and MA(2) model. Both models was selected by using SCAN option in SAS software. The autocorrelation check of residuals for AR(1) model and MA(2) model are illustrated in Table 5 and 6, respectively. It appears from both tables that the Pr > ChiSq values are greater than 0.05 for all lag order, which implying that we cannot reject the null hypothesis of white noise and the estimated residuals are white noise. Therefore the identified model AR(1) and MA(2) passed the diagnostic test.

Table 5. Autocorrelation Check For Residuals of AR(1) Model

To Lag	Chi-Square	DF	Pr > ChiSq	Autocorrelations					
6	7.98	5	0.1571	0.020	-0.047	-0.027	0.047	0.015	-0.007
12	13.05	11	0.2903	-0.039	0.007	-0.015	0.042	0.002	-0.006
18	18.11	17	0.3817	0.014	0.041	0.032	-0.009	-0.006	-0.025
24	24.70	23	0.3658	-0.040	-0.048	-0.006	0.020	-0.012	-0.015
30	30.01	29	0.4137	0.017	-0.007	-0.032	0.043	-0.007	0.024
36	34.05	35	0.5139	0.000	-0.010	-0.008	-0.011	0.013	0.049
42	42.26	41	0.4164	0.027	0.010	-0.047	0.043	0.021	-0.024
48	45.33	47	0.5419	0.014	0.003	-0.003	0.027	0.018	-0.030

Source: Author's calculation using data S.M.R 20

Table 6. Autocorrelation Check For Residuals of MA(2) Model

To Lag	Chi-Square	DF	Pr > ChiSq	Autocorrelations					
6	3.39	4	0.4954	-0.000	0.004	0.016	0.045	0.014	-0.004
12	8.26	10	0.6031	-0.036	0.011	-0.019	0.042	0.004	-0.002
18	13.19	16	0.6590	0.016	0.040	0.030	-0.009	-0.006	-0.026
24	19.72	22	0.6005	-0.039	-0.049	-0.009	0.016	-0.014	-0.016
30	25.53	28	0.5991	0.018	-0.006	-0.032	0.046	-0.011	0.024
36	29.06	34	0.7083	0.001	-0.009	-0.004	-0.008	0.013	0.047
42	37.39	40	0.5884	0.025	0.016	-0.045	0.045	0.020	-0.024
48	40.42	46	0.7045	0.018	0.004	-0.004	0.025	0.019	-0.028

Source: Author's calculation using data S.M.R 20

Table 7. In-Sample Analysis of Testing an ARCH Effect For S.M.R 20

Order	Q	Pr > Q	LM	Pr > LM	LK	Pr > |LK|	WL	Pr > WL
1	166.1245	<.0001	165.9043	<.0001	25.0250	<.0001	175.8346	<.0001
2	227.7188	<.0001	178.7377	<.0001	24.5244	<.0001	243.5460	<.0001
3	422.5273	<.0001	297.4051	<.0001	30.6907	<.0001	311.7386	<.0001
4	518.0371	<.0001	301.3239	<.0001	31.1285	<.0001	352.9029	<.0001
5	528.9433	<.0001	308.9287	<.0001	27.8513	<.0001	370.6685	<.0001
6	556.9039	<.0001	309.1092	<.0001	26.9007	<.0001	415.0158	<.0001
7	589.8222	<.0001	309.7842	<.0001	26.5255	<.0001	443.5837	<.0001
8	594.4754	<.0001	309.9478	<.0001	24.8546	<.0001	494.9318	<.0001
9	594.9220	<.0001	311.7765	<.0001	23.0565	<.0001	535.6105	<.0001
10	596.8305	<.0001	312.4006	<.0001	21.8958	<.0001	572.0933	<.0001
11	597.1771	<.0001	312.4701	<.0001	20.7063	<.0001	624.4816	<.0001
12	597.5316	<.0001	313.8788	<.0001	19.7216	<.0001	635.2950	<.0001

Source: Author's calculation using data S.M.R 20

Table 7 provides the result of ARCH effect using four tests which are Q test, LM test, LK test and WL test. The p-value of four tests indicate that we rejected the null hypothesis of no ARCH effect at the 5 percent level of significance. From Table 7, the results show that the presence of ARCH effect in S.M.R 20. This can be explained that heteroscedastic appeared in residuals.

Table 8. Comparison Model of Selection Criteria ARMA-GARCH Model

Models	AIC	AICC	SBC	HQC
ARMA(0,1)	4124.7460	NA	4135.1850	NA
ARMA(0,2)	4108.7360	NA	4124.3950	NA
ARMA(1,0)	4110.6460	4110.6550	4121.0850	4114.5530
ARMA(1,1)	4109.1410	NA	4124.8000	NA
ARMA(1,2)	4110.7340	NA	4131.6130	NA
ARMA(2,0)	4109.2020	4109.2190	4124.8610	4115.0620
ARMA(2,1)	4111.1220	NA	4132.0000	NA
ARMA(2,2)	4112.7330	NA	4138.8310	NA
GARCH(1,1)	4034.7170	4034.7460	4055.5960	4042.5310
GARCH(1,2)	4016.4990	4016.5430	4042.5970	4026.2670
GARCH(2,1)	4329.4060	4329.4150	4339.8450	4333.3130
GARCH(2,2)	4327.1800	4327.1890	4337.6200	4331.0880
ARMA(1,0)-GARCH(1,1)	3836.9930	3837.0370	3863.0910	3846.7610
ARMA(1,0)-GARCH(1,2)	3831.8580	3831.9200	3863.1760	3843.5800
ARMA(1,0)-GARCH(2,1)	4109.9150	4109.9330	4125.5740	4115.7760
ARMA(1,0)-GARCH(2,2)	4107.7620	4107.7790	4123.4210	4113.6230
ARMA(2,0)-GARCH(1,1)	3837.7330	3837.7950	3869.0510	3849.4540
ARMA(2,0)-GARCH(1,2)	3831.7290	3831.8120	3868.2670	3845.4050
ARMA(2,0)-GARCH(2,1)	4110.9660	4110.9960	4131.8450	4118.7810
ARMA(2,0)-GARCH(2,2)	4110.7770	4110.8060	4131.6550	4118.5910

NA = Not applicable

After confirmed heteroscedastic exist in residuals, then we proceed to specify the model. We generates 20 models from different conditional mean ad conditional variance specification in ARMA(m,n)-GARCH(p,q) models where m and n were either 0,1 or 2 while p and q were either 1 or 2. The 20 different specification models was compared based on criteria such as AIC, AICC, SBC and HQC to select the best model. The comparison among 20 models are highlighted in Table 8. The ARMA(2,0)-GARCH(1,2) model presents the minimum value of AIC and AICC. While

ARMA(1,0)-GARCH(1,1) and ARMA(1,0)-GARCH(1,2) model shows minimum in SBC and HQC value, respectively. As Table 9 indicates, the ARMA(1,0)-GARCH(1,2) model and ARMA(2,0)-GARCH(1,2) model are the lowest in total rank. From both models, we decided ARMA(1,0)-GARCH(1,2) model as the best model in-sample part.

Table 9. Comparison of ARMA-GARCH Model Based on Ranking

Models	Rank AIC	Rank AICC	Rank SBC	Rank HQC	Total rank
ARMA(0,1)	18	20	17	19	74
ARMA(0,2)	8	15	9	15	47
ARMA(1,0)	12	10	7	8	37
ARMA(1,1)	9	16	10	16	51
ARMA(1,2)	13	17	13	17	60
ARMA(2,0)	10	8	11	9	38
ARMA(2,1)	16	18	16	18	68
ARMA(2,2)	17	19	18	20	74
GARCH(1,1)	6	6	6	6	24
GARCH(1,2)	5	5	5	5	20
GARCH(2,1)	20	14	20	14	68
GARCH(2,2)	19	13	19	13	64
ARMA(1,0)-GARCH(1,1)	3	3	1	3	10
ARMA(1,0)-GARCH(1,2)	2	2	2	1	7
ARMA(1,0)-GARCH(2,1)	11	9	12	10	42
ARMA(1,0)-GARCH(2,2)	7	7	8	7	29
ARMA(2,0)-GARCH(1,1)	4	4	4	4	16
ARMA(2,0)-GARCH(1,2)	1	1	3	2	7
ARMA(2,0)-GARCH(2,1)	15	12	15	12	54
ARMA(2,0)-GARCH(2,2)	14	11	14	11	50

3.2. Out-of-Sample Part

In Fig. 2 (left panel) there is a clear trend of 586 daily observations prices of the S.M.R 20 from 5th August 2015 to 29th December 2017. The plot in Fig. 2 (right panel) illustrates there are large negative values especially on 28th September 2017. Both Figures can be explained that there exists volatility clustering in daily prices and returns on S.M.R 20.

In out-of-sample part, the performance of forecasting was evaluated. The evaluation of ARMA-GACRH models in out-of-sample are summarized in Table 10. As illustrated in Table 10, the accurate forecast for MSE and RMSE statistic are ARMA(2,0)-GARCH(2,1) model and ARMA(2,0)-GARCH(2,2) model. While ARMA(1,0)-GARCH(1,2) model present more accurate in MAPE statistic. Therefore among 14 models, we decided ARMA(1,0)-GARCH(1,2) model as the best accurate model in out-of-sample part. This is supported by[25] which suggested that the selection of MAPE statistic is the best if there are different results between three statistics. In addition, based on the SCAN option during in-sample part analysis, only AR(1) model and MA(2) model selected for forecasting.

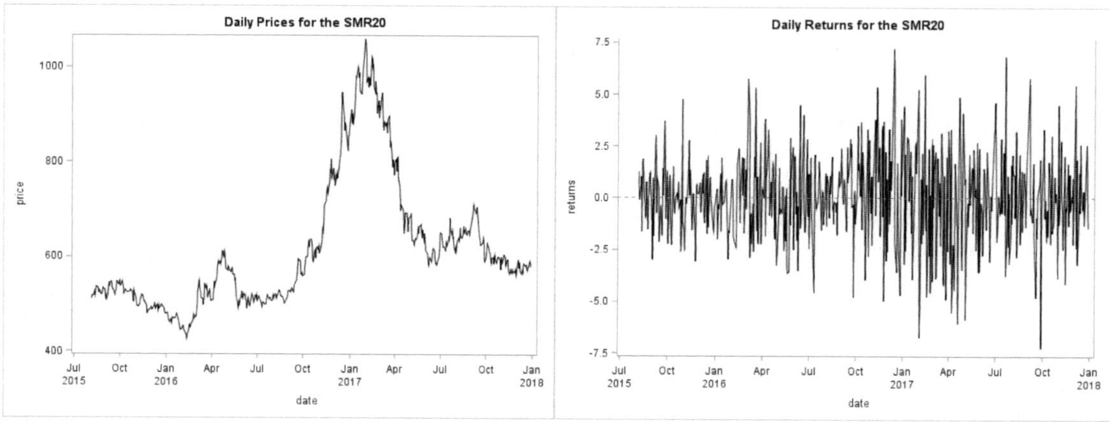

Fig. 2: Out-of-sample results of daily prices (left panel) and daily returns (right panel) of S.M.R 20

Table 10: Comparison Model Evaluation of ARMA-GARCH Models

Models	MSE	RMSE	MAPE
ARMA(1,0)	4.4771	2.1159	99.7326
ARMA(2,0)	4.4845	2.1177	99.7914
GARCH(1,1)	4.4650	2.1131	99.9117
GARCH(1,2)	4.4727	2.1149	100.4053
GARCH(2,1)	4.4650	2.1131	99.8665
GARCH(2,2)	4.4650	2.1131	99.8665
ARMA(1,0)-GARCH(1,1)	4.4620	2.1123	99.8477
ARMA(1,0)-GARCH(1,2)	4.4620	2.1123	99.6798
ARMA(1,0)-GARCH(2,1)	4.4618	2.1123	99.7328
ARMA(1,0)-GARCH(2,2)	4.4618	2.1123	99.7328
ARMA(2,0)-GARCH(1,1)	4.4638	2.1128	100.3370
ARMA(2,0)-GARCH(1,2)	4.4664	2.1134	100.2446
ARMA(2,0)-GARCH(2,1)	4.4615	2.1122	99.7919
ARMA(2,0)-GARCH(2,2)	4.4615	2.1122	99.7919

Source: Author's calculation using SAS

3.3. Overall Part

For overall part, there is a clear trend of 1953 daily observations prices of the rubber S.M.R 20 in Malaysia from 4th January 2010to 29th December 2017 depicted in Fig. 3. When daily prices converted to log returns, plotted of daily returns of rubber S.M.R 20 clearly show exhibit volatility clustering.

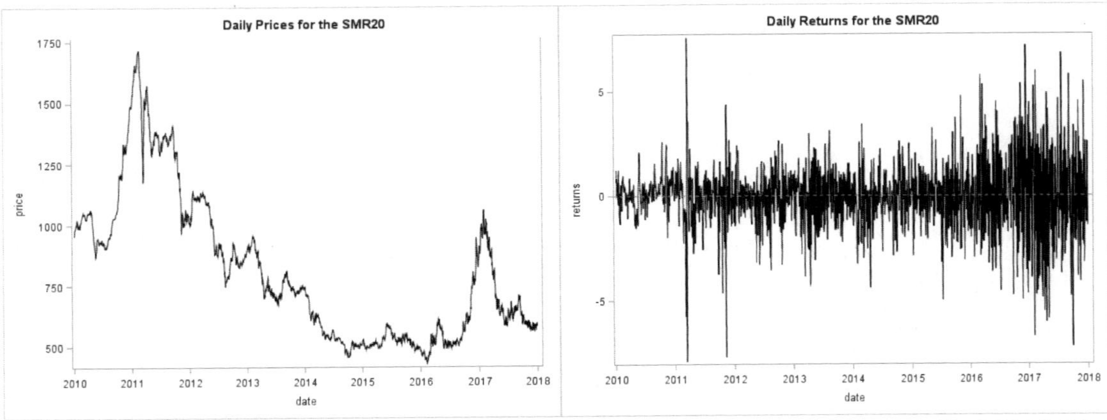

Fig. 3: Overall results of daily prices (left panel) and daily returns (right panel) of S.M.R 20

Table 11 provides the result of ARCH effect using four tests which are Q test, LM test, LK test and WL test. The p-value of four tests indicate that we reject the null hypothesis of no ARCH effect at the 5 percent level of significance. Table 11 shows that the presence of ARCH effect in S.M.R 20. This can be explained that heteroscedastic appeared in residuals.

Table 11. Overall Results of Testing an ARCH Effect For S.M.R 20

Order	Q	Pr > Q	LM	Pr > LM	LK	Pr > |LK|	WL	Pr > WL
1	150.624	<.0001	150.410	<.0001	19.838	<.0001	6800.491	<.0001
2	217.600	<.0001	175.112	<.0001	20.691	<.0001	10690.557	<.0001
3	335.227	<.0001	235.213	<.0001	23.911	<.0001	14325.307	<.0001
4	416.569	<.0001	251.162	<.0001	24.831	<.0001	16981.035	<.0001
5	451.437	<.0001	251.959	<.0001	23.986	<.0001	19081.184	<.0001
6	489.096	<.0001	254.964	<.0001	23.673	<.0001	21779.230	<.0001
7	567.360	<.0001	276.011	<.0001	24.629	<.0001	24279.598	<.0001
8	590.339	<.0001	276.195	<.0001	23.927	<.0001	27379.125	<.0001
9	654.523	<.0001	296.258	<.0001	24.501	<.0001	30696.639	<.0001
10	680.825	<.0001	296.770	<.0001	24.111	<.0001	33706.437	<.0001
11	701.056	<.0001	296.961	<.0001	23.648	<.0001	36433.951	<.0001
12	733.562	<.0001	299.649	<.0001	23.607	<.0001	38825.800	<.0001

Source: Author's calculation using data S.M.R 20

Since the best model in estimation part and forecasting part are consistent, then the parameter estimation of ARMA(1,0)-GARCH(1,2) model summarized in Table 12. From Table 12, the estimated conditional mean and conditional variance of ARMA(1,0)-GARCH(1,2) model can expressed as

$$\hat{y}_t = 0.043 - 0.312y_{t-1}$$

$$\hat{\sigma}_t^2 = 0.011 + 0.913\sigma_{t-1}^2 + 0.224\varepsilon_{t-1}^2 - 0.136\varepsilon_{t-2}^2$$

Finally we used ARMA(1,0)-GARCH(1,2) model to forecast the daily S.M.R 20 rubber price for 20 days ahead (2nd January 2018 to 30th January 2018) which presented in Table 13. From this data there is a clear trend forecasting of rubber price and returns, visible in Fig. 4.

IOP Conf. Series: Materials Science and Engineering **548** (2019) 012023 doi:10.1088/1757-899X/548/1/012023

Table 12. Estimation Result of ARMA(1,0)-GARCH(1,2) Model For Rubber S.M.R 20

Parameter	Estimate	Standard error	t-value	Approx. Pr > |t|
Intercept	0.043	0.035	1.24	0.2158
ϕ_1	-0.312	0.025	-12.49	<.0001
η	0.011	0.003	4.18	<.0001
α_1	0.224	0.030	7.4	<.0001
α_2	-0.136	0.030	-4.49	<.0001
β_1	0.913	0.008	109.26	<.0001
Forecast performance				
AIC	6396.60874			
AICC	6396.65193			
SBC	6430.0684			
HQC	6408.90953			
MSE	2.28525			
MAPE	122.27325			

Table 13. Forecast Prices for Rubber S.M.R 20 (sen/kg)

Date	Actual Price RM (sen/kg)	Forecast Price RM (sen/kg)	L95	U95	Returns
2/1/2018	582.00	580.85	543.581	618.122	-0.29172
3/1/2018	576.00	573.75	527.962	619.538	-0.07991
4/1/2018	571.00	572.77	504.148	641.392	-0.04130
5/1/2018	572.00	567.49	485.423	649.55	-0.03427
8/1/2018	572.50	565.55	464.742	666.359	-0.03299
9/1/2018	580.00	561.39	446.294	676.489	-0.03276
10/1/2018	588.50	558.99	427.973	690.012	-0.03273
11/1/2018	590.50	555.49	410.962	700.009	-0.03273
12/1/2018	595.50	552.91	394.67	711.139	-0.03273
15/1/2018	603.00	549.79	379.34	720.233	-0.03274
16/1/2018	603.00	547.17	364.827	729.504	-0.03274
17/1/2018	597.50	544.30	351.104	737.488	-0.03275
18/1/2018	602.00	541.70	338.142	745.263	-0.03276
19/1/2018	595.00	539.01	325.857	752.154	-0.03277
22/1/2018	600.50	536.47	314.241	758.702	-0.03277
23/1/2018	590.00	533.91	303.211	764.601	-0.03278
24/1/2018	584.50	531.44	292.756	770.132	-0.03279
25/1/2018	585.50	528.99	282.812	775.16	-0.03279
26/1/2018	591.50	526.60	273.362	779.839	-0.03280
29/1/2018	580.50	524.24	264.357	784.116	-0.03281
30/1/2018	580.50	521.93	255.778	788.077	-0.03281

L95 = Lower 95, U95 = Upper 95

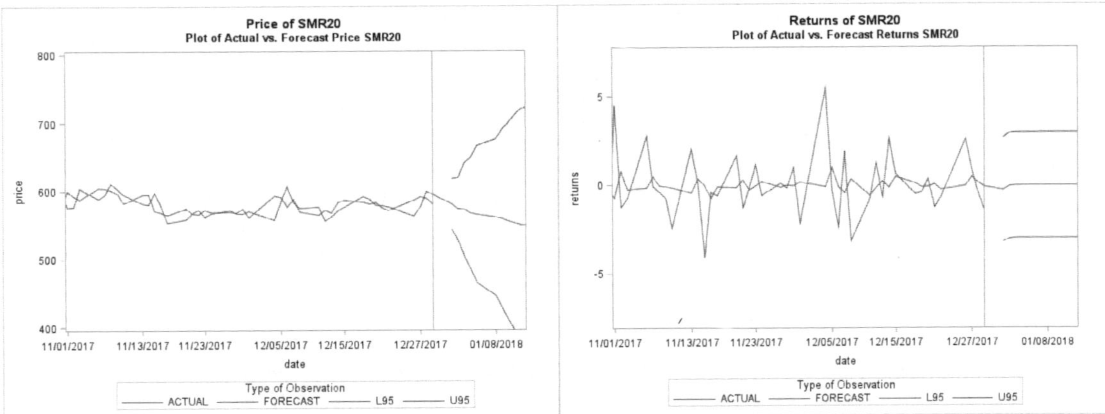

Fig. 4. Forecast of daily prices of S.M.R 20 (left panel) and Forecast of daily returns of S.M.R 20 (right panel)

4. Conclusion

In this paper, the aim was to model the S.M.R. 20 rubber price using 20 different specification of ARMA-GARCH models. The results of this paper shows that ARMA(1,0)-GARCH(1,2) model is the best model for the period of in-sample part and out-of-sample sections. The ARMA(1,0)-GARCH(1,2) model was used to estimate and forecast the daily S.M.R 20 rubber price for 20 days ahead in the future market effectively. The findings of this study suggest that GARCH(1,2) model is suitable used to model and forecast the current S.M.R 20 rubber price compared to GARCH(1,1) model. Future research should therefore concentrate on the investigation of excess kurtosis on S.M.R 20 rubber price which caused by outlier effect.

5. References

[1] Goh H H, Tan K L, Khor C Y and Ng S L 2016 Volatility and Market Risk of Rubber Price in Malaysia: Pre- and Post-Global Financial Crisis *Journal of Quantitative Economics* **14**(2) p 323–344

[2] Salisu A A and Fasanya I O 2013 Modelling oil price volatility with structural breaks. *Energy Policy* p 554–562

[3] Liu H and Shi J 2013 Applying ARMA-GARCH approaches to forecasting short-term electricity prices *Energy Economics* **37** p 152–166

[4] Kang S H and Yoon S M 2013 Modeling and forecasting the volatility of petroleum futures prices *Energy Economics* **36**(July 2015) p 354–362

[5] Daníelsson J 2011 *Financial risk forecasting: The theory and practice of forecasting market risk with implementation in R and MATLAB* (United Kingdom: John Wiley and Sons) chapter 1 pp 9

[6] Engle R F 1982 Autoregressive conditional heteroscedasticity with estimates of the variance of United Kingdom inflation. *Econometrica* **50** p 987–1007

[7] Hill R C, Griffiths W E and Lim G C 2011 *Principles of Econometrics* ed Fourth (United States of America: John Wiley & Sons, Inc) chapter 14 pp 526

[8] Brooks C 2014 *Introductory Econometrics for Finance* ed Third (New York: Cambridge University Press) chapter 9 pp 430

[9] Olson D L and Wu D 2017 *Predictive Data Mining Models* (Singapore: Springer) chapter 6 pp 63

[10] Isa Z and Jamil N A 2004 Tabiat Kemeruapan Perubahan Harga Getah Asli Malaysia *Jurnal Ekonomi Malaysia* **38** p 63–79

[11] Bollerslev T 1986 Generalized autoregressive conditional heteroskedasticity *Journal of Econometrics* **31** p 307–327

[12] Epaphra M 2017 Modeling Exchange Rate Volatility: Application of the GARCH and EGARCH Models *Journal of Mathematical Finance* **7**(1) p 121–143

[13] Joukar A and Nahmens I 2015 Volatility Forecast of Construction Cost Index Using General Autoregressive Conditional Heteroskedastic Method *J. Constr. Div., Am. Soc. Civ. Eng.* **142**(1) 4015051

[14] Huq M M, Rahman M M, Rahman M S, Shahin A and Ali A 2013 Analysis of Volatility and Forecasting General Index of Dhaka Stock Exchange *American Journal of Economics* **3**(5) p 229–242

[15] Hickey E, Loomis D G and Mohammadi H 2012 Forecasting hourly electricity prices using ARMAX-GARCH models: An application to MISO hubs *Energy Economics* **34**(1) p 307–315

[16] Wang Y and Wu C 2012 Forecasting energy market volatility using GARCH models: Can multivariate models beat univariate model *Energy Economics* **34**(6) p 2167–2181

[17] Pham H T and Yang B S 2010 Estimation and forecasting of machine health condition using ARMA/GARCH model *Mechanical Systems and Signal Processing* **24**(2) p 546–558

[18] Box G E P and Jenkins G M 1976 Time Series Analysis: Forecasting and Control ed Second (San Francisco: Holden-Day)

[19] Poon S H and Granger C W 2003 Forecasting Volatility in Financial Markets: A Review *Journal of Economic Literature* **41**(2) p 478–539

[20] Akaike H 1974 A New Look at the Statistical Model Identification *IEEE Trans. Autom. Control* **19**(6) p 716–723

[21] Schwarz G 1978 Estimating the Dimension of a Model *The Annals of Statistics* **6**(2) p 461–464

[22] Phillips P C B and Perron P 1988 Testing for a Unit Root in Time Series Regression *Biometrika* **75**(2) p 335–346

[23] Dickey D A and Fuller W A 1979 Distribution of the Estimators for Autoregressive Time Series With a Unit Root *J. Am. Stat. Assoc.* **74**(366) p 427–431

[24] Kwiatkowski D, Phillips P C B, Schmidt P and Shin Y 1992 Testing the null hypothesis of stationarity against the alternative of a unit root *Journal of Econometrics* **54**(1–3) p 159–178

[25] Makridakis S 1993 Accuracy measures: theoretical and practical concerns *International Journal of Forecasting* **9** p 527–529

Challenges of the Export for Natural Rubber Latex in the ASEAN Market

Aye Aye Khin [1], Raymond Ling Leh Bin [2], Sia Bik Kai [3], Kevin Low Lock Teng [4], Fong Yi Chiun [5]

[1] Assistant Professor, Faculty of Accountancy & Management (FAM), Universiti Tunku Abdul Rahman (UTAR), Jalan Sungai Long, Bandar Sungai Long, Cheras, 43000 Kajang, Selangor, MALAYSIA
[2] Lecturer, Faculty of Accountancy & Management (FAM), Universiti Tunku Abdul Rahman (UTAR), Jalan Sungai Long, Bandar Sungai Long, Cheras, 43000 Kajang, Selangor, MALAYSIA
[3] Senior Lecturer, Faculty of Accountancy & Management (FAM), Universiti Tunku Abdul Rahman (UTAR), Jalan Sungai Long, Bandar Sungai Long, Cheras, 43000 Kajang, Selangor, MALAYSIA
[4] Associate Professor, Faculty of Accountancy & Management (FAM), Universiti Tunku Abdul Rahman (UTAR), Jalan Sungai Long, Bandar Sungai Long, Cheras, 43000 Kajang, Selangor, MALAYSIA
[5] M.Phil Student, Faculty of Accountancy & Management (FAM), Universiti Tunku Abdul Rahman (UTAR), Jalan Sungai Long, Bandar Sungai Long, Cheras, 43000 Kajang, Selangor, MALAYSIA

E-mail: ayekhin@utar.edu.my

Abstract. There are approximately 90% of the world natural rubber (NR) supplies are produced by the top 4 NR exporting countries, particularly Thailand, Indonesia, Malaysia and Vietnam. The export challenges and uncertainties of the NR latex in the ASEAN market are influenced by many factors. All these issues have driven the motivation of this study that aims to investigate the factors affecting the export of the NR latex, to analyse the relationship among the export price, production and exchange rate with the export, and to predict the export of the NR latex of the four countries. The data from 1999 to 2016 are utilized as the panel data analysis and granger homogeneously causality test for estimations. Estimation reveals that the independent variables, namely export price and production are exhibiting significant positively and the exchange rate shows a negative relationship with export. To conclude, Indonesia and Malaysia are estimated to have decreasing trends of NR latex export. Inversely, Thailand and Vietnam are predicted to exhibit increasing trends of NR latex export. The novelty of this study concerns effectively on the enhancement and stability of NR production in the ASEAN markets despite of the challenges arising from the global market integration.

1. Introduction

With the utilization of econometric panel data analysis, the factors that affecting the NR export and challenges arise on natural rubber latex in ASEAN market will be examined extensively in this paper. The export of Natural rubber (NR) plays a vital role as one of the major contributions to the economic growth in ASEAN developing countries. Concurrently, the rubber latex has been considered as one of the crucial natural rubber commodities and contributing to the rise of exports in the emerging ASEAN economies. Latex [1] is better known as white liquid sticky form that has been extracted and collected from "tapping" process which is a long cut made on rubber trees (Fig. 1). Generally, the exports of rubber products are namely the dry and latex natural rubber, compound rubber, tyres, inner tubes, catheters, gloves, footwear, rubber bands, rubber sheets, condoms and so forth. On the other hand, the gloves, condoms, catheters, latex thread, and many more are the usual exports of rubber latex products.

The export challenges and uncertainties of the rubber latex in the ASEAN market are influenced by many factors. According to the past study of rubbers' supply and the export challenges [2], the supply of rubber was determined by the expected price in the market place, in line with production capacity of rubber such as planted area, new planting, replanting, and the age of the rubber tree, input costs, and underlying technological progress. In other circumstances, the rubber production was also influenced by the public (government) policy and private contributors that involved in the rubber production and development. In retrospect, in the case if the rubber productions have been increased in a particular country, the rubber latex export would be increased simultaneously, holding other variables as constant [3]. Similarly, for the NR exporting countries, if there is a continuous rise in NR export price, the rubber export will be increased as well.

Fig. 1. The Rubber Plantation and Collection of Latex from a Tapped Rubber Tree
Source: [1]

The past literatures were later interacted the rubber production with dynamic and recursive manner of demand. Demand was set by the expected rubber price, as well as the income level in the overall economy, prices of rubber substitutes, and prices of final goods, technology, consumer preferences, stocks, and utilization of manufacturing capacity [2]. The organizational structure of rubber production, marketing, consumption, and the government measures were also significant [2], however, the emphasis of rubber framework must examine both the supply and demand factors. It is possible to examine these relationships by constructing the rubber export model to capture the interdependencies between these factors in the emerging ASEAN market.

There are other past literatures [3, 4] elucidated the potential challenges of the export model utilization. Numerous challenges namely the favourable weather, political stability in the leading rubber producing countries, revival of abandoned rubber plantings due to long run high prices, instability of the demand side, slowing down of the GDP growth following sharp rises in oil prices and numerous factors are bound to have a direct impact on the long-term stability of rubber prices. The rise of NR price may also lead the great concern of some NR consuming countries and producers to acquire for more synthetic rubber (SR) in place of NR demand for more reclaimed NR and the hunt for more alternatives to NR products. Fundamentally, the NR prices were strongly influenced by numbers of external factors, namely the integration of global economic situations, supply-demand gap, export-import situation, and primarily the changes in crude petroleum oil prices, exchange rate volatility, time-lag, slowing growth in agricultural productivity, as well as government policies, futures markets interventions, rubber stock and unstable demand [5].

All these factors and challenges will provide a good starting point in constructing a solid theoretical framework in this study, in line with the general rubber economy perspectives, the export challenges and uncertainties of rubber latex.

Table 1 portrays the top 10 largest rubber exporters and importers in the world in 2016 and exhibits a huge opportunity for ASEAN developing countries as the main exporters and of NR in the world economy. There are about 65% of global supplies are contributed by Thailand and Indonesia. Diversely, there are approximately 90% of the world NR supplies are produced by the top 4 NR exporting countries, particularly Thailand, Indonesia, Malaysia and Vietnam. Currently, Malaysia stands in the lists of both the top 10 rubber exporters and importers; and is the country has grows to be the 1st exporter of rubber latex glove products (in Fig. 2) and 3rd largest producers of the world rubber producing countries in 2017.

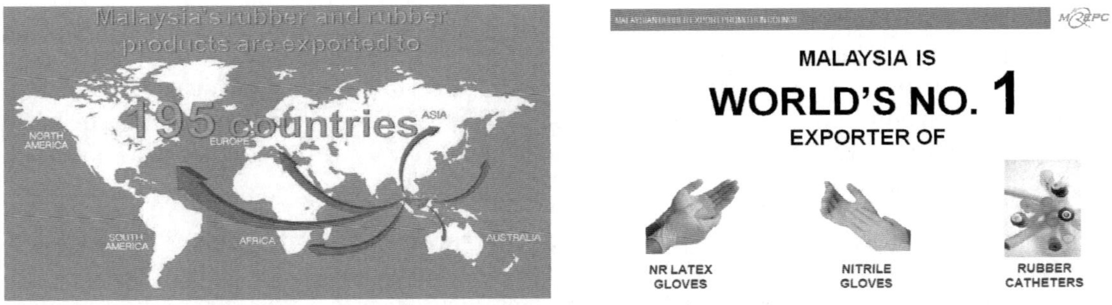

Fig. 2. Malaysia's Rubber and Rubber Products Export Countries [6]

In Malaysia, for the rubber latex glove products have been increased from approximately 12,000 million pairs to 37,000 million pairs in 2001 to 2017. Moreover, the rubber latex products' export volumes have been increased continuously from RM 4,000 million to RM 17,000 million in 2001 to 2017. In addition, the rubber latex import volumes has raised from approximately RM 200 million to RM 700 million in between 2001 to 2017 [7]. Conversely, Thailand was the largest exporter and amounting to 36.8% of the world rubber export, followed by Indonesia (28.1%), Vietnam (7.5%), Malaysia (7.3%) and Cote d' Ivoire (6.1%) in 2016. China was the largest importer, amounting to 26.7% of the world rubber import, followed by United States (11.7%), Malaysia (9.3%), Japan (7.4%) and India (5.2%) in 2016 (Table 1).

Table 1: Top 10 Largest Rubber Exporters & Importers in the World in 2016 [8]

Rank	Country	Export Value (USD)
1.	**Thailand**	**4.4 billion (36.8%)**
2.	**Indonesia**	**3.4 billion (28.1%)**
3.	**Vietnam**	**904.1 million (7.5%)**
4.	**Malaysia**	**871.1 million (7.3%)**
5.	**Côte d'Ivoire**	**726.3 million (6.1%)**
6.	Myanmar (Burma)	194.1 million (1.6%)
7.	Germany	153.4 million (1.3%)
8.	Belgium	126.4 million (1.1%)
9.	Guatemala	120.9 million (1.0%)
10.	Liberia	106.3 million (0.9%)

Rank	Country	Import Value (USD)
1.	**China**	**3.4 billion (26.7%)**
2.	**United States**	**1.5 billion (11.7%)**
3.	**Malaysia**	**1.2 billion (9.3%)**
4.	**Japan**	**927.2 million (7.4%)**
5.	**India**	**656 million (5.2%)**
6.	Rep. of Korea	537.1 million (4.3%)
7.	Germany	478.9 million (3.8%)
8.	Brazil	321.7 million (2.6%)
9.	Spain	241.8 million (1.9%)
10.	France	239.1 million (1.9%)

2. Literature Reviews

The research on "Factor Affecting Thailand's Major Agricultural Exports Using Panel Co-integration Method" illustrates the export volumes of Thailand's agricultural products that were affected by number of variables [9]. The unit root tests, panel co-integration test, and ordinary least squares (OLS) have been tested on annual panel data from 2001 to 2010. This research concludes a negative relationship between export volume and rubber's domestic price. Also, this research summarizes that when there is a decrease in domestic price of natural rubber, it will lead to the decrease in export of the rubber. Inversely, if the NR is highly demanded from other countries, there will still be a high exportation of the nature rubber.

Another research on "An Econometric Analysis of Natural Rubber Market in Malaysia" aims to investigate how NR industry in Malaysia is affected by essential factors [10]. The unit root test, co-integration test and two stages least squares (2SLS) have been utilized in this research. The outcomes of the export equation of this research show a relationship between world industrial production index and time trend at 5% significance level. The results revealed that there is a positive relationship between export and production of natural rubber, but a negative relationship between production and time trend of rubber trees. This research also concludes that in the case if the time trend of the rubber trees were increased above 30 years, the production of NR product will be reduced year by year.

In the perspective of the neighbourhood country, a research on "Determinant Analysis for Rubber Export in Indonesia" was executed to evaluate the factors that affect the rubber export in Indonesia [11]. Robust regression is used in conducting this research. The researchers revealed that there is a negative relationship between exchange rate and export quantity of rubber and there is also a positive relation between production and export quantities of rubber. The novelty of this research explains the connection among the exchange rate, production level, and export of natural rubber in Indonesia. Within the same ASEAN region, a research on "Exchange Rate Variations and Agricultural Exports in Vietnam" aims to investigate on how exchange rate variations will affect the agricultural export market in Vietnam [12]. The results revealed that the depreciation of exchange rate will lead to the decline of agricultural products' exports in Vietnam. The data utilized the agricultural market in Vietnam, however, the specification of specific agricultural products are not clearly defined.

3. Methodology

In retrospect, the export challenges and uncertainties of the NR latex in the ASEAN market were affected by many factors. This study (Fig. 3) assessed the breakdowns of both the supply and demand factors in line with the modifications made on the functional form of NR export model and some key factors such as the export price, production and exchange rate. All these key factors have driven the motivation of this study that aims 1) to investigate the factors affecting the export of the rubber latex, 2) to analyse the relationship among the export price, production and exchange rate with the export, and 3) to predict the export of the rubber latex of the selected ASEAN countries. The data from 1999 to 2016 are utilizing the panel data analysis and granger homogeneously causality test for estimations. All the secondary data of four selected countries are gathered from Malaysian Rubber Board (MRB), Malaysian Rubber Export Promotion Council (MREPC) and The Association of Natural Rubber Producing Countries (ANRPC).

Fig. 3. Conceptual Framework of the Export of the Rubber Latex of the 4 ASEAN countries

Thailand: $\mathbf{nrexpTH_{it}} = \alpha_1 + \beta_1 nrstr20_{it} + \beta_2 exTHB_{it} + \beta_3 nrpth_{it} + e_{1it}$

Indonesia: $\mathbf{nrexpIN}_{it} = \alpha_2 + \beta_4 nrsir20_{it} + \beta_5 exIDR_{it} + \beta_6 nrpin_{it} + e_{2it}$
Malaysia: $\mathbf{nrexpMS}_{it} = \alpha_3 + \beta_7 nrsmr20_{it} + \beta_8 exRM_{it} + \beta_9 nrpms_{it} + e_{3it}$
Vietnam: $\mathbf{nrexpVN}_{it} = \alpha_4 + \beta_{10} nrsvr20_{it} + \beta_{11} exVN_{it} + \beta_{12} nrpvt_{it} + e_{4it}$

where:
nrexpTH = NR export of the rubber latex in Thailand ('000 tonnes)
nrexpIN = NR export of the rubber latex in Indonesia ('000 tonnes)
nrexpMS = NR export of the rubber latex in Malaysia ('000 tonnes)
nrexpVN = NR export of the rubber latex in Vietnam ('000 tonnes)
nrstr20 = STR20 NR Thailand export price (USD/ton) deflated by the CPI
nrsir20 = SIR20 NR Indonesia export price (USD/ton) deflated by the CPI
nrsmr20 = SMR20 NR Malaysia export price (USD/ton) deflated by the CPI
nrsvr20 = SVR20 NR Vietnam export price (USD/ton) deflated by the CPI
exTHB = Exchange rate (THB100/USD) (THB = Thai Baht)
exIDR = Exchange rate (IDR100/USD) (IDR = Indonesian Rupiah)
exRM = Exchange rate (RM/USD) (RM = Malaysian Ringgit)
exVN = Exchange rate (VND/USD) (VND = Vietnamese Dong)
nrpth = NR latex production of Thailand ('000 tonnes)
nrpin = NR latex production of Indonesia ('000 tonnes)
nrpms = NR latex production of Malaysia ('000 tonnes)
nrpvt = NR latex production of Vietnam ('000 tonnes)
it = Time trend 1999 to 2016 yearly panel data of Thailand, Indonesia, Malaysia and Vietnam
e_{it} = Error terms respectively

3.1 Hypothesis Development

H_o1: There is no positive relationship between export price and export of the rubber latex of Thailand, Indonesia, Malaysia and Vietnam.
H_A1: There is a positive relationship between export price and export of the rubber latex of Thailand, Indonesia, Malaysia and Vietnam.

H_o2: There is no negative relationship between exchange rate and export of the rubber latex of Thailand, Indonesia, Malaysia and Vietnam.
H_A2: There is a negative relationship between exchange rate and export of the rubber latex of Thailand, Indonesia, Malaysia and Vietnam.

H_o3: There is no positive relationship between NR latex production and export of the rubber latex of Thailand, Indonesia, Malaysia and Vietnam.
H_A3: There is a positive relationship between NR latex production and export of the rubber latex of Thailand, Indonesia, Malaysia and Vietnam.

3.2. Panel Data Analysis

Panel data is a set of data that consists of the combination of time series and cross-sectional data in the dataset, or better known as longitudinal data which is a study over times of a group of variables [16]. There are 72 observations (18 years x 4 countries) in total consists of Thailand, Indonesia, Malaysia and Vietnam from 1999 to 2016. The panel data are used to analyse the issues that are hardly to be solved by time series data or cross-sectional data, and to avoid the issues of omitted variables that would cause bias in cross-sectional data [16, 17]. There are four different kinds of variables used in this research: 1) variables that vary among individuals but do not change over time (country code); 2)variables that vary over times but are similar for all individuals in a specific time period (price of natural rubber and exchange rate); 3) variables that changed on both over times and among individuals (export volume and export price); and 4) tendency variables that vary in predictable ways (exchange rate) [17,18].

3.2.1 Selection of Panel Data Models

The observations of panel data (*it*) will be indicated by both cross-section dimension (indicated by subscript *i*) and time series dimension (indicated by subscript *t*). Panel data models will examine the individual-specific effects, time effects, or both in order to deal with heterogeneity or better known as individual effect (cross-sectional or time specific effect). There are three panel data models will be utilized in this study, namely Pooled OLS (POLS), Fixed Effect Model (FEM) and Random Effect Model (REM). POLS is preferred if the individual effect does not exist, due to the reason that ordinary least square method (OLS) can produce more efficient and consistent parameters estimates. If utilising both POLS vs FEM, the testing of the Redundant Fixed Effects can be done. If comparing both POLS vs REM, the Breusch-Pagan LM test can be done in line with the past researches and testing by [16], [17] and [18]. In order to test the presence of individual effects, both fixed and random effects can be examined. A FEM model will examine in the case if the intercepts vary across groups or time periods, whereas a REM explores differences in error variance components across individuals or time periods. Hausman test [19] needs to be carried out to confirm which of the models (REM vs FEM) will be the most appropriate method for analysis. In short,

POLS vs FEM : **Redundant Fixed Effects test** will examine the assumptions of variances for individual effect

POLS vs REM : **Breusch-Pagan LM test** will examine the assumptions of variances for individual effect

REM vs FEM : **Hausman test** will detect the significant differences between the two estimators

Finally, the Hausman test will be appraised in order to select between both **REM and FEM**, and the results outlines in Table 2. H_0: REM is preferred; and H_A: FEM is preferred.

Table 2: Results of the Correlated Random Effects - Hausman Test

Equation: REM			
Test period random effects			
Test Summary	Chi-Sq. Statistic	Chi-Sq. d.f.	Prob.
Period random	2.5359	3	**0.4688**

Source: Own Data Analysis

Table 2 shows that p-value (0.4688) is larger than α 0.05, and therefore, the null hypothesis will be accepted. In this case, the **REM** is more preferred rather than FEM. Eventually, **REM** serves as the more appropriate panel data analysis to be employed for the export model of the ASEAN market.

4. Results & Discussion

4.1 REM of the Natural Rubber Latex Export Model

$$\text{Export}_{it} = 1.4322 + 0.0323\,\text{Price}_{it} + 0.0512\,\text{Production}_{it} - 0.0221\,\text{Exr}_{it} + 0.0129\,e_{it}$$

$$[2.7661^{**}] \qquad [4.2039^{**}] \qquad [-11.7833^{***}]$$

$R^2 = 0.2369$; Adjusted $R^2 = 0.2033$

Residual Diagnosis: Jarque-Bera Statistics = 2.4517; sig-p value = 0.2935

As refer to the export model among the ASEAN 4 countries, NR latex export is set as the dependent variable while the independent variables in this model include export price, NR production and the exchange rate of the individual country. R-squared means that there is approximately 23.69 percent of total variation in the export equation is well explained by the explanatory variables, namely the export price, NR production and exchange rate. Amongst the 3 explanatory variables, the NR production portrays as the most important variable in the NR latex export model. Every 1 unit increase in export price, in average, has a positive effect by increasing 0.0323 unit in NR latex export with statistically significance at α=0.05 level. In addition, every 1 unit increase in NR production, in average, has a positive impact by increasing 0.0512 unit in NR latex export with statistically significance at α=0.05 level. Lastly, every 1 unit increase in exchange rate, in average, will have a negative impact by decreasing 0.0221 unit in NR latex export with statistically significance at α=0.01 level. Besides, based on the residual analysis of the normality test, the Jarque-Bera Statistics = 2.4517; and sig-p value = 0.2935 is more than α=0.05. Therefore, the residuals are proven to be normally distributed.

Based on the outcomes of the Granger Homogeneously Causality Test [20], the PRODUCTION is homogeneously granger causes the EXPORT. Therefore, PRODUCTION and EXPORT are having long term equilibrium and co-integrated with statistically significance at α 5% level (sig-p value = 0.0107 < α 0.05). PRODUCTION → EXPORT, it is proven to have a unidirectional relationship. The PRICE is homogeneously granger causes the PRODUCTION. Therefore, PRICE and PRODUCTION are having long term equilibrium and co-integrated with statistically significance at α 5% level (sig-p value = 0.0292 < α 0.05). PRICE → PRODUCTION, it is also proven to have a unidirectional relationship. Lastly, the PRODUCTION is homogeneously granger causes the EXCHANGE_RATE. Therefore, PRODUCTION and EXCHANGE_RATE are having long term equilibrium and co-integrated with statistically significance at α 5% level (sig-p value = 0.0397 < α 0.05). PRODUCTION → EXCHANGE_RATE, it is also proven to have a unidirectional relationship.

4.2 Model Simulation of the Prediction of the NR Latex Export Trends

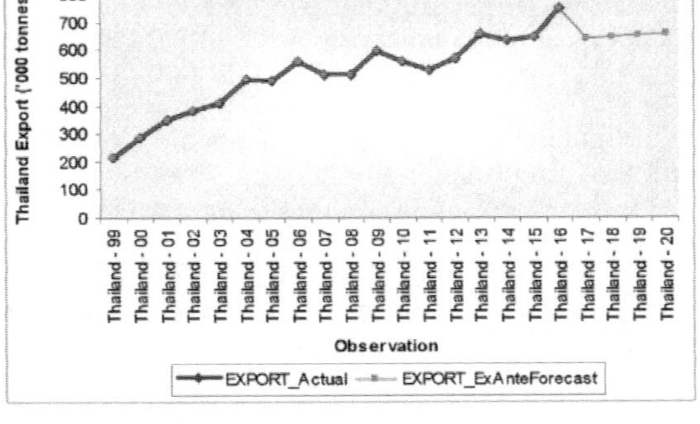

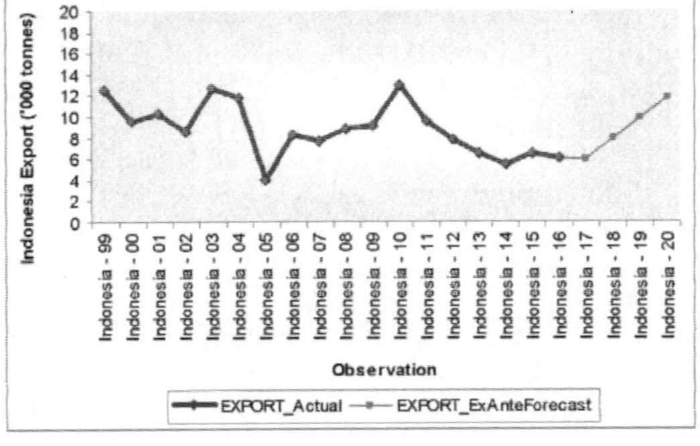

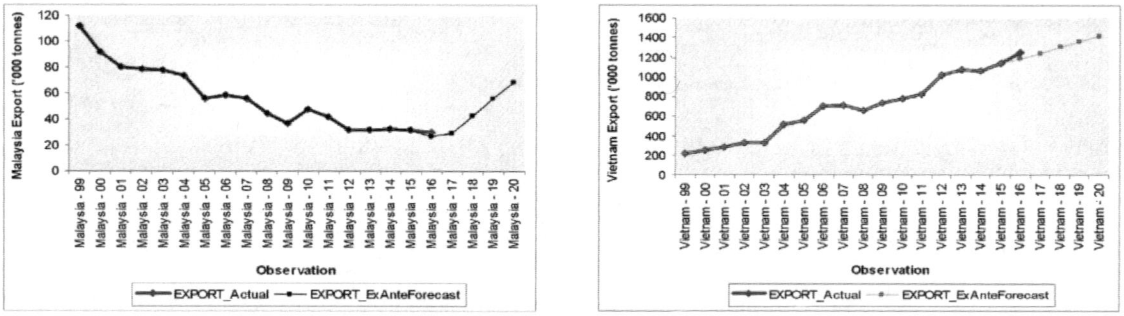

Fig. 4. Model Simulation of the Export of the NR Latex of the 4 ASEAN countries
Source: Own Development

Fig. 4 illustrates that Indonesia and Malaysia are estimated to have decreasing trends of NR latex export. Inversely, Thailand and Vietnam are predicted to exhibit increasing trends of NR latex export from 1999 to 2016. In general, all the NR latex export trends on the 4 countries are predicted to be increased from 2017 to 2020.

5. Conclusion

Based on the Random Effect Model (REM) of the NR latex export model equation, there is a positive relationship between export price and export NR latex of Thailand, Indonesia, Malaysia and Vietnam. In summary, the results shown that exchange rate exhibits a negative relationship with export NR latex among the four selected ASEAN countries. In addition, NR production illustrates a positive relationship with the export NR latex among Thailand, Indonesia, Malaysia and Vietnam. This study examines some significance of variables that are beneficial for prediction and forecasting of the export of latex in the rubber sector (export price, exchange rate and NR latex production). All the significant variables within the REM equation utilize the correct functional form and a strong long-term relationship exhibits among the variables. In addition, the important identified variables in this study may benefit the authorities, businesses and other related sectors in different countries. The novelty of this study concerns effectively on the enhancement and stability of NR production in the ASEAN markets despite of the challenges arising from the global market integration.

References

[1] Global Rubber Markets. (2017). Retrieved from: https://globalrubbermarkets.com/
[2] Barlow, C., Jayasuriya, S. and Tan, C. S. (1994). The world rubber industry. Routledge, London and New York.
[3] Hassan, N., Hamzah, H. H. M., & Zain, S. M. M. (2013). A goal programming approach for rubber production in Malaysia. *Am-Eur. J. Sustain. Agri.* **7(2)**, pp 50-53.
[4] Yogaratnam, N. (2012). Natural rubber production policies to meet the challengers. Retrieved from: http://www.dailymirror.lk/19044/natural-rubber-production-policies-to-meet-the-challengers-.
[5] Burger, K., Smit, H., and Vogelvang, B. (2002), Exchange rates and natural rubber prices, the effect of the Asian crisis. *Eur. Assoc. Agri. Econo.*, 2002 International Congress, August 28-31, Zaragoza, Spain, http://purl.umn.edu/24958, pp.1-24.
[6] MREPC. (2017). Malaysian Rubber Export Promotion Council Statistics, Retrieved from:http://www.mrepc.com/ mrepc_new/industry/world_production.php
[7] DOSM. (2018). Department of Statistics Malaysia (DOSM), Retrieved from: https://www.dosm.gov.my/v1/
[8] IRSG. (2017). Latest rubber statistical bulletin. International Rubber Study Group (IRSG), 111 North Bridge Road, #23-01/06 Peninsula Plaza, Singapore 179098.
[9] Leelawattanapan, T., & Chaiboonsri, C. (2012). Factors Affecting Thailand's Major Agricultural Exports Using Panel Cointegration Method, Proceedings-Sufficiency Economy and Community Enterprise-007, 4th International Conference on Humanities and Social Sciences, April 21st , 2012, Faculty of Liberal Arts, at Prince of Songkla University, Thailand, pp 1-10.
[10] MdLudin, N., Applanaidu, S., & Abdullah, H. (2012). An econometric analysis of natural rubber market in Malaysia. *Int. J. Environ. Agri. Res.* **2(6)**, pp 2454–1850.

[11] Yanita, M., Yazid, M., Alamsyah, Z., & Mulyana, A. (2016). Determinant analysis for rubber export in Indonesia. *International Journal of Scientific and Research Publications*, ***6*(9)**, pp 478–2250. Retrieved from www.ijsrp.org.

[12] Dinh, C. X., & Nguyen, T. (2017). Exchange rate variations and agricultural exports in Vietnam. *Accounting and Finance Research*, ***5*(2)**, pp 54. Retrieved from: https://doi.org/10.5430/afr.v5n2p54

[13] Raju, K. V. (2016). Instability in Natural Rubber Prices in India: An Empirical Analysis. *IOSR Journal of Economics and Finance Ver. III*, *7*(3), 2321–5933. https://doi.org/10.9790/5933-0703032428Salvatore.

[14] Abolagba, E. O., Abolagba, O. O., Omokhafe, K. O., Musa, E., Mesike, S. C., & Osazuwa, S. (2016). The Performance of Rubber Exports in Nigeria, *5*(10), 347–353.

[15] Yanita, M., Yazid, M., Alamsyah, Z., & Mulyana, A. (2016). Determinant Analysis for Rubber Export in Indonesia. *International Journal of Scientific and Research Publications*, *6*(9), 478–2250. Retrieved from www.ijsrp.org

[16] Studenmund, A. H. (2014). *Using econometrics : a practical guide. Addison-Wesley series in economics.* https://doi.org/10.1017/CBO9781107415324.004

[17] Wooldridge, J. M. (2015). *Introductory econometrics: A modern approach*. Nelson Education.

[18] Gujarati, D.N., & Porter, D.C. (2009). Basic Econometrics, 5th ed. USA: McGraw-Hill. ISBN: 978-007-127625-2

[19] Hausman, J. A. (1978). Specification tests in econometrics. *Econometrica: Journal of the econometric society*, pp 1251-1271.

[20] Dumitrescu, E. I., & Hurlin, C. (2012). Testing for Granger non-causality in heterogeneous panels. *Economic Modelling*, ***29*(4)**, pp 1450-1460.

Institute of Physics
Dirac House, Temple Back
Bristol BS1 6BE UK

ISSN: 1757-8981
ISBN 978-1-5108-9383-2

9 781510 893832

CURRAN ASSOCIATES INC.
proceedings.com